Tribology of Magnetic Storage Systems

Tribology of Magnetic Storage Systems

A collection of significant papers originally published in the *Journal of Engineering Tribology*, Part J of the Proceedings of the Institution of Mechanical Engineers.

Edited by Bharat Bhushan

Professional Engineering Publishing

Published by Professional Engineering Publishing Limited,
Bury St Edmunds and London, UK

First Published 1998

The papers included in this volume were first published in the following publications:
Proc. Instn Mech. Engrs, Part J, J. Engineering Tribology, 1997, **211** (J4), 249–372
Proc. Instn Mech. Engrs, Part J, J. Engineering Tribology, 1998, **212** (J1), 1–32
Tribology Letters, 1998, **4**, 1–35 (J. C. Baltzer AG, Science Publishers)

ISBN 1 86058 174 9

A CIP catalogue record for this book is available from the British Library.

Printed by The Ipswich Book Company, Suffolk, UK.

Contents

Contact modelling

Microtribology

Related Titles of Interest

Title	Author	ISBN
New Directions in Tribology – First World Tribology Congress – Plenary Papers	Ed. I. M. Hutchings	1 86058 099 8
First World Tribology Congress – Abstracts of Papers	Ed. W. Roberts *et al.*	1 86058 109 9
History of Tribology (2nd Edition)	D. Dowson	1 86058 070 X
A Tribology Casebook	J. D. Summers-Smith	1 86058 041 6
Introductory Guide to Industrial Tribology	J. D. Summers Smith	0 85298 896 6
Journal of Engineering Tribology	Proceedings of the Institution of Mechanical Engineers, Part J	ISSN 1350-6501

For the full range of titles published by Professional Engineering Publishing contact:

Sales Department
Professional Engineering Publishing Limited
Northgate Avenue
Bury St Edmunds
Suffolk
IP32 6BW
UK

Tel: +44 (0) 1284 763 277
Fax: +44 (0) 1284 704 006

About the Editor

Professor Bharat Bhushan is an internationally recognized expert in the fields of conventional tribology and micro/nanotribology. He is presently an Ohio Eminent Scholar and Howard D. Winbigler Professor in the Department of Mechanical Engineering, and Director of the Computer Microtribology and Contamination Laboratory, at the Ohio State University, Columbus, Ohio. His wide experience encompasses both academic and industrial working environments, including the Department of Mechanical Engineering, Massachusetts Institute of Technology (Cambridge, Massachusetts), Automotive Specialists (Denver, Colorado), the Research and Development Division of Mechanical Technology Incorporated (Latham, New York), the Technology Services Division of SKF Industries Incorporated (Pennsylvania), the General Products Division Laboratory of IBM Corporation (Tucson, Arizona), the Almaden Research Centre of IBM Corporation (San José, California) and the Department of Mechanical Engineering at the University of California (Berkeley, California).

He is a prolific author, having written five technical books, 22 handbook chapters and more than 325 technical papers for refereed journals; in addition he has authored over 60 technical reports, edited more than 24 books and holds seven US patents. He has a particular research interest in the tribology of magnetic storage systems and organized the first symposium on 'Tribology and Mechanics of Magnetic Storage Systems' in 1984 and the first international symposium on 'Advances in Information Storage Systems' in 1990. He is founding editor of the World Scientific 'Advances in Information Storage Systems' Series and CRC Press 'Mechanics and Materials Science' Series and the Journal of Information Storage and Processing Systems.

Editorial

The magnetic recording process is accomplished by relative motion between a magnetic medium against a stationary or rotating read/write magnetic head. Under steady operating conditions, a load-carrying air film is formed. There is physical contact between the medium and the head during starting and stopping. In modern high-end computer tape and disk drives, the surfaces have roughnesses ranging from 1–5 nm r.m.s. and head-to-medium separations ranging from 25–150 nm. The need for higher and higher recording densities requires that surfaces be as smooth as possible and flying height be as small as possible. The very smooth surfaces and ultra-low flying heights required present challenges and opportunities unparalleled in any other industry.

Theories of conventional friction, stiction, interface temperatures and wear relevant to head–medium interface are well developed (**1, 2**). The advent of new techniques, namely atomic force microscopy and friction force microscopy, has led to the development of the new field of microtribology (**3**). The microtribological studies are commonly conducted to develop fundamental understanding of interfacial phenomena in magnetic storage devices.

This book entitled *Tribology of Magnetic Storage Systems* is dedicated to the modern developments in the macro- and microtribology and mechanics of magnetic storage systems. It contains ten papers detailing the latest research from the group of Professor Bharat Bhushan, Ohio Eminent Scholar and Howard D. Winbigler Professor and Director of the Computer Microtribology and Contamination Laboratory of the Ohio State University, a paper from the group of Professor John L. Sullivan of Aston University and two papers from the group of Professor Frank E. Talke of the University of California, San Diego. Twelve of the papers were originally published in the *Journal of Engineering Tribology, Proceedings of the Institution of Mechanical Engineers*, 1997, Volume 211 (Issue J4) and 1998, Volume 212 (Issue J1). One of the papers is reprinted with permission of J. C. Baltzer Science Publishers, from *Tribology Letters*, 1998, Volume 4.

References

1 **Bhushan, B.** *Tribology and Mechanics of Magnetic Storage Devices*, 2nd edition, 1996 (Springer, New York).

2 **Bhushan, B.** *Mechanics and Reliability of Flexible Magnetic Media*, 1992 (Springer, New York).

3 **Bhushan, B.** *Handbook of Micro/Nanotribology*, 2nd edition, 1999 (CRC Press, Boca Raton, Florida).

B. Bhushan
Editor

Head–disk interface tribology

In situ instrumentation for localized wear studies of magnetic thin-film disks

B Bhushan and **S M Forehand**
Computer Microtribology and Contamination Laboratory, Department of Mechanical Engineering, The Ohio State University, Columbus, Ohio, USA

Abstract: An experimental set-up and a procedure to make *in situ* localized measurements of various wear processes on magnetic thin-film disks are developed. The coefficient of friction, acoustic emission and optical reflectance are monitored simultaneously. Results are analysed to determine 'precursors to failure' and to compare the effects of lubricant, surface roughness, overcoat and normal load on disk wear. Disk wear is observed to follow a pattern of lubricant depletion followed by head slider–overcoat asperity contact. Wear tracks are formed after asperity contact. Lubricant extends disk wear life, as does a polished surface texture, and an SiO_2 overcoat wears more slowly than a diamond-like carbon overcoat. The experimental methodology developed in this study should be of interest in studying precursors to failure of any moving interface.

Keywords: magnetic disk, head, acoustic emission, optical sensor, strain gauge, friction and wear

1 INTRODUCTION

Rigid disks are the principal on-line magnetic storage media in computers, and the tribological study of disk durability is of significant interest to the computer industry. Specifically, as the demand for larger storage density grows, the need for lower head flying height increases. Lower flying height, however, increases the potential for asperity contacts at the head–disk interface, which results in higher wear rates (**1**). Figure 1 shows a general cross-section of a thin film disk (**1**). A hard overcoat and liquid lubricant film are applied to provide protection against wear and corrosion and to maintain low friction and stiction.

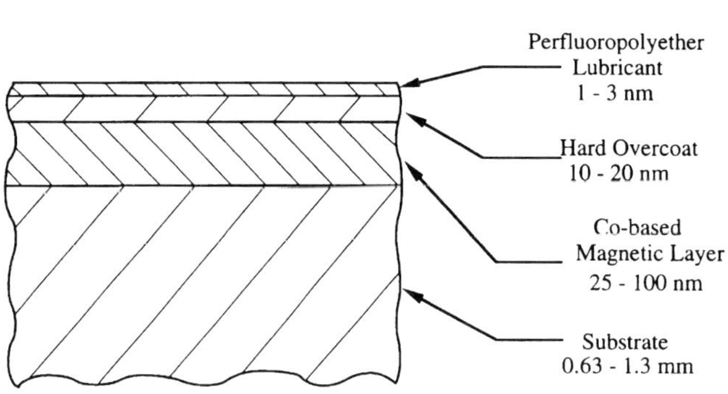

Fig. 1 Cross-section of a typical thin-film disk

The MS was received on 30 April 1996 and was accepted for publication on 13 November 1996.

A typical durability test is an accelerated sliding wear test in which the head slider contacts the spinning disk under a normal load higher than the typical slider load (e.g. the slider load for a microslider is 7 g). Load and disk speeds are chosen to ensure that a hydrodynamic lift does not occur at the head–disk interface. Durability tests in the past have focused primarily on *in situ* measurements of interface friction and post-failure evaluation of wear tracks. These approaches are limited in that friction and wear are related but independent, and post-failure measurements give only end results. To design machine elements with low friction and wear, 'precursors to failure' need to be understood. One approach is a test apparatus that provides simultaneous *in situ* localized information about specific wear phenomena, such as interface friction, acoustic emission, formation of wear debris and wear track progression, during a sliding wear test.

Friction measurements (**1**) and acoustic emission measurements at the sliding interface (**2**–**8**) and optical measurements of worn interfaces (**9**–**12**) have been used separately to study failure mechanisms. The purpose of this study is to develop a test apparatus that utilizes several measurement instruments to monitor the wear phenomena mentioned earlier. Experimental results are used to form a definitive characterization of wear progression and to compare the effects of various disk properties, such as lubricant, roughness and overcoat. A review of the literature reveals that a comprehensive, *in situ* wear test such as this has not been reported. It is important to note that the experimental methodologies developed in this study will be applicable to the wear characterization of not only computer disk drives but also to any material pair that may undergo a sliding and/or rolling durability test.

2 EXPERIMENTAL APPARATUS AND PROCEDURE

2.1 Description of apparatus

Figure 2 shows a schematic of the experimental set-up. A modified commercial disk drive with an in-line head slider suspension is used. The hub is modified to adapt both 95 and 65 mm diameter disks. Instruments used for *in situ* wear measurements are a dual-axis strain gauge beam, an acoustic emission (AE) transducer and an optical reflectance sensor. The strain arm is mounted on an x–z positioner to move it in the horizontal plane and along the vertical axis, and it has the head slider–suspension assembly attached to one end. It measures both normal and friction forces and thus allows calculation of the coefficient of friction. The AE transducer is mounted on the strain arm and measures the high-frequency acoustic emission generated by head slider–overcoat contact. The optical reflectance sensor is attached to a motorized linear stage. It scans radially across the wear track and registers changes in disk surface reflectance. In addition, a vertically mounted linear stage is used to change the probe-to-sample spacing.

2.1.1 *Strain gauge beam*

An increase in the coefficient of friction detects lubricant depletion and damage to the interface. The coefficient of friction is measured with a dual-axis strain gauge beam.

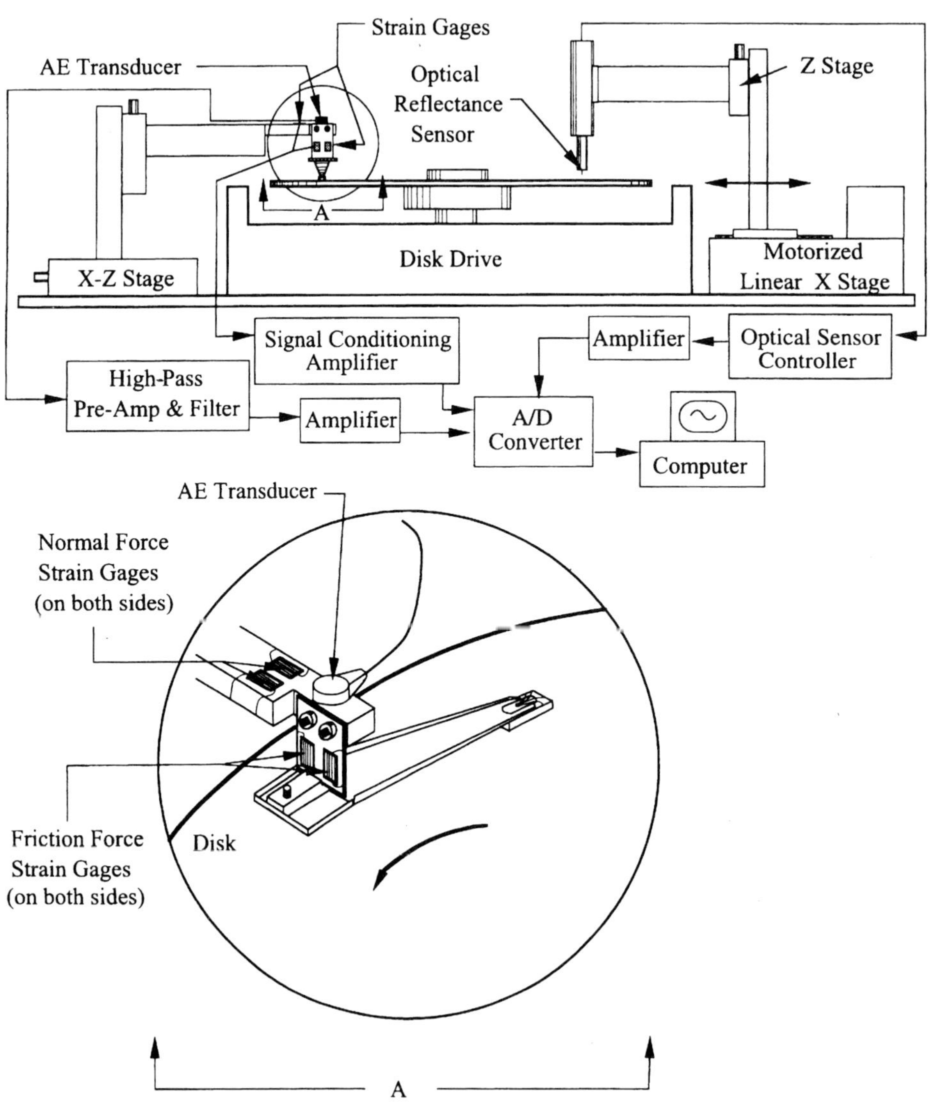

Fig. 2 Schematic of experimental set-up and close-up of the slider–suspension assembly over a disk surface

The beam used is a model HF 650 with a 7075 aluminium (with a gold alodine finish) strain arm and resistive gauges from Lotus Technologies Inc. in Los Gatos, California. Four active gauges (two on each side) are mounted on each axis for high signal output and temperature compensation. The beam has a natural frequency of 6.5 kHz and is mounted on an in-line suspension. It is rated at 30 g full scale with non-linearity of less than 1 per cent and resolution of 20 mg. Strain signals are output from a standard Wheatstone bridge configuration with an excitation voltage of 8 V. The strain arm is attached to a manual x–z positioner, which allows the head slider to be placed at a particular radial position on the disk, and also allows the head slider to be set at a specific normal load. The z positioner is used to lower the head slider on to the disk until the desired normal load is reached. Then the z positioner is locked into place for the duration of the test.

Continuous monitoring of the strain gauge signals allows the normal load to be constantly observed. In this experimental set-up, the working normal load range was about 0–24 g. The strain gauge beam is calibrated by applying known loads to a head slider and recording the change in voltage output. Loads are applied separately in both friction and normal directions. Recording of both friction and normal forces leads to the calculation of a coefficient of friction. The force signals, after suitable amplification using a signal conditioning amplifier, are fed to an analogue-to-digital (A/D) converter and digitized at a frequency of 1 kHz. Digitized data are recorded on a PC and averaged.

2.1.2 Acoustic emission transducer

When the mechanical state of a material body changes, part of the energy that is subsequently released goes into the formation of transient elastic stress waves called acoustic emission (AE). The surface displacements that occur when the AE interacts with material surfaces are easy to measure using a displacement sensor mounted on the surface. A common displacement sensor used to monitor stress waves is a piezoelectric transducer (also called an acoustic emission transducer). Piezoelectric material possesses the property of generating electric charge when under mechanical deformation. Piezoelectric charge output is proportional to the intensity of vibration which it experiences (**13**). Asperity impact and/or particles (wear debris or particulate contaminates) at the interface generate high-frequency elastic stress waves which vibrate surrounding surfaces. These vibrations are measured by an acoustic emission transducer (**3**). Acoustic emission measurement has shown serious promise for monitoring the contacts at the head–disk interface (**2**, **3**, **6**, **7**). Many investigators have correlated acoustic emission with surface wear and head–disk contact. Kita *et al.* (**2**) used an AE sensor to study slider take-off profiles, and Briggs *et al.* (**6**) and Peck *et al.* (**7**) used an AE sensor to detect asperity contact. Klamecki and Hanchi (**5**) proposed a relationship between the AE count rate (number of times the signal exceeds a threshold) and wear rate and Matsuoka *et al.* (**8**) related AE to the energy required to remove material during abrasion. Jiaa and Dornfeld (**4**) identified three regions of wear that could be determined based on AE signals.

The acoustic emission detection sensor used in this work is a model MC500 LT (500 kHz) from On-Line Digital Technologies in Carmichael, California. The transducer is glued to the strain arm with epoxy bond. The AE signal is amplified with a charge pre-amp of 60 dB gain and filtered through a high-pass filter (FL12Y, 125–1000 kHz) with a usable frequency range above 500 kHz before it is digitized in an A/D converter and recorded on a PC, where an r.m.s. is taken. This allows measurement of the high-frequency elastic stress waves associated with asperity impact to pass through, while it attenuates lower-frequency machine vibration and noise (the first bending mode of the Al_2O_3–TiC slider is about 300 kHz). Since an r.m.s. of the AE signal is recorded at a sampling rate of 1 kHz, the actual limiting frequency is lowered, but this rate still accurately preserves the high-intensity AE signals. An interesting observation is that the detail of an AE r.m.s. plot is significantly affected by the size of the averaging window in data acquistion. Figure 3 shows that when the AE signal is averaged every 0.3 s, much more detail is obtained than when the signal is averaged every 0.6 s. In this work, the averaging rate used is 0.3 s. One factor that must be kept in mind is that the AE transducer is separated from the stress wave source by the strain arm and slider suspension assembly. As pointed out by Klamecki and Hanchi (**5**), this separation can possibly result in signal changes or attenuation, but an assumption must be made that certain AE characteristics survive interactions in the transmission path.

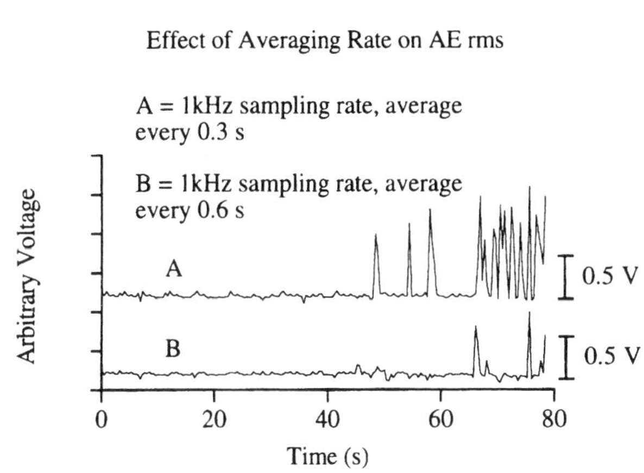

Fig. 3 Plot of the AE r.m.s. signal for a standard texture, lubed Al–Mg disk after about 0.75 km sliding distance, with averaging rates of 0.3 and 0.6 s

2.1.3 *Optical sensor*

The concept of using optical methods to monitor wear track progression arose from the need to determine wear volume *in situ*, without disturbing track generation. Nunnelley *et al.* (**9**) successfully developed a high-speed scanning microellipsometer that makes on-line wear measurements of thin-film disks. Though this method was accurate, a microellipsometer is both expensive and bulky and does not lend itself well to test stand use. The latest generation of optical on-line wear inspection tools appears to contain an optical surface analyser developed by Meeks *et al.* (**12**), which utilizes measurement of a laser diode beam reflected off the disk surface. Sensitivity to 1 Å of carbon wear of lubricant depletion, a lateral resolution of 5 μm and a 2 MHz bandwidth were reported using this device. In 1992, Jen and Gillis (**10**) demonstrated the effectiveness of the Fotonic™ sensor as an *in situ* wear measurement device for thin-film disks. The Fotonic probe is compact, inexpensive and highly sensitive to changes in local reflectance, with fair lateral resolution (**14**–**16**). Jen and Gillis (**10**) show that increases in reflectance (and thus increases in Fotonic voltage output) correspond to localized thinning of the carbon overcoat, whereas scratching (roughening) of the surface is expected to reduce the reflectance.

The optical reflectance sensor employed is a dual-channel, fibre-optic system that performs non-contact displacement and vibration measurements. It is the MTI 2000 Fotonic sensor mainframe with MTI 2020R probe plug-in module (tip diameter = 508 μm, active diameter = 178 μm, maximum frequency responses = 90 kHz, resolution = 30 nm) from Mechanical Technology, Inc. in Latham, New York. The probe transmits and receives light through a 178 μm diameter fibre-optic bundle, as shown in Fig. 4a. The transmitting and receiving fibres are 50 μm each in diameter and are arranged randomly within the probe. Light from an incandescent halogen lamp inside the mainframe is sent through the transmitting fibres, while reflected light from the test surface is captured by the receiving fibres. The reflected light is then sent to a photodetector in the mainframe which converts the light intensity into an electrical signal (voltage). The signal is amplified and digitized and sent to a PC. The probe is scanned radially across the disk surface by means of a stepper motor.

Although wear tests are run at 200 r/min, the sensor's 10 kHz bandwidth allows for use in high-speed measurement applications (~ 1000 r/min). For all wear tests data are collected at a 1 kHz sampling rate and averaged every 0.3 s in order to eliminate the effects due to disk run-out at 200 r/min. To improve upon the lateral resolution obtained by Jen and Gillis (**10**), who scanned across tracks rather quickly in 100 μm steps, the probe is scanned at approximately 15 μm per step. This increases scan time significantly, but also produces profile plots of much higher detail. The trade-off, obviously, is the

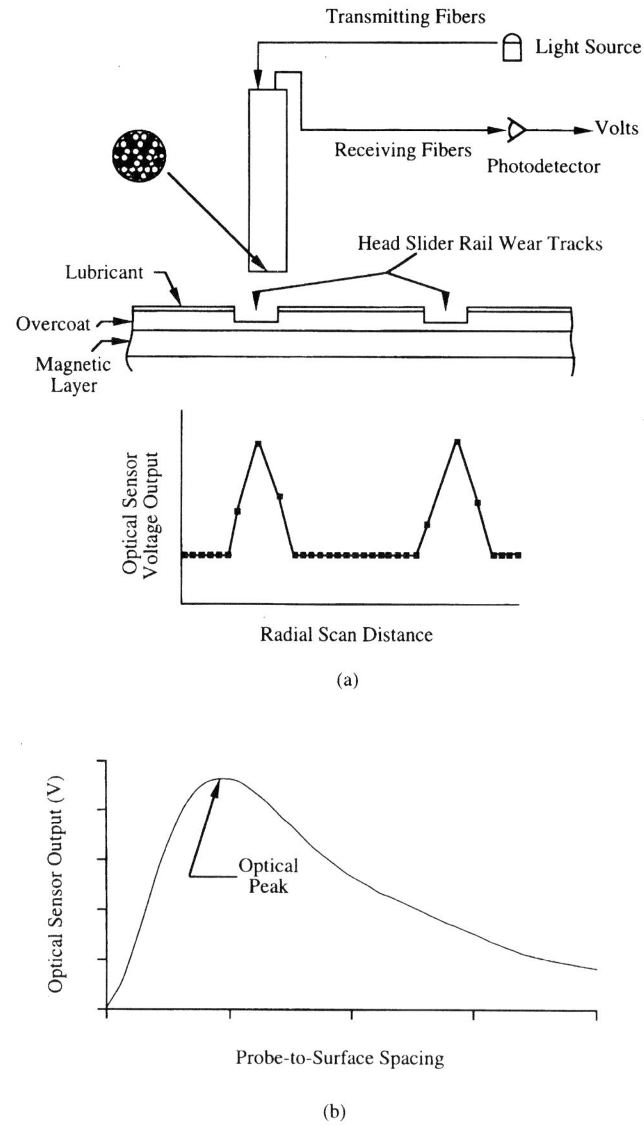

Fig. 4 (a) Operating principle of the optical reflectance sensor and schematic of optical sensor output while scanning across wear tracks and (b) optical sensor output as a function of probe-to-surface spacing

increased possibility that the optical profiles may change during a scan, but at typical scan times of less than 2 min, this risk is acceptable.

The intensity of reflected light as seen by the sensor changes as a function of probe-to-surface spacing. Specifically, the intensity increases from zero, peaks and then decreases as the probe-to-surface spacing is increased from zero. Figure 4b shows the probe response characteristics. A vertical positioner allows the probe to be set at the optical peak (of the order of 0.25 mm), which is the chosen operating point. This point was found by Jen and Gillis (**10**) and Karis and Lyn (**11**) to correspond to the lowest sensitivity to disk run-out, as well as to the highest signal-to-noise ratio. Additionally, a d.c. bias offset and the incandescent light source intensity can be adjusted to set a standard reflectance level on each disk before commencing a wear test. An increase in sensor output corresponds to a thinning of the overcoat, while a decrease in sensor output corresponds to the total removal of the overcoat (scratching).

2.2 Optical sensor comparison to stylus profiler

To obtain a relationship between the optical sensor and wear track depth, optical sensor output is compared to stylus profiler depth measurement for a variety of cases. The stylus profiler used is a Tencor alpha-step 200 with a 5 μm radius diamond tip stylus.

Figure 5 shows measurements taken of four artificially created scratches on a disk surface. Scratches were produced by loading a fine scalpal blade on a spinning disk. Several stylus profiles of each scratch were taken, and a representative profile is shown in Fig. 5. Optical sensor data are taken at a 30 mm radius on a disk spinning at 200 r/min in order to accurately mimic test conditions, but the stylus profiles are taken while the disk is stationary. As a result, optical sensor output is circumferentially averaged around the entire wear track, while stylus profiles represent a specific local line across a wear track. Curve shapes from the optical sensor output generally follow that of the stylus profiler, suggesting good correlation between the two. The optical sensor tends to exaggerate the width of the scratches, due in part to lateral resolution limitations and also to the eccentricity of the spinning disk. The first scratch extends down 20 nm, which is well into the carbon overcoat layer. This scratch is clearly picked up by the optical sensor, peaking at about 1 V. The two peaks present in the optical sensor output are probably due to non-uniformities in the scratch depth, which would be picked up in the track average, but not in the specific line picked by the stylus profiler. The second scratch illustrates the optical sensor's sensitivity to small changes

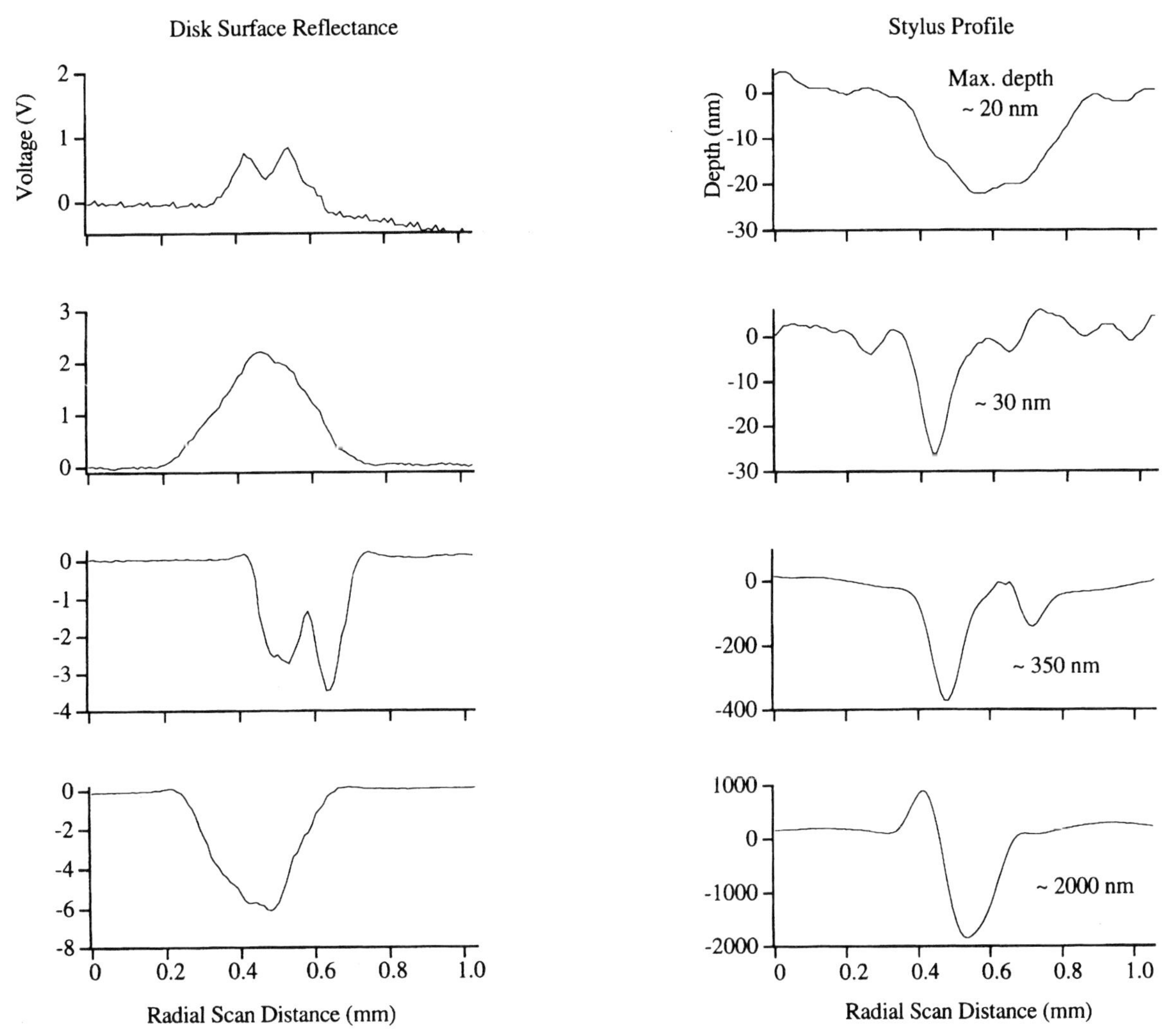

Fig. 5 Comparison of optical sensor output to stylus profile for 20, 30, 350 and 2000 nm deep scratches made artificially with a fine scalpel blade

in track depth. This scratch is 30 nm deep, and the voltage peaks near 2.5 V. One of the important principles of the Fotonic probe is that the reflectance should decrease in scratches that extend beyond the overcoat layer, as light is scattered and the scratch surface appears darker. In Fig. 5 the third and fourth scratches are deep enough to penetrate into the layers underneath the overcoat. The optical sensor returns a strong decrease in reflectance at the scratch locations.

2.3 Experimental procedure

To ensure that a hydrodynamic lift does not occur at the head–disk interface, the tangential speed of the disk at the interface location must be kept below about 1 m/s. For a disk speed of 200 r/min and a radial distance of 30 mm, the tangential speed is 0.63 m/s, which is safely below the critical speed. Tests are conducted using Al_2O_3–TiC microsliders with a standard normal load of 7 g. As polished disks and glass disks show no wear at 7 g, they are tested at 21 g. Data are analysed with SnapMaster^{TM} data acquisition software.

Due to the location of components in the experimental set-up, care is taken to ensure that the optical probe is aligned over the wear tracks. This is accomplished by positioning a tiny piece of lint-free paper on the disk at the intended location of the head slider–disk contact, and then manually rotating the disk until the paper is underneath the probe tip. The probe is aligned to scan across the paper marker. To compensate for the error in this positioning method, a radial scan distance of 4 mm is chosen. This gives a generous positioning tolerance of about ± 1 mm to scan over wear tracks made by the 2.1 mm wide head slider. The paper is carefully removed and the disk is cleaned with an air line before testing begins.

Wear tests are conducted on various disks in a continuous sliding mode until interface failure occurs, and the tests are stopped when significant carbon wear has occurred, with some exceptions, as will be seen later. Failure is defined by a significant or sudden increase in any of the instrument signals.

2.4 Test disks and sliders

Tests are run on three types of Al–Mg disks (standard texture/lubed, standard texture/unlubed and polished texture/lubed) and one type of glass disk (standard texture/lubed). Disks with mechanically textured Al–Mg substrates are 95 mm in outer diameter, 25 mm in inner diameter and 0.800 mm in thickness. Thin-film construction consists of a 75 nm thick Co–Pt–Cr magnetic coating and a 20 nm thick hydrogenated diamond-like carbon (DLC) coating, both deposited by sputtering. Lubed disks are lubricated with about 2.6 nm thick perfluoropolyether (Demnum) lubricant film. Standard texture disks are commercial and were tested to 37 nm glide over the surface, while polished disks are specially ordered. Commercial disks with chemically strengthened alkali-aluminosilicate glass substrates are 65 mm in outer diameter, 20 mm in inner diameter and 0.635 mm in thickness. Thin-film construction consists of a 120 nm thick sputtered chromium underlayer, a 50 nm thick sputtered Co-based magnetic alloy film and a 20 nm thick spun-coated silicon dioxide (SiO_2) overcoat with fine ceramic particles to provide surface texture. The overcoat is lubricated with a 2 nm thick perfluoropolyether (Z-DOL) lubricant. Commercial Al_2O_3–TiC microsliders mounted on the disk drive's in-line suspension are used. Typical three-dimensional surface roughness maps of the various disks and the slider are shown in Fig. 6.

3 RESULTS AND DISCUSSION

For each disk several plots describe the various wear phenomena. A continuous time history of the coefficient of friction and acoustic emission are given. Coefficient of friction data are sampled at 1 kHz and averaged every 0.2 s for the standard texture lubed and unlubed Al–Mg disks, whereas the data signal is averaged once per minute for the polished Al–Mg and glass disks (due to longer test durations). The AE signal is recorded similarly, except that an r.m.s. of the data is taken instead of an average. Because of the larger averaging windows in the longer tests, the coefficient of friction and AE graphs for these disks have less detail than the graphs for the standard texture Al–Mg disks. Also included is a wear track progression over time from the optical sensor. In some cases, a close-up of a particular wear track is provided. The coefficient of friction, AE and optical sensor data are plotted as functions of sliding distance in kilometres. Additionally, a stylus profile (with a 5 μm radius diamond tip) and an optical micrograph of the final wear track give post-failure information.

Figure 7a shows extreme optical sensor output values for scans taken during a wear test on a standard texture, lubed Al–Mg disk which lasted until a deep scratch occurred. Included also are stylus profiles of four wear regions: lubricant depletion, overcoat burnishing, overcoat wear and total overcoat penetration. Optical micrograph examples of these four regions are included. Disk wear is characterized by a gradual increase in optical sensor output until the overcoat is fully penetrated. At this point the reflectance drops sharply as the disk sputtered layers fall apart quickly. This behaviour has also been reported by Jen and Gillis (**10**). The optical micrographs further illustrate the wear regions. A track cannot be discerned during the lubricant depletion phase. A barely visible track can be observed during burnishing, while overcoat wear produces a track clearly brighter than its surroundings. Finally, penetration to subsurface layers produces large dark streaks in the disk, thus decreasing the reflectance (**17**, **18**).

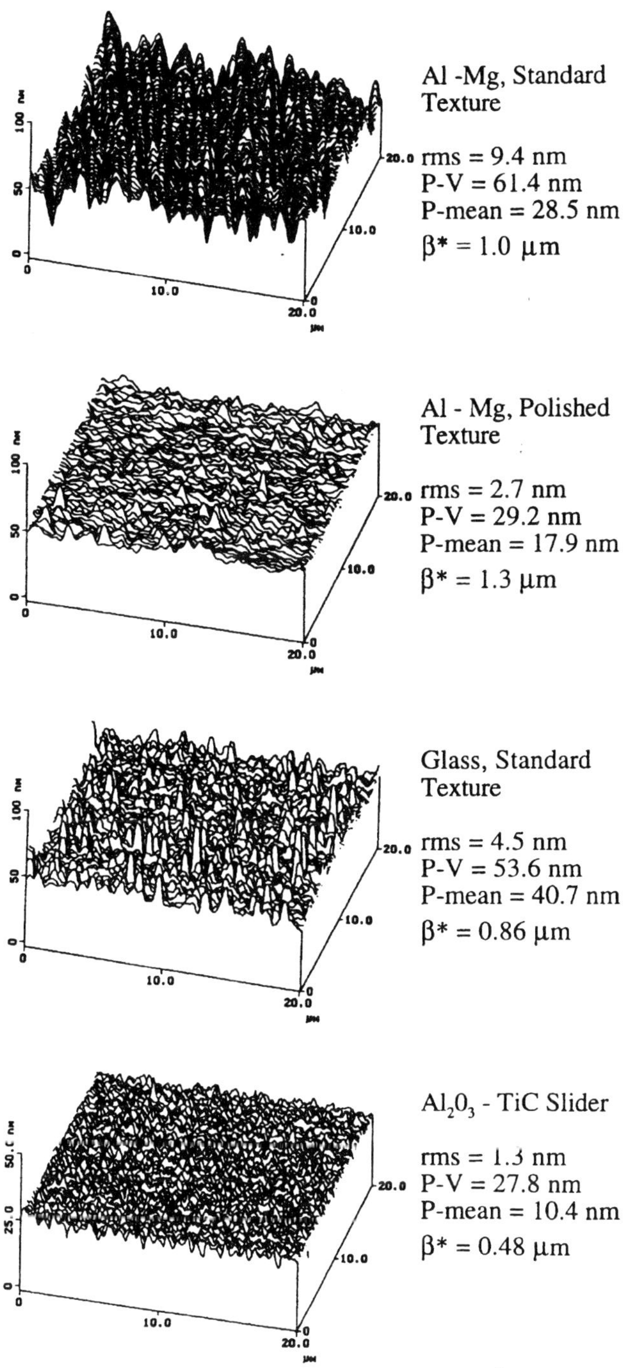

Fig. 6 Three-dimensional AFM surface maps of three thin-film disks and a head slider

One observation worth noting is the tendency of wear to occur under the head slider's hub rail before the rim rail. This is due mostly to head tilt and non-uniformity in the applied load, but also to the slightly lower tangential speed at the hub rail, which is closer to the disk centre. Optical sensor data from the tests show this tendency clearly. In addition, microscopic examination of the disks reveals a tendency of the disks to not wear uniformly under an individual rail.

Figure 7b, which contains optical micrographs of a head slider hub rail during a standard texture Al–Mg disk test, shows the wear progression on the head slider. The head slider shows no visual evidence of damage or debris transfer in the burnishing phase, but the overcoat wear phase produces a narrow dark band along the rail surface. The dark streaks are lubricant and wear debris which adhere to the rail during head slider–overcoat contact.

Figure 8 further shows results in detail for a standard texture, lubed Al–Mg disk. The coefficient of friction rises quickly as the lubricant is depleted, until a peak is reached. The coefficient of friction then decreases slightly and begins a slower rise until another peak is reached. The friction peaks occur at the same time as high-intensity AE signals are produced. The initial onset of high-intensity AE corresponds to head slider–overcoat contact under the hub rail, while the AE burst during the second friction peak corresponds to contact under the rim rail. AE activity in the regions between these two are the result of further carbon wear under the hub rail and generation and entrapment of wear particles at the interface. Several researchers (**1**) concluded that the rise in coefficient of friction during a sliding test on an unlubed carbon disk was due to the increased real area of contact associated with abrasive surface smoothing. This smoothing under the hub rail, along with the continuing lubricant depletion under the rim rail, accounts for the increase in the coefficient of friction above its initial peak. Optical sensor data provide convincing evidence to complement the coefficient of friction and AE data. The first significant measurement of a wear track occurs at about 380 m. Since the AE signal is still quiet at this point, the reflectance change is due to overcoat burnishing. Scans taken at 1900 and 3750 m show increasing reflectance, which is expected, because the high-intensity AE output during this time corresponds to debris generation and asperity contacts. The rim rail wear track can be seen beginning to form in the full scan graph, appearing as a small bump. A stylus profile of the final hub rail wear track reveals the track depth to be about 30 nm, which corresponds well to the +2 V optical sensor peak signal. An optical micrograph of the final wear track shows a bright streak clearly distinguishable from the unworn surface. Valleys in the optical sensor output are probably due to bits of dark debris present in the track. For this test, the coefficient of friction and AE data were sampled at 1 kHz, and an average (or r.m.s. in the case of AE) was taken every 0.2 s.

Figure 9 shows results for a standard texture, unlubed Al–Mg disk. The coefficient of friction again rises to a peak quickly and then slowly tapers off. AE growth is steady from the beginning, but does not reach high intensity until about 450 m. This corresponds well to the optical sensor close-up, which shows a small reflectance increase (burnishing) by 380 m and then a significant reflectance increase (heavy carbon wear) by 750 m. The

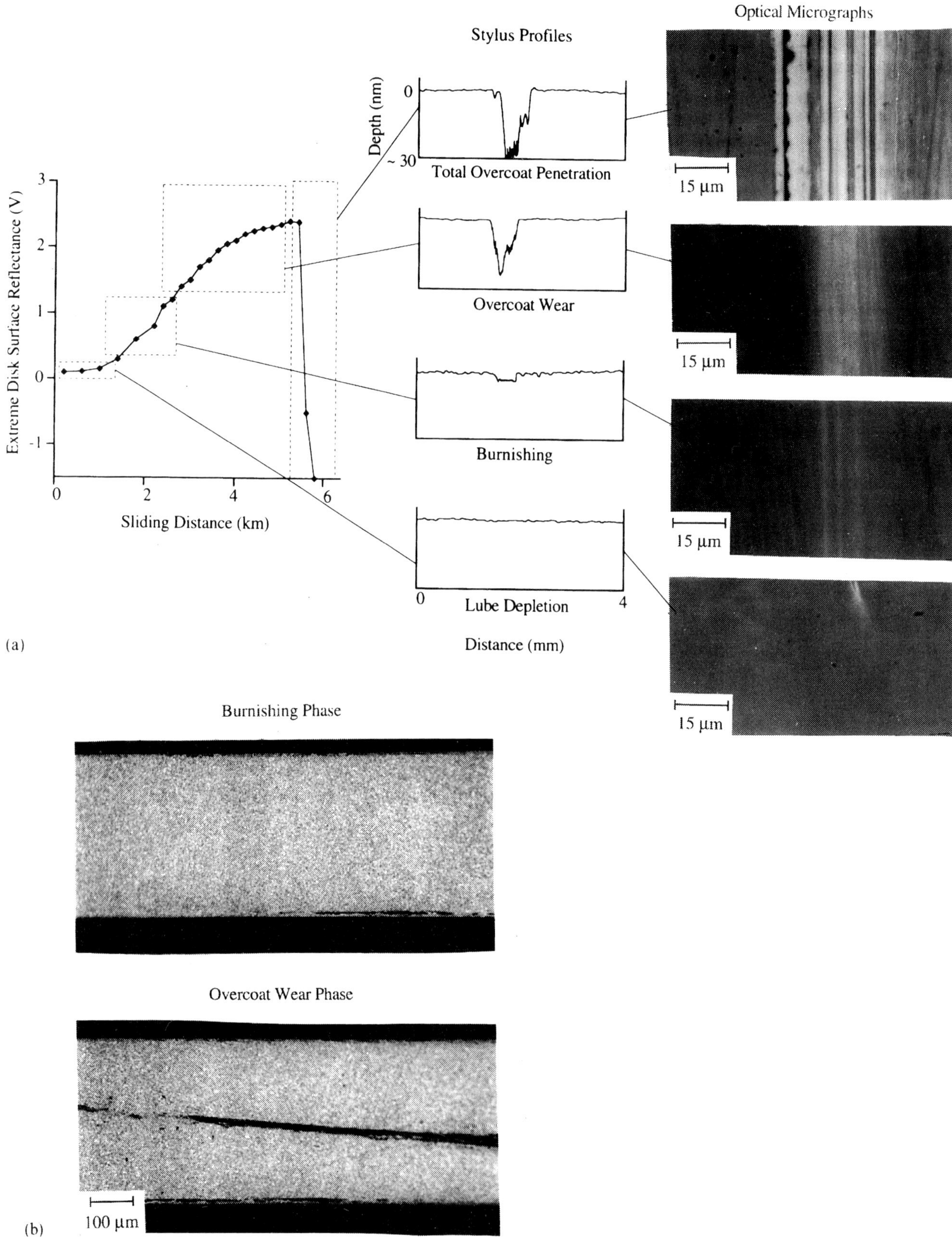

Fig. 7 (a) Extreme optical sensor voltage as a function of sliding distance, corresponding wear track profiles and optical micrographs of the wear tracks for a standard texture, lubed Al–Mg disk during a sliding test conducted at 0.63 m/s and a normal load of 7 g and (b) optical micrographs of a head slider hub rail during a sliding wear test on a standard texture, Al–Mg disk

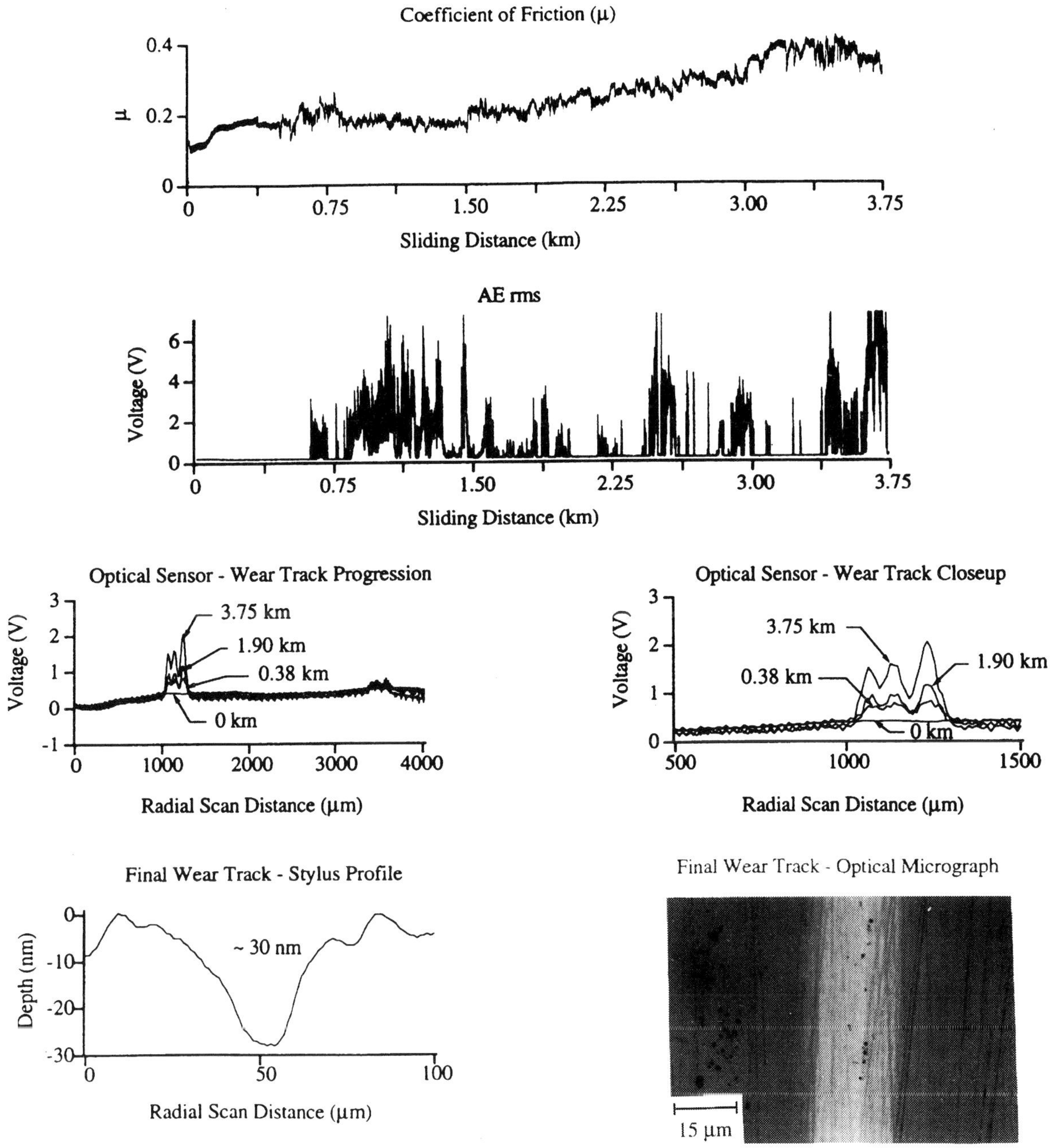

Fig. 8 Test results for a standard texture, lubed Al–Mg disk run at 0.63 m/s and 7 g normal load: coefficient of friction, AE r.m.s., optical sensor—full scan distance; optical sensor—wear track close-up; final wear track profile, optical micrograph—final wear track

AE signal abruptly stops at about 910 m as the last of the overcoat asperities is worn away, and the coefficient of friction significantly drops. Optical sensor output shows a significant increase in reflectance between 750 and 1130 m, but little between 1130 and 1500 m, further suggesting a slower rate of carbon wear after 910 m. The decrease in the coefficient of friction can be explained in the photomicrograph, which reveals the wear track to be striated, non-uniform and relatively wide. This roughness prevents a large real area of contact from developing. Stylus profile of the final wear track shows a track depth of about 30 nm, which corresponds well to the +2.5 V peak optical sensor output. For this test, the coefficient of friction and AE data were sampled at 1 kHz, and an average (or r.m.s. in the case of AE) was taken every 0.2 s.

Figure 10 shows results for a polished, lubed Al–Mg disk run at the same normal load of 7 g as the previous tests. Again, data graphs for this and the following disks are of less detail than the previous ones, because of the long test durations. Approximately 30 000 m of testing fails to produce a wear track, and the test is ended. As

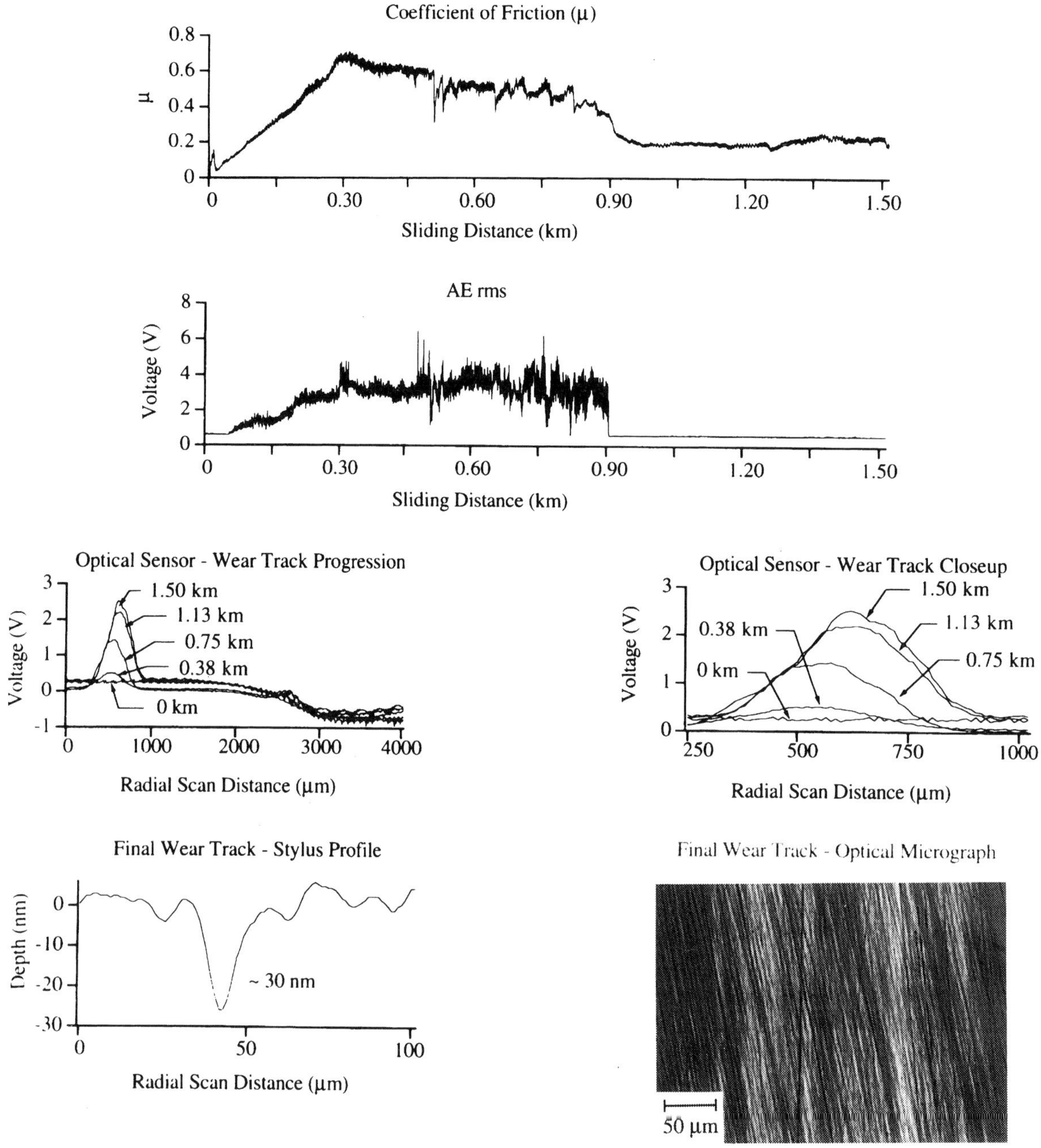

Fig. 9 Test results for a standard texture, unlubed Al–Mg disk run at 0.63 m/s and 7 g normal load: coefficient of friction, AE r.m.s., optical sensor—full scan distance; optical sensor—wear track close-up; final wear track profile, optical micrograph—final wear track

expected, the AE signal is quiet throughout the test. The coefficient of friction rises slowly as lubricant is depleted, but optical scans reveal that a burnishing wear phase is still not obtained. Tests were next conducted at a higher normal load of 21 g (Fig. 11). A wear track is created in this disk after about 25 000 m. The same rise-and-peak behaviour of the friction coefficient is observed in this disk as it was in the standard texture disks. The AE plot shows a pattern of high intensity during carbon wear that is similar to the standard texture disks. Optical sensor data show that no wear track can be observed after 18 900 m, which agrees with the AE plot at that point. A scan taken at 24 900 m, approximately 2300 m after the onset of carbon wear, shows a clear increase in reflectance. Further wear does not increase the reflectance much beyond that point. The final wear track is measured to be slightly less than 20 nm by the stylus profiler, which corresponds well to the peak optical sensor voltage of just under +2 V. The valley seen in the middle of the track in the optical scan is clearly seen as a dark stripe in the photomicrograph, and the stylus profile reveals it to be an unworn area. For this test, the

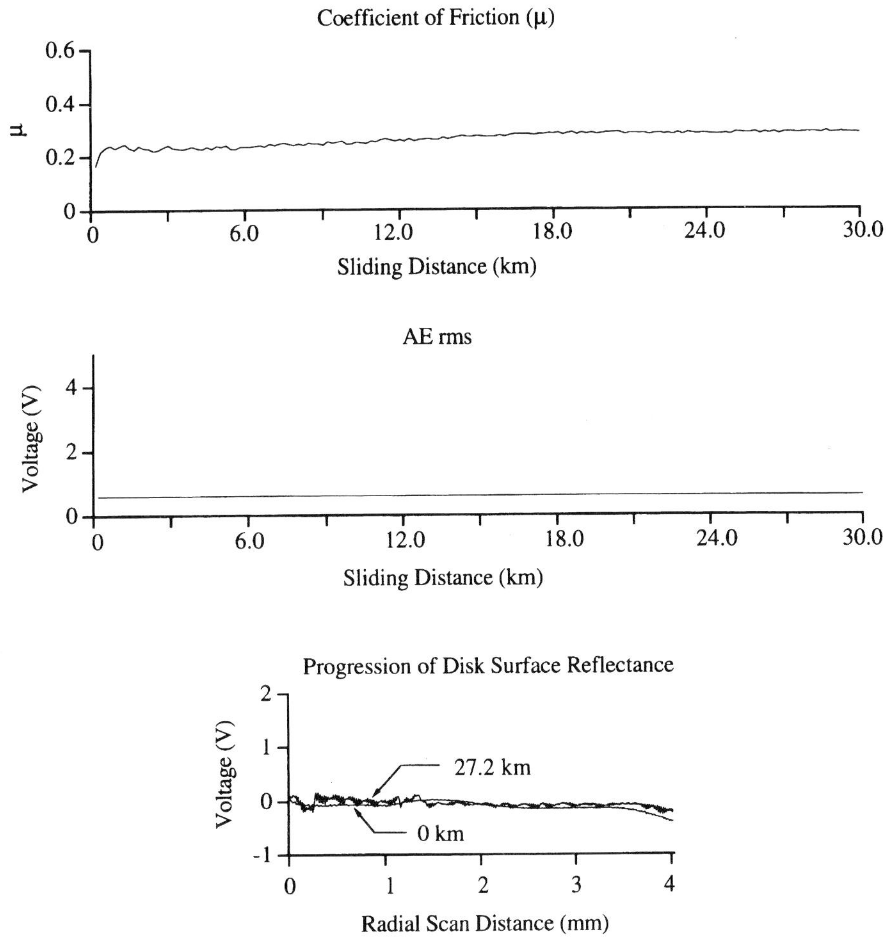

Fig. 10 Test results for a polished, lubed Al–Mg disk run at 0.63 m/s and 7 g normal load: coefficient of friction, AE r.m.s., optical sensor—full scan distance

coefficient of friction and AE data were sampled at 1 kHz, and an average (or r.m.s. in the case of AE) was taken once per minute.

No significant wear was observed for a standard texture, lubed glass disk at a normal load of 7 g, which agrees with the findings of Bhushan *et al.* (**17**) that these disks have a lower wear rate than standard texture, lubed Al–Mg disks. Figure 12 shows results for a standard texture, lubed glass disk at a normal load of 21 g. Wear progression closely follows that of the polished disk at 21 g, with overcoat wear occurring slightly later. This test continues until the overcoat is fully penetrated, because wear tracks are not discernible against the near-white background colour of the glass disk. This explains the absence of a reflectance increase well after the AE data suggest that overcoat wear has begun. The strong negative reflectance at 22 500 m is due to the penetration to layers underneath the SiO_2 overcoat. This penetration produces the strong dark streaks seen in the optical micrograph. The stylus profile shows the final track to be over 200 nm in depth. For this test, the coefficient of friction and AE data were sampled at 1 kHz, and an average (or r.m.s. in the case of AE) was taken once per minute.

4 SUMMARY

Several physical phenomena contribute to the wear process in thin-film disks. The first phase of wear is lubricant depletion. This phase is followed by a short period of overcoat burnishing. The phase that produces visible wear tracks is overcoat wear, which generally lasts the longest of all the wear phases. Total overcoat penetration leads to deep gouging of the disk. Lubricant depletion is seen as an initial increase in the coefficient of friction. Overcoat wear and the presence of wear particles at the interface are characterized by high-intensity acoustic emission. Wear tracks can be picked up clearly by an optical reflectance sensor. Sensors that measure the coefficient of friction, acoustic emission and surface reflectance during a sliding wear test are useful individually, but they are most effective when used simultaneously. A strain gauge beam measures the

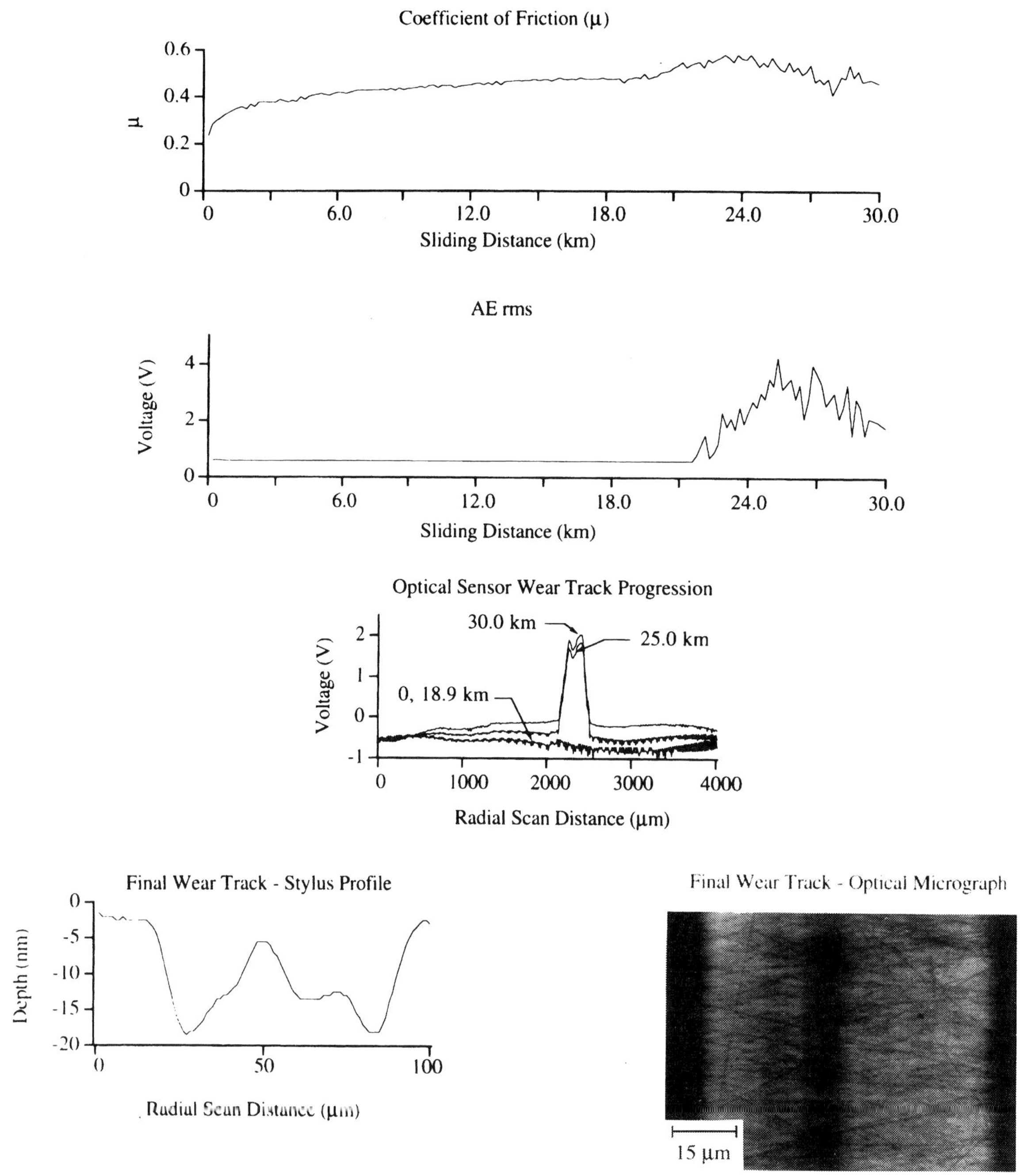

Fig. 11 Test results for a polished, lubed Al–Mg disk run at 0.63 m/s and 21 g normal load: coefficient of friction, AE r.m.s., optical sensor—full scan distance; final wear track profile, optical micrograph—final wear track

coefficient of friction, and thus gives some insight into lubricant depletion, but its relevance beyond the onset of carbon wear is questionable. An acoustic emission transducer monitors stress waves in the head–disk interface, which accurately reveals head slider–overcoat contact and the presence of interface particles, but it cannot give a direct indication of wear volume. An optical reflectance sensor monitors the progression of wear tracks (carbon overcoat thinning versus scratching) with good sensitivity, but it has no capability to reveal the causes of wear. When three such instruments are used in concert, however, on-line information about both wear volume and precursors to failure is gained. The effectiveness of sensor co-ordination thus established opens numerous possibilities for such a system in sliding wear tests of any material surface.

The particular disks chosen for the study allow the effects of lubricant, roughness, overcoat and normal load on disk wear to be observed. A polished texture disk wears much more slowly than a standard texture one.

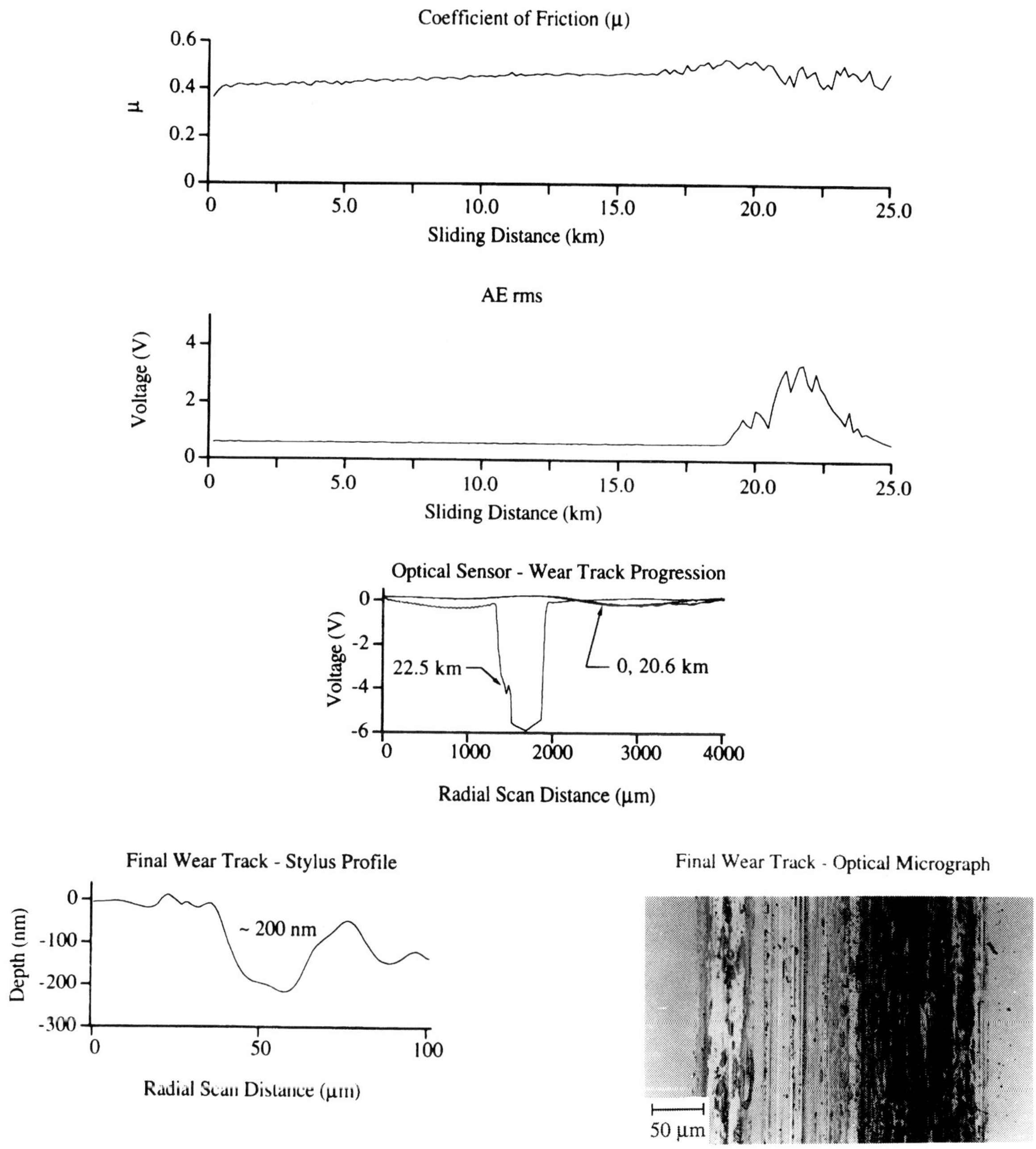

Fig. 12 Test results for a standard texture, lubed glass disk run at 0.52 m/s and 21 g normal load: coefficient of friction, AE r.m.s., optical sensor—full scan distance; final wear track profile, optical micrograph—final wear track

Even at 3 times the load, a polished Al–Mg disk takes about 8 times longer to wear than a standard Al–Mg disk. Lubricant plays a major role in disk wear. An unlubed Al–Mg disk experiences head slider–overcoat contact virtually from the outset and thus wears faster than a lubed disk. As for overcoat, diamond-like carbon wears faster than SiO_2, because a standard texture SiO_2 disk experiences roughly the same wear rate as a polished DLC disk. Finally, as expected, higher load contributes to faster wear. A polished Al–Mg disk under 7 g normal load has no discernible wear after the same amount of time that it takes the same disk to fail under 21 g normal load.

ACKNOWLEDGEMENTS

The authors would like to thank the following individuals for their contributions during this study: David Jen and Don Gillis of the IBM Storage Systems Products Division in San Jose, California, for their insight into the workings of the Fotonic sensor; Joe West at the Department of Mechanical Engineering electronics shop at The Ohio State University, for his patient work on nearly every piece of electronics in the experimental set-up; Curtis D. Kissinger of Mechanical Technology, Inc. in Latham, New York, for technical support. Al–Mg disks were supplied by Komag Corp. in Milpitas, California, glass disks were supplied by Hoya Electronics Corp. in San Jose, California, and microsliders were supplied by Applied Magnetics Corp. in Goleta, California. Financial support for this research was provided by the industrial membership of the Computer Microtribology and Contamination Laboratory.

REFERENCES

1 **Bhushan, B.** *Tribology and Mechanics of Magnetic Storage Devices,* 2nd edition, 1996 (Springer-Verlag, New York).

2 **Kita, T., Kogrue, K., Mitsuya, Y.** and **Nakanishi, T.** New method of detecting contact between floating-head and disk. *IEEE Trans. Magn.,* 1980, **MAG-16**, 873–875.

3 **Yeack-Scranton, C. E.** Novel piezoelectric transducer to monitor head–disk interactions. *IEEE Trans. Magn.,* 1986, **MAG-22**, 1011–1016.

4 **Jiaa, C. L.** and **Dornfeld, D. A.** Experimental studies of sliding friction and wear via acoustic emission signal analysis. *Wear,* 1990, **139**, 403–420.

5 **Klamecki, B. E.** and **Hanchi, J.** Wear process description based on acoustic emission. *Trans. ASME, J. Tribology,* 1990, **112**, 469–476.

6 **Briggs, J. C., Chang, M. K.** and **Tse, M.-K.** High frequency slider vibrations during asperity impacts on rigid magnetic disk systems. *Adv. Inf. Stor. Systems,* 1992, **4**, 181–194.

7 **Peck, P. R., Kono, R. N., Jhon, M. S.** and **Karis, T. E.** A novel accelerated wear test for magnetic recording disks. *IEEE Trans. Magn.,* 1993, **29**, 3969–3971.

8 **Matsuoka, K., Forrest, D.** and **Tse, M.-K.** On-line wear monitoring using acoustic emission. *Wear,* 1993, **162–164**, 605–610.

9 **Nunnelley, L., Burleson, M.** and **Fuller, G.** On-line tribology measurements on lubricated rigid disks. *IEEE Trans. Magn.,* 1990, **MAG-26**, 2079–2081.

10 **Jen, D.** and **Gillis, D.** Optical profiler for thin film disk wear measurement. *Adv. Inf. Stor. Systems,* 1992, **4**, 219–230.

11 **Karis, T. E.** and **Lyn, D.** A novel high-speed optical inspection tool for thin-film disks. *Adv. Inf. Stor. Systems,* 1993, **5**, 363–373.

12 **Meeks, S. W., Weresin, W. E.** and **Rosen, H. J.** Optical surface analysis of the head–disk interface of thin film disks. *Trans. ASME, J. Tribology,* 1995, **117**, 112–118.

13 **Doebelin, E.** *Measurement Systems Application and Design,* 4th edition, 1990 (McGraw-Hill, New York).

14 **Menadier, C., Kissinger, C.** and **Adkins, H.** The Fotonic sensor. *Instruments and Control Systems,* 1967, **40**, 114–120.

15 **Kissinger, C. D.** and **Smith, R. L.** Improved noncontact fiber optics/lens displacement measuring system. Proceedings of Conference on *Technical Program Electro-Optical Systems Design,* 1973, pp. 372–377.

16 **Lagace, L. J.** and **Kissinger, C. D.** Non-contact displacement and vibration measurement systems employing fiber optic and capacitance transducers. Presented at the Twenty-third International Instrumentation Symposium, Las Vegas, Nevada, 1–5 May 1977.

17 **Bhushan, B., Yang, L., Gao, C., Suri, S., Miller, R. A.** and **Marchon, B.** Friction and wear studies of magnetic thin-film rigid disks with glass–ceramic, glass and aluminum–magnesium substrates. *Wear,* 1995, **190**, 44–59.

18 **Bhushan, B.** and **Li, Y.** Wear studies of contact recording interface with a microfabricated head. *J. Appl. Phys.,* 1996, **79**, 5791–5793.

Degradation of perfluoropolyether lubricants on magnetic recording disks during sliding

Y Li and **B Bhushan**
Computer Microtribology and Contamination Laboratory, Department of Mechanical Engineering, The Ohio State University, Columbus, Ohio, USA

Abstract: Degradation of perfluoropolyether lubricants (Fomblin Z-Dol and Fomblin AM 2001) on the magnetic recording thin-films disks are studied during sliding with an Al_2O_3–TiC slider under high vacuum. Gaseous products generated from the head/disk interfaces are detected and monitored as a function of sliding distance using a quadrupole mass spectrometer. Effects of overcoats and surface roughnesses of disks on degradation of lubricants and coefficient of friction were also studied. Experimental results show that the gaseous products generated from Z-Dol and AM 2001 lubricants during sliding are quite similar. Degradation of both lubricants is found to take place from the beginning of sliding, and the evolution rate of decomposition fragments decreases with the sliding distance. Removal of the lubricants is accompanied by a rapid rise in the coefficient of friction of the interfaces, followed by gaseous products from the tribochemical reaction of the overcoat material. Surface roughness of the disk has an effect on the sliding distance to failure, but little effect on the decomposition of lubricant.

Keywords: magnetic disk, slider, magnetic storage devices, perfluoropolyether lubricants, lubricant degradation

1 INTRODUCTION

To reduce friction and wear, magnetic recording thin-film rigid disks have a hard overcoat lubricated with a thin layer of a liquid lubricant. The thickness of this lubricant layer must be chosen carefully to provide maximum wear protection of slider and disk and to guarantee acceptable life performance. If the lubricant film is too thick, excessive stiction and mechanical failure of the head/disk interface is observed during start-up. On the other hand, if the lubricant film is too thin, insufficient protection of the head/disk interface is provided, and tribological failure of the head/disk interface can occur early on. In general, the thickness of liquid lubricant on thin-film disks ranges from 1 to 3 nm (**1**).

The lubricants, in general, used on today's magnetic recording disks are perfluoropolyethers. These lubricants are chemically inert and non-reactive. To improve their adhesion and lubrication performance, bifunctionally terminated perfluorocarbons are commonly used. Fomblin Z-Dol (80 cSt) with an —OCF_2—CH_2—OH end group and Fomblin AM 2001 (80 cSt) with a piperonyl end group have been commonly used as functional (or polar) lubricants with molecular weights ranging from 2000 to 3000 Da (daltons) on the basis of number distribution. Polar lubricants bond to the surface on which they are applied without any post-treatment. The ability to bond a polar lubricant depends on the cleanliness of the overcoat surface and chemical reactivity of the overcoat. For increased bonding and reproducible results, the lubricants are generally partially or fully bonded by heating the lubricated disk surface, for example, at 150 °C for 1 hour for partial bonding. The polar lubricants are generally found to have better friction and wear performance.

During disk drive operation, the head slider slides on the disk surface during start-up and stop operations. Lubricant film on the disk surface may be depleted by mechanical contact of the slider with the disk, which has been shown to affect their lubricating performance. Ohno *et al.* (**2**) measured lubricant thickness using polarized micro-FTIR (Fourier transform infrared spectroscopy) and found localized loss of disk lubricant in the path of the slider rails after continued sliding at 0.6 m/s. This change in lubricant distribution was accompanied by a rise in both friction coefficient and photoelectron emission intensity at the sliding interface. This type of localized lubricant loss has also been detected during oscillating wear experiments using ellipsometry (**3**). More recent investigations using mass spectrometry have shown that while some of the displaced lubricant was transferred to the slider or to the surrounding lubricant film, a

The MS was received on 4 June 1996 and was accepted for publication on 20 December 1996.

portion of the lubricant was removed from the head/disk interface entirely (**4–7**). Novotny *et al.* (**6**) compared the mass spectra obtained during sliding experiments with those obtained in thermal desorption and found that the spectrum during sliding at 0.5 m/s resembled the thermal spectrum at 300 °C rather than that at 90 °C. They speculated that the loss mechanism of lubricants could be of two types: either transient interfacial temperatures were high enough to evaporate lubricant from the surface or polymer scission produced fragments small enough to be gaseous. Strom *et al.* (**4**) ran similar experiments at a lower sliding speed and found rapid increases in partial pressure of degradation products of lubricants. Based on the work by Bair *et al.* (**8**) and Bhushan (**9**), local increases in temperature at contact spots is of the order of 10 °C for test conditions. Thus Strom *et al.* suggested that degradation of lubricants was probably triboelectrically activated. Based on the mass spectrum of wear fragments and thermal and electron decomposition products of perfluoropolyethers (PFPEs), Vurens *et al.* (**10**) concluded that decomposition of PFPE lubricants in sliding is an electron mediated process, since most of the product fragment distribution is very similar to the electron decomposition fragment distribution.

In this study, degradation of lubricants on the magnetic recording disks during sliding, as well as its effect on the frictional and wear behaviour of the head/disk interface, are studied. Effects of two types of lubricants, two disk roughnesses and two disk overcoats on the lubricant degradation process and friction and wear are studied in detail.

2 EXPERIMENTAL

2.1 Test apparatus

The sliding experiments were carried out in an ultra-high vacuum (UHV) environmental tribotester, which has been described elsewhere (**11–13**). It features a mass spectrometer with its ionization chamber located in the line of sight of the slider/disk interface so that emission from the interface can be monitored *in situ* during the sliding under high vacuum. Figure 1 shows the overall schematic of the UHV pin-on-disk apparatus, which can be used in continuous sliding or reciprocating sliding mode. The vacuum system includes a mechanical rotary pump and a turbomolecular pump, capable of reaching to 1×10^{-8} torr with mild baking (150 °C). The system pressure is monitored by a vacuum ionization gauge (843 Ratiomatic ionization gauge, manufactured by Varian), which consists of a thermocouple gauge for the pressure range of 1 atm (760 torr)–1×10^{-3} torr and an ionization gauge for the pressure range of 1×10^{-3}–1×10^{-8} torr. A Dycor quadrupole gas analyser or mass spectrometer (M200M) with a Faraday cup detector, electron multiplier and video display is connected to a port on the vacuum chamber. It has a mass range of 1–200 atomic mass units (amu). The operating pressure range is 10^{-4}–10^{-11} torr with a minimum detectable partial pressure of 5×10^{-14} torr.

The disk was rotated by a d.c. motor which is controlled with a closed-loop controller through a high-speed

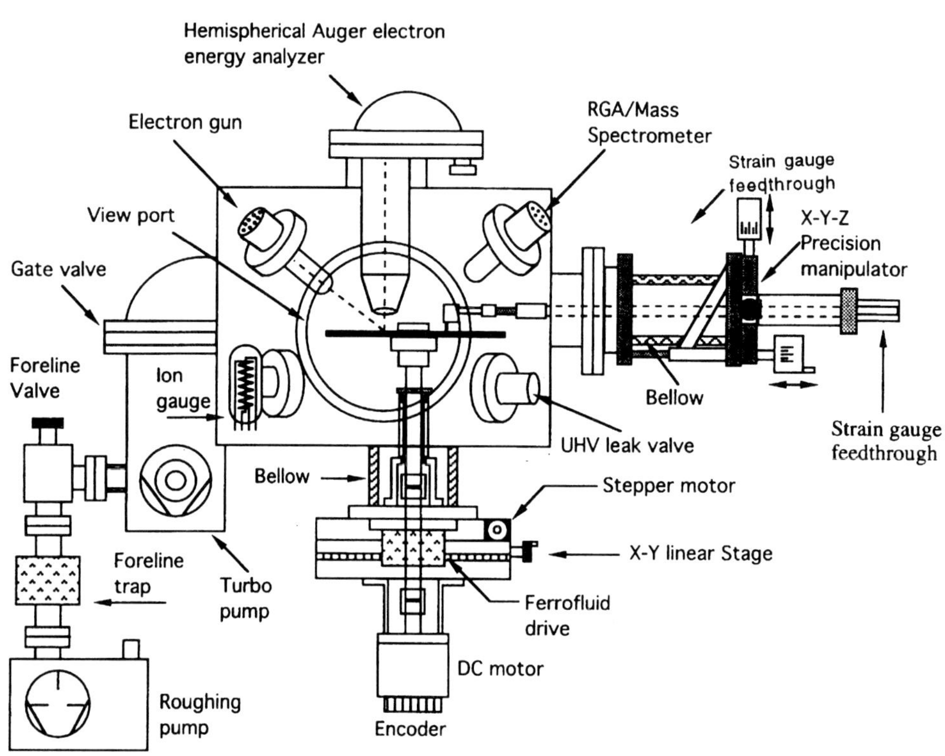

Fig. 1 Overall schematic of the UHV environmentally controlled tribotest apparatus

ultra-high vacuum compatible ferrofluidic rotational feedthrough. The motor rotational speed can be varied from 0 to 3600 r/min. The slider was mounted on a head suspension assembly, which was mounted on a strain gauge cantilever beam. The desired normal load was applied to the head by adjusting the vertical position of the cantilever beam relative to the test disk surface using an $X-Y-Z$ precision position manipulator. Both friction and normal forces at the head/disk interface were measured using resistive strain gauge transducers at a sampling frequency of 1 Hz. A lower sampling frequency was used to match with the data acquisition rate of the mass spectrometer. The sensitivity for both normal and friction forces is about 1 mN.

Before the slider and disk were mounted, the vacuum chamber was baked at 150 °C for at least 15 hours. The system was then allowed to cool down to room temperature to achieve a pressure of about 5×10^{-8} torr. All of the filaments and spindle feedthroughs were completely outgassed during the bake-out to make sure that no contaminants were detected by the mass spectrometer. After cooling to room temperature, the vacuum was filled with Ar gas for installation of the magnetic head (slider) and disk. Then the system was again evacuated to a base pressure of 5×10^{-7} torr. Since the degradation of lubricants on the disk surfaces is sensitive to the environment, the vacuum chamber was kept at room temperature once the slider and disk were installed. Mass spectra were obtained using an electron energy of 70 eV. Besides taking the scans in the range of 1–150 amu during the experiments, the time dependence of certain characteristic mass peaks was also monitored during sliding tests. They are H_2 (2), $CO + N_2$ (28), CO_2 (44), CFO (47) and CF_3 (69). For all the sliding tests, the applied load was 7 *g* and the rotating speed of the disk was 200 r/min, which corresponds to a linear speed of about 0.5 m/s at a disk diameter of 50 mm. Tests were conducted until all of the lubricant was removed. This was found to be of the order of 500 s (or 250 m of sliding distance).

2.2 Test materials

The thin-film disks used in this work include Al–Mg substrate disks with a hard carbon overcoat, with and without partially bonded perfluoropolyether lubricant, and commercially available glass substrate disks with an SiO_2 overcoat, with and without partially bonded perfluoropolyether lubricant. The commercial Al–Mg disks were made of an Al–Mg substrate, with a layer of about 10 μm thick Ni–P coating. The hard hydrogenated diamond-like carbon coating was about 20 nm thick. These disks were 95 mm in outer diameter and 25 mm in inner diameter, with a thickness of 0.800 mm. Commercial disks had a standard texture and some of the disks had a polished texture. Some of the disks were lubricated with a bonded Fomblin Z-Dol perfluoropolyether lubricant. The chemical structure of Fomblin Z-Dol is as follows:

$$HO{-}CH_2{-}CF_2{-}O{-}(CF_2{-}CF_2{-}O)_m{-}(CF_2{-}O)_n{-}CF_2{-}CH_2{-}OH$$

with a molecular weight of 2000 Da and kinematic viscosity of 80 cSt (mm^2/s).

The commercially textured glass disks were 65 mm in outer diameter and 20 mm in inner diameter, with a thickness of 0.635 mm. Some of the glass disks were bonded with Fomblin AM 2001 perfluoropolyether lubricant. The chemical structure of AM 2001 is as follows:

$$\text{Piperonyl}{-}O{-}CH_2{-}CF_2{-}O{-}(CF_2{-}CF_2{-}O)_m{-}(CF_2{-}O)_n{-}CF_2{-}CH_2{-}O{-}\text{piperonyl}$$

with a molecular weight of 2300 Da and kinematic viscosity of 80 cSt (mm^2/s).

The sliders used were microsliders fabricated from Al_2O_3–TiC (overall dimensions: 2.85 mm × 2.24 mm, rail length = 2.59 mm and rail width = 0.33 mm). Surface roughness profiles of these disks and of the Al_2O_3–TiC slider were measured using an atomic force microscope (AFM) with a scan size of 20 μm × 20 μm, and are shown in Fig. 2.

3 RESULTS AND DISCUSSION

The experimental results obtained using Al–Mg disks with and without Z-Dol lubricant will be described and discussed first, followed by the results obtained using glass disks with and without AM 2001 lubricant.

During sliding of the slider on the lubricated, textured Al–Mg disk, decomposition of the lubricant took place. The produced fluorinated compounds entered the gas phase and were detected using the quadrupole mass spectroscope. The mass spectrum of gaseous species generated at the sliding interface after 45 m of sliding, in the mass range 40–100 atomic mass units (amu), is shown in Fig. 3b. Figure 3a shows the mass spectrum in the same amu range obtained before sliding began. In trace (a), no peaks were detected above the base level in the amu range 40–100, except the peaks at 55–58 amu, which may be attributed to hydrocarbons emanating from the spindle bearings, since these hydrocarbons were produced when the spindle was rotated without a slider loaded on the disk, consistent with other researchers' results (**4**). In trace (b), taken over 1.5 min after 45 m of sliding, the peaks at 44, 47, 51, 66, 69 and 70 amu may be attributed to the fragments CO_2, CFO, HCF_2, CF_2O, CF_3 and HCF_3 respectively. No peaks were observed in the mass spectrum beyond 100 amu. In the mass range 1–40 amu, shown in Fig. 4, there were peaks at 2, 16, 18, 28, 40 amu, which are attributed to the compounds H_2, O, H_2O, $CO + N_2$ and Ar respectively.

Since the reaction products from the fluorinated

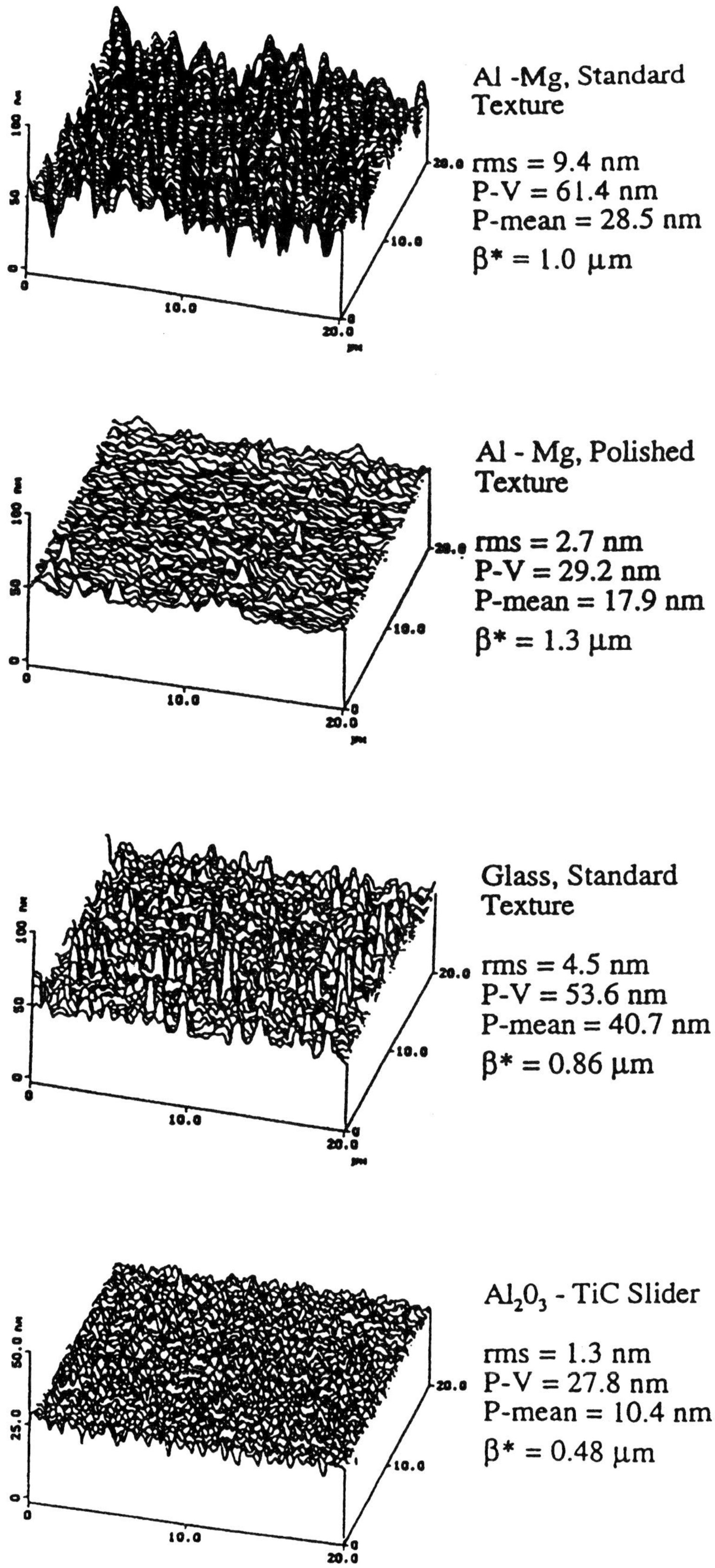

Fig. 2 Surface roughness profiles measured using an atomic force microscope with a scan size of 20 μm × 20 μm for (a) textured Al–Mg disk, (b) polished Al–Mg disk, (c) textured glass disk and (d) Al_2O_3–TiC microslider

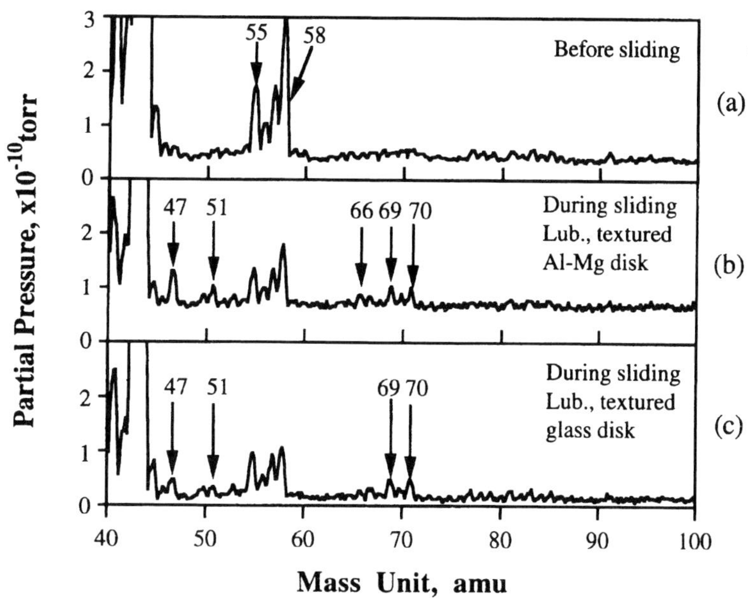

Fig. 3 Mass spectra ranging from 40 to 100 amu. Trace (a) is background, obtained before sliding begins. Trace (b) was obtained while sliding on a textured Al–Mg disk with Fomblin Z-Dol lubricant. Trace (c) was obtained while sliding on a textured glass disk with Fomblin AM 2001 lubricant. Traces were taken after 45 m of sliding

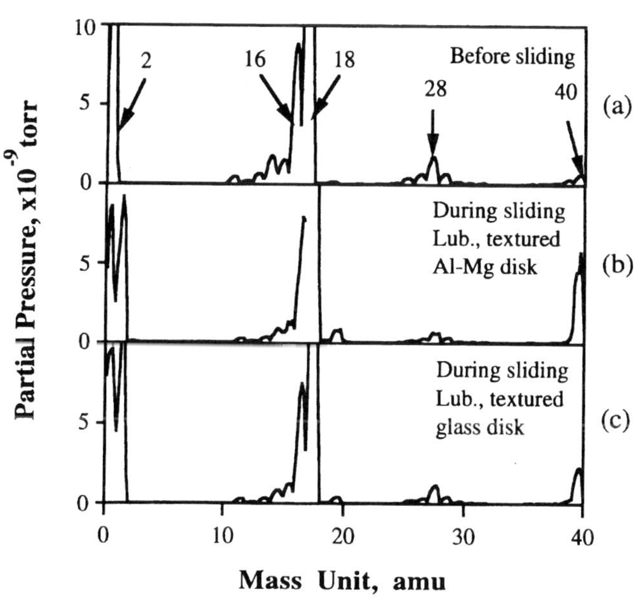

Fig. 4 Mass spectra ranging from 1 to 40 amu. Trace (a) is background, obtained before sliding begins. Trace (b) was obtained while sliding on a textured Al–Mg disk with Fomblin Z-Dol lubricant. Trace (c) was obtained while sliding on a textured glass disk with Fomblin AM 2001 lubricant. Traces were taken after 45 m of sliding

lubricant during sliding contain fluorine atoms, the partial pressure of CFO (47 amu) and CF_3 (69 amu) fragments were monitored as a function of sliding time. The partial pressure of H_2 (2 amu), CO + N_2 (28 amu) and CO_2 (44 amu) were also monitored during sliding. Figure 5 shows the time history of the partial pressure of these gaseous species generated at the sliding interfaces between the slider and unlubricated, textured Al–Mg disk and the slider and lubricated, textured Al–Mg disk. The coefficients of friction (COF) measured at a frequency of 1 Hz for both interfaces are also shown in Fig. 5.

As shown in Fig. 5a, for the sliding interface between the slider and unlubricated, textured Al–Mg disk, the coefficient of friction starts at a high value of about 0.75 at the beginning of sliding and then increases slightly as the sliding continues. Evolution of H_2, CO + N_2 and CO_2 is observed from the beginning of sliding and continues to the end of sliding. No significant changes in the partial pressure of H_2, CO + N_2 and CO_2 were detected during the entire duration. The evolution of H_2, CO + N_2 and CO_2 from the interface is likely due to the desorption of organic adsorbates and water from the disk surface and the removal of surface groups from the hydrogenated carbon overcoat initially (**6**), followed by reaction of water molecules with the overcoat and/or evolution of hydrogen within the overcoat. During sliding, the organic adsorbates and water on the overcoat surface are desorbed due to the sliding. The disk surface is cleaned when the slider passes by and fresh surface is exposed to the environment. Water molecules, which are the main compounds in the chamber at high vacuum, re-adsorbs polarly on the fresh overcoat surface. Subsequent sliding may break the bonds between the hydrogen and oxygen atoms, causing evolution of hydrogen. The oxygen atoms react with the carbon overcoat, producing CO and CO_2 gases. Another source of hydrogen may be the

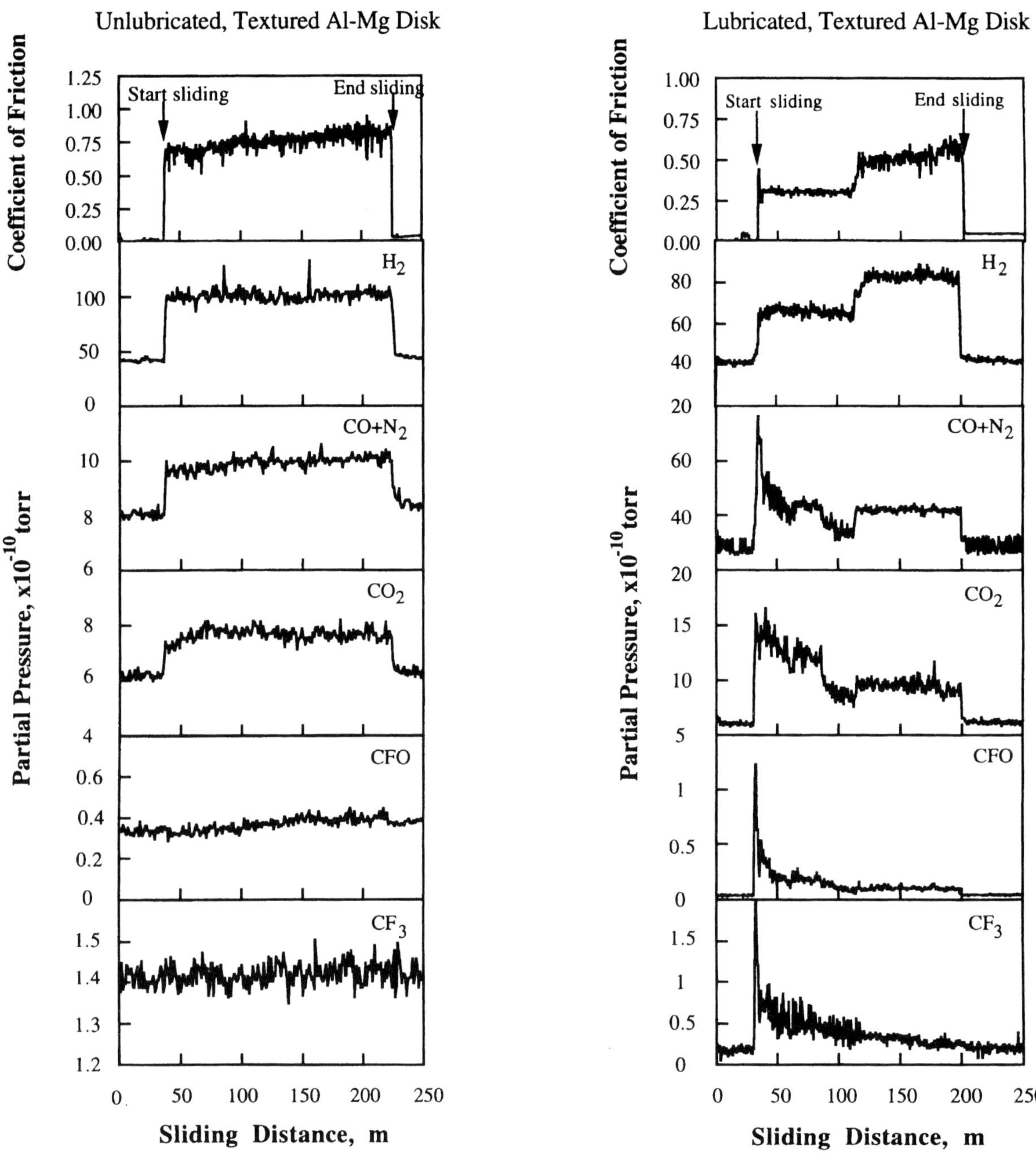

Fig. 5 Mass spectra of H_2, $CO + N_2$, CO_2, CFO, CF_3 and the coefficient of friction as a function of sliding distance on the textured Al–Mg disks (a) without and (b) with Fomblin Z-Dol lubricant

hydrogenated carbon overcoat, which may be activated by the work done during sliding. No evolution of CFO and CF_3 was observed during sliding as this disk was unlubricated.

During sliding on the lubricated, textured disk, lubricant fragments and desorbed adsorbates are detected, as shown in Fig. 5b. A transient in the coefficient of friction is observed at the beginning of sliding, which may be due to the stiction of the interface. As sliding continues, the coefficient of friction remains at a steady value of about 0.26. The partial pressures of $CO + N_2$, CO_2 CF_3 and CFO go through their maxima at the beginning of sliding, which can be as much as two orders of magnitude above the previous background, and then decrease. For the H_2 signal, it increases to about 7×10^{-9} torr from a previous background of about 4×10^{-9} torr at the beginning and remains constant with small fluctuations around the mean value as the sliding continues. The initial evolution of gases is likely due to the desorption of organic adsorbates and water from the disk surface. As the sliding continues, decomposition of the lubricant begins due to the mechanical action of sliding.

Decomposition of the lubricant may be caused by thermal desorption, mechanical or triboelectric process. Novotny *et al.* (**6**) found that the mass spectrum from the sliding experiment at a sliding speed of 0.5 m/s resembles the mass spectrum obtained in thermal desorption at 300 °C, rather than that obtained at 90 °C. The transient (flash) temperature at the head/disk interface is very difficult to measure and varies with the load and sliding speed (**8**). Bhushan (**9**) calculated the average and maximum transient temperature at the head/disk interface at a load of 10 *g* and sliding speed of about 20 m/s (disk rotating speed of 4320 r/min) to be about 56 and 81 °C respectively, and found that the temperature rise is directly proportional to normal pressure and the square root of the sliding speed. Compared with the conditions used in his calculation, the load (7*g*) and sliding speed (0.5 m/s) used in this study were smaller. The average and maximum transient temperature at the head/disk interface in this study are thus expected to be lower than those calculated earlier. This rise in temperature is thus expected to be not high enough to generate the rapid increase in partial pressures of gases observed. Vurens *et al.* (**10**) found that most of the product fragment distribution of Krytox lubricant from the sliding test is very similar to the electron decomposition fragment distribution, and suggested that the decomposition of the perfluoropolyethers is an electron mediated process. Electric potential differences of up to 0.1 V have been reported to develop between slider tracks and reference areas (**14**). Shear rates of the lubricant films in these experiments are quite high (**1**) and the energy dissipated during sliding may be sufficient to activate the lubricant to decompose. It is more likely that the decomposition reaction of the lubricant at the head/disk interface is either mechanically or triboelectrically activated, rather than thermally activated.

After a sliding of 100 m, the H_2 signal increases suddenly to a higher value and so do the $CO + N_2$ and CO_2 signals. These are accompanied by a sudden increase in the coefficient of friction to about 0.5. At the same time, CFO and CF_3 signals decrease to lower levels, which are slightly higher than the background levels and become less noisy. After some delay, appreciable wear track is observed, as shown in Fig. 6b. As can be seen, there are some regions where the lubricant has been removed and/or desorbed in the wear track, which suggests that at these regions the slider surface comes into contact with the hydrogenated carbon overcoat directly. The sudden increase in the coefficient of friction may be due to the removal of the lubricant at contact regions. As the sliding continues, the lubricant layer on the overcoat surface becomes thinner due to the sliding. At some contact regions, the lubricant has been removed and/or desorbed completely, and the slider surface contacts with the overcoat surface directly. The coefficient of friction rises quickly, and CFO and CF_3 signals, which are lubricant fragments, decrease to lower levels and become

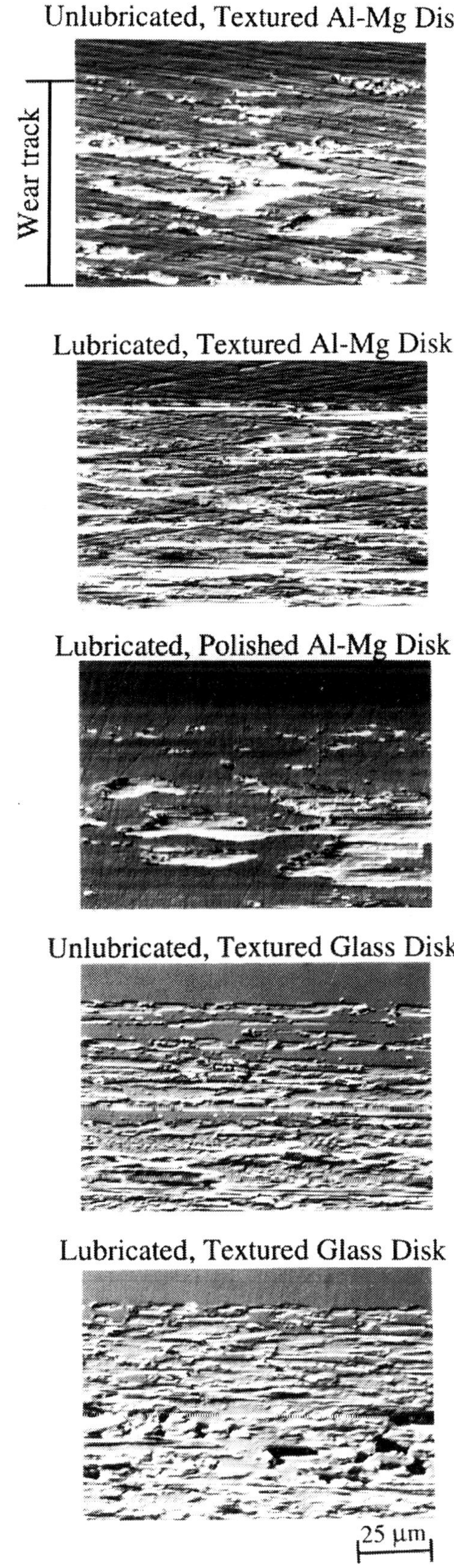

Fig. 6 Optical micrographs of wear tracks on (a) unlubricated, textured Al–Mg disk, (b) lubricated, textured Al–Mg disk, (c) lubricated, polished Al–Mg disk, (d) unlubricated, textured glass disk and (e) lubricated, textured glass disk after failure

less noisy. The increase in H_2, $CO + N_2$ and CO_2 may be attributed to evolution of hydrogen within the overcoat and/or the reaction of water vapour with the overcoat. After the slider slides directly on the overcoat, the work done during sliding 'cleans' the overcoat surface and may activate the evolution of hydrogen within the hydrogenated carbon overcoat. The fresh overcoat surface exposures to the environment and the water molecules in the vacuum chamber will adsorb on the fresh overcoat surface. Subsequent sliding may break the bonds between the hydrogen and oxygen atoms, and cause the evolution of hydrogen. The oxygen atoms react with the overcoat surface, producing CO and/or CO_2, causing the increase in $CO + N_2$ and CO_2 signals.

Figure 7 shows the time dependence of the friction coefficient and the partial pressure of gases generated at the interface between the slider and the lubricated, polished Al–Mg disk. They follow a pattern similar to the results with the lubricated, textured Al–Mg disk, except that the sliding time to failure is shorter than that of the polished disk. The pattern of $CO + N_2$ and CO_2 signals from the polished disk is slightly different from that from the textured disk. The signals of $CO + N_2$ and CO_2 do not increase after the coefficient of friction rises rapidly. The reason is unclear. One possibility is that, for the polished disk, the contact area with the slider surface may be larger than that for textured disk under the same load, and thus the contact pressure is smaller, which may affect the adsorption of water molecules on the overcoat surface.

The mass spectrum of chamber gases obtained while the slider is sliding on the textured glass disk with AM 2001 lubricant, as shown in Fig. 3c, is similar to that obtained while sliding on the Fomblin Z-Dol lubricated Al–Mg disk, except the peak at 66 is not appreciable. The peaks at 44, 47, 51, 55–58, 69 and 70 were all detected. There are no peaks observed in the mass spectrum beyond 100 amu. In the mass range 1–40 amu, shown in Fig. 4c, there are peaks at 2, 16, 18, 28 and 40 amu, which are attributed to the compounds H_2, O, H_2O, $CO + N_2$ and Ar respectively.

Friction and the partial pressure of compounds at mass numbers 2, 28, 44, 47 and 69 were monitored during sliding on the glass disks with and without AM 2001 lubricant, and the results are shown in Fig. 8. For the unlubricated glass disk (Fig. 8a), the coefficient of friction starts at a high value of 0.5 and then increases to about 1.0 at the end of sliding. The increase in the coefficient of friction may be due to the three-body abrasive wear, caused by the debris generated at the interface during sliding. Evolution of H_2 was observed at the beginning of sliding and then decreases slightly. No significant change in the $CO + N_2$, CO_2, CF_3 and CFO signals was observed before, during and after sliding. This may be because there is no lubricant on the overcoat surface and the overcoat is made of SiO_2.

For the AM 2001 lubricated glass disk (Fig. 8b),

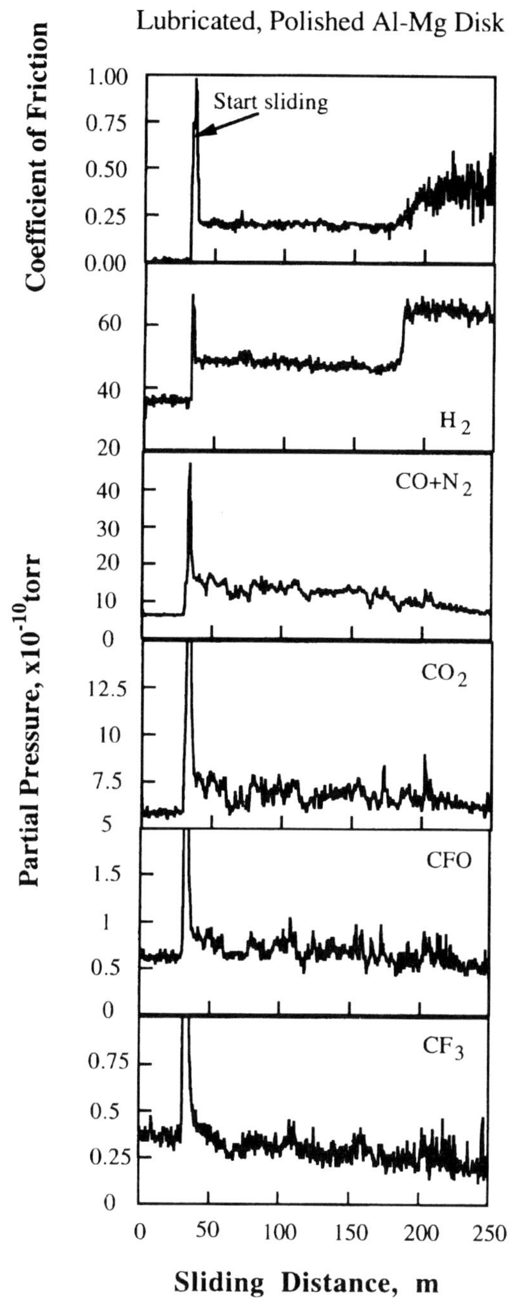

Fig. 7 Mass spectra of H_2, $CO + N_2$, CO_2, CFO, CF_3 and the coefficient of friction as a function of sliding distance on the textured Al–Mg disk with Fomblin Z-Dol lubricant

stiction was again observed at the beginning of sliding and was followed by a low coefficient of friction of about 0.15. Partial pressures of all gases monitored reach a maximum at the beginning of sliding and then decrease non-linearly. The initial evolution of gases is likely due to the desorption of organic adsorbates and water from the disk surface, similar to that of sliding on the lubricated Al–Mg disk, and is then followed by decomposition of the lubricant. The decrease in the partial pressure

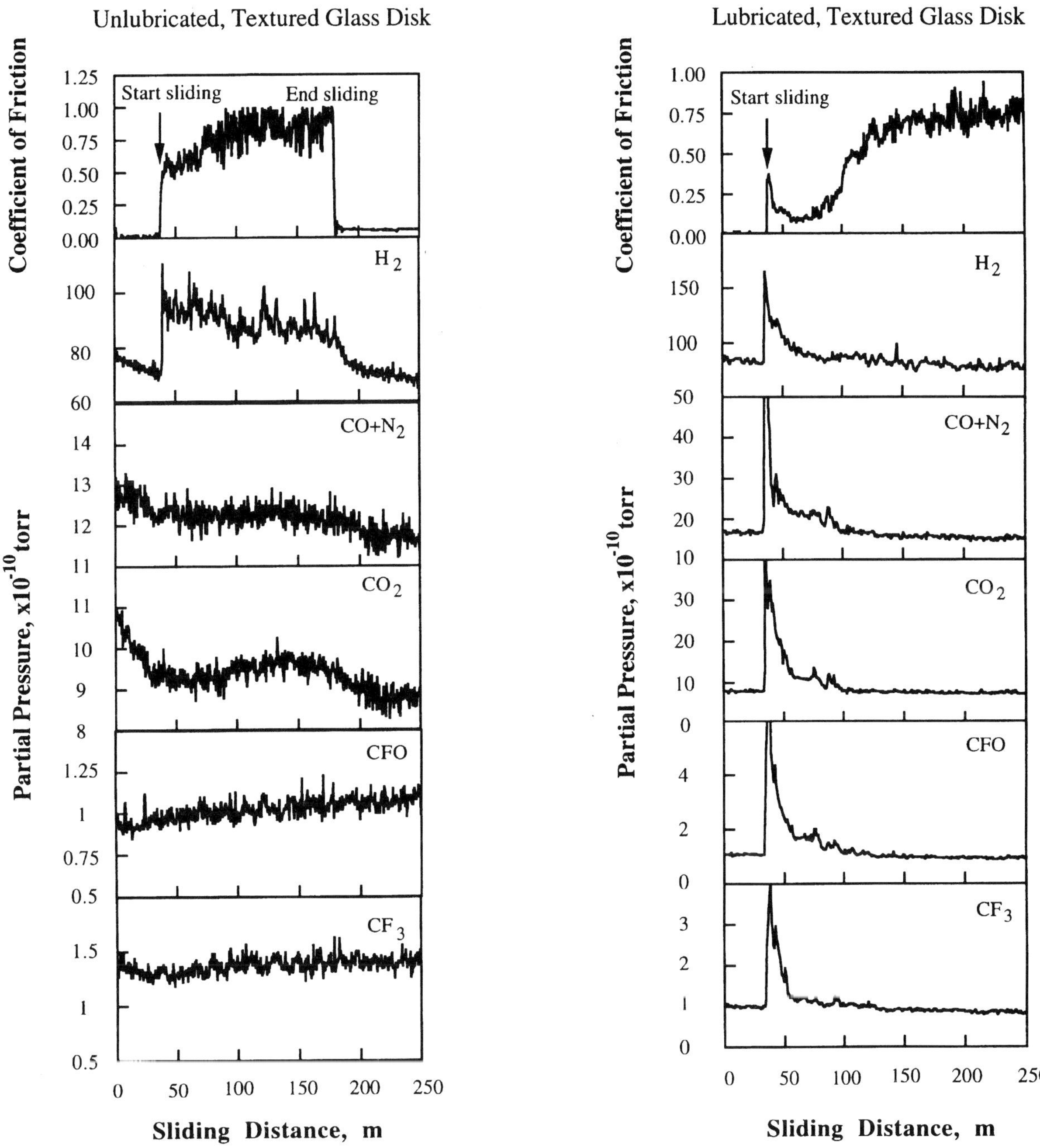

Fig. 8 Mass spectra of H_2, $CO + N_2$, CO_2, CFO, CF_3 and the coefficient of friction as a function of sliding distance on the textured glass disks (a) without and (b) with Fomblin AM 2001 lubricant

of these gases with the sliding time suggests that the evolution rate of these gases decreases with the sliding time. After about 100 m of sliding, the partial pressures of all gases decrease down to the background levels, and are accompanied by a rapid increase in the coefficient of friction. Appreciable wear track was observed and at some regions lubricant has been removed, as shown in Fig. 6e, suggesting that the increase in the coefficient of friction is due to the removal of lubricant from the interface and direct contact between the slider surface and the SiO_2 overcoat surface.

4 CONCLUSIONS

Degradation products of both Fomblin Z-Dol and Fomblin AM 2001 lubricants on the thin-film disk surfaces were detected during sliding using a mass spectrometer. The gaseous products generated from these two lubricants are quite similar. Decomposition of both lubricants is found to take place from the beginning of sliding, and the evolution rate of these gases decreases with the sliding time. The rapid rise in the coefficient of friction of the interfaces between the slider and lubricated disk during sliding is attributed to the removal of lubricant from the interface and direct contact between the slider surface and overcoat surface. Surface roughness has an effect on the sliding time to failure, but little effect on the decomposition of lubricant.

ACKNOWLEDGEMENTS

Thanks are given to Yuping Cheng for assistance in the tests. Financial support for this research was provided by the industrial membership of the Computer Microtribology and Contamination Laboratory.

REFERENCES

1 **Bhushan, B.** *Tribology and Mechanics of Magnetic Storage Devices*, 2nd edition, 1996 (Springer-Verlag, New York).

2 **Ohno, T., Shiroishi, Y., Saitoh, S., Yahisa, Y.** and **Nakano, F.** In-situ analysis of lubricant loss during sliding tests using polarized micro-FTIR and photoelectron emission spectroscopy. In Proceedings of Japan International Tribology Conference, 1990, pp. 605–610.

3 **Hu, Y.** and **Talke, F. E.** A study of lubricant loss in the rail region of a magnetic recording slider using ellipsometry. In *Tribology and Mechanics of Magnetic Storage Systems* (Eds B. Bhushan and N. S. Eiss), Vol. 5, 1988, SP-25, pp. 43–48 (STLE, Park Ridge, Illinois).

4 **Strom, B. D., Bogy, D. B.** and **Bhatia, C. S.** Gaseous wear products from lubricated thin film disks. *IEEE Trans. Magn.*, 1993, **29**, 253–258.

5 **Mori, S., Onodera, N.** and **Itoh, M.** Tribochemical reactions of very thin layers on magnetic recording disks. *Wear*, 1993, **168**, 85–90.

6 **Novotny, V. J., Pan, X.** and **Bhatia, C. S.** Tribochemistry at lubricated interfaces. *J. Vac. Sci. Technol. A*, 1994, **12**, 2879–2886.

7 **Pan, X.** and **Novotny, V. J.** Head material effects on interface tribochemistry. *IEEE Trans. Magn.*, 1994, **MAG-30**, 433–439.

8 **Bair, S., Green, I.** and **Bhushan, B.** Measurement of asperity temperatures of a read/write head slider bearing in hard magnetic recording disks. *Trans. ASME, J. Tribology*, 1991, **113**, 547–554.

9 **Bhushan, B.** Magnetic head–media interface temperatures—Part 3: application to rigid disks. *Trans. ASME, J. Tribology*, 1992, **114**, 420–430.

10 **Vurens, G., Zehringer, R.** and **Saperstein, D.** The decomposition mechanism of perfluoropolyether lubricants during wear. In *Surface Science Investigation in Tribology* (Eds Y. W. Chung, A. M. Homola and G. B. Street), 1992, Symposium Series 485, pp. 169–180 (ACS, Washington, D.C.).

11 **Bhushan, B.** and **Ruan, J.** Tribological performance of thin film amorphous carbon overcoats for magnetic recording rigid disks in various environments. *Surf. and Coat. Technol.*, 1994, **68–69**, 644–650.

12 **Bhushan, B., Yang, L., Gao, C., Suri, S., Miller, R. A.** and **Marchon, B.** Friction and wear studies of magnetic thin-film rigid disks with glass-ceramic, glass and aluminum–magnesium substrates. *Wear*, 1995, **190**, 44–59.

13 **Li, Y.** and **Bhushan, B.** Wear and friction studies of contact recording interface with microfabricated heads. *Wear*, 1996, **202**, 60–67.

14 **Novotny, V. J.** and **Baldwinson, M. A.** Lubricant dynamics in sliding and flying. *J. Appl. Phys.*, 1991, **70**, 5467–5652.

Effect of environmental humidity on the friction/stiction and durability of lubricated magnetic thin-film disks

Z Zhao and **B Bhushan**
Computer Microtribology and Contamination Laboratory, Department of Mechanical Engineering, The Ohio State University, Columbus, Ohio, USA

Abstract: Drag and CSS (contact start and stop) tests were conducted to study the effects of the degree of chemical bonding of the lubricant and environmental humidity on the friction/stiction and durability of the head–disk interface. Disks with untreated, partially bonded and fully bonded films of a polar perfluoropolyether lubricant were tested at different levels of humidity. Static friction is low for unlubricated disks and disks with a fully bonded lubricant film as compared to disks with untreated and partially bonded lubricant films. Static friction gradually increases with an increase in the relative humidity and the effect of humidity at high humidity values is more pronounced in the case of disks with untreated and partially bonded lubricant films. At high humidities, durability is low for all lubricated disks but is high for unlubricated disks, as compared to that at low humidities. At low humidities, durability in drag tests is more sensitive to the degree of bonded fraction than that in the CSS tests. In general, partially bonded lubricant films are desirable for low friction/stiction and high durability and lubricated disks perform best at low to moderate humidities. Operation at high humidities must be avoided.

Keywords: environmental humidity, friction/stiction, durability, lubricated disks

NOTATION

$K\alpha$ energy level
P-V peak-to-valley
XPS X-ray photoelectron spectroscopy

1 INTRODUCTION

Environmental humidity plays a significant role in the tribological performance of the head–disk interface (HDI) (**1**, **2**). It is well known that friction/stiction increases with an increase in humidity and it increases rapidly above a certain humidity. Liu and Mee (**3**) gave an early report about stiction occurrence at the HDI at high humidities. The effect of humidity on the friction and stiction has been studied by several researchers (**4**–**7**). There are very few recent studies on the effect of humidity on the durability (**7**, **8**).

High stiction and friction arises as a result of the development of menisci at the asperity contacts. Bonded lubricant film can moderate stiction and friction because it does not develop menisci at the interface. However, some mobile fraction is required for long durability (**1**, **7**). Partially bonded films of polar perfluoropolyether (PFPE) lubricants are commonly used on commercial thin-film rigid disks. Recent studies by Zhao and Bhushan (**9**) and Bhushan and Zhao (**7**) have been conducted to systematically understand the effects of the degree of chemical bonding of the lubricant and humidity on friction/stiction and durability in drag tests. No studies have been reported to date that examine the effects of degree of bonded fraction on tribological performance at various levels of humidity in contact start and stop (CSS) tests. The purpose of this research is to study these effects in CSS tests and compare the tribological performance with that in drag tests.

2 EXPERIMENTAL

The single-disk tester was set up in a humidity-controlled chamber (**7**). All experiments were conducted at a

The MS was received on 6 January 1997 and was accepted for publication on 9 July 1997.

temperature of 22 ± 1 °C in a class-100 clean room environment. Static friction on virgin disks was measured after a 30 s rest time, with an acceleration to 3600 r/min in 10 s (disk acceleration = $1.4\ m/s^2$) and a normal load of 7 g. A sample rate of 5000 Hz was chosen for friction force measurements. A slow-speed drag test at 0.0016 m/s and a normal load of 7 g was also performed on virgin disks to better examine the variation in kinetic friction around the disk circumference. A sample rate of 50 Hz was chosen for kinetic friction force measurements. In drag tests for durability, a constant disk velocity of 0.75 m/s at 200 r/min and a normal load of 14 g were used. A sample rate of 720 Hz was chosen for kinetic friction force measurements. The failure point was determined from the advent of significant increase in the coefficient of friction to 0.5, the appearance of wear tracks on the disk surface or when the coefficient of friction showed a sharp decrease (> 20 per cent) after having reached a well-defined maximum, whichever occurred first. The coefficient of kinetic friction was recorded once every 20 revolutions and obtained through averaging the time trace over 2 revolutions. In CSS tests, one CSS cycle time of 60 s consisted of 10 s at acceleration, 10 s at a speed of 3600 r/min, 10 s at deceleration and 30 s at rest. The suspension normal load used was 7 g. A sample rate of 5000 Hz was chosen for friction force measurements. The failure point was determined from the advent of a significant increase in the coefficient of static friction (to about 1) or the appearance of wear tracks on the disk surface, whichever occurred first. The CSS test is closer to drive operating conditions than the drag test but it takes a much longer time.

Thin-film Al–Mg disks used in this study had a diamond-like carbon overcoat with an r.m.s. roughness of 10.4 nm, a P-V (peak-to-valley) distance of 73.0 nm and P-mean (peak-to-mean) distance of 39.1 nm. The disks were lubricated with a PFPE lubricant (Z-DOL) with reactive end groups with the following chemical structure: $HO—CH_2—CF_2—O—(CF_2—CF_2—O)_m—(CF_2—O)_n—CF_2—CH_2—OH$. The two rail taper-flat microsliders fabricated from Al_2O_3–TiC (70–30 wt%) were chosen for tests with an r.m.s. roughness of 1.3 nm.

Tribological tests were performed on unlubricated and lubricated disks. A total of 1.7 nm thick lubricant film was applied on the lubricated disks. Lubricant film is called (chemically) untreated if no post-thermal treatment is followed after the disk is dip-lubricated, partially bonded if thermal treatment at 150 °C for 30 min is followed and fully bonded if thermal treatment at 150 °C for 90 min is followed and then the mobile fraction of the lubricant film is washed off with a solvent (FC-72). The immobile fraction of the partially bonded lubricant film was measured to be about 1.1 nm with a mobile fraction of remaining 0.6 nm thick film. The total film thickness was measured using an ellipsometer (**9**).

3 RESULTS AND DISCUSSION

3.1 Coefficient of static friction

Figure 1 shows the profiles of the coefficient of static friction during the first cycle of CSS tests on unlubricated and lubricated disks at 5, 50 and 80 per cent levels of relative humidity (RH). Unlubricated disks and disks lubricated with a fully bonded lubricant film give the low static friction because only the water film adsorbed on the disk surface is mobile and contributes to the meniscus force, with little difference among the values of static friction. For disks lubricated with untreated and partially bonded lubricant films, the coefficient of static friction is higher than those on unlubricated disks and disks with a fully bonded lubricant film. The static friction generally

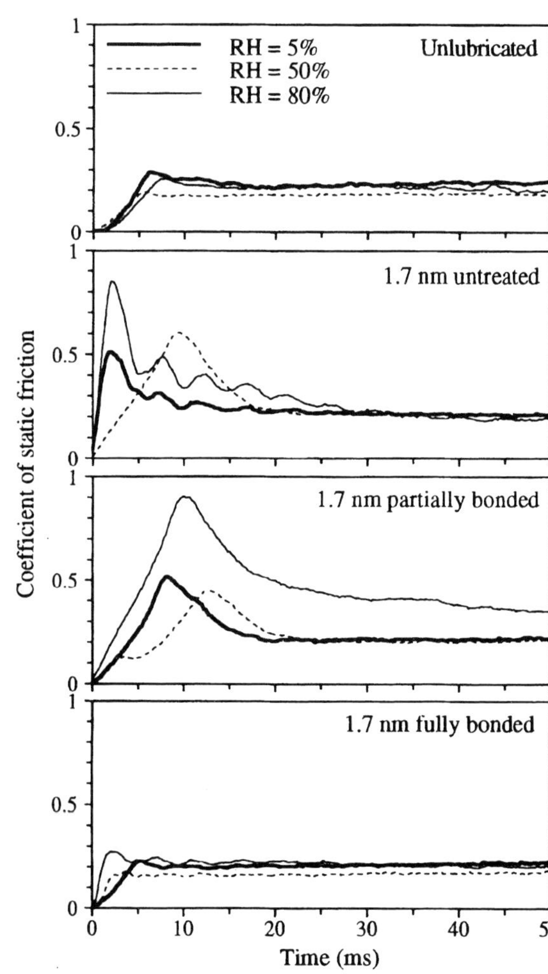

Fig. 1 Profiles of the coefficient of static friction on virgin unlubricated and lubricated disks, measured after a rest time of 30 s, acceleration to 3600 r/min in 10 s (disk acceleration = $1.4\ m/s^2$) and at a normal load of 7 g

increases with an increase in the relative humidity. The effect of humidity at high humidity values is more pronounced in the case of untreated and partially bonded lubricated films.

3.2 Coefficient of kinetic friction

Disk surfaces exposed to a humid environment may adsorb a layer of water film. The total liquid film on the disk surface consists of the lubricant and adsorbed water. The uniformity of the liquid film is necessary for a uniform value of friction. Variation in the coefficient of kinetic friction over one revolution can be used as an indicator of the uniformity of liquid film thickness on the disk surface at different levels of humidities. The results of a low-speed drag test on virgin disks at 0.0016 m/s and a normal load of 7 g are presented in Fig. 2. For unlubricated disks (Fig. 2a), a negligible effect on the coefficient of kinetic friction exists at 5 and 50 per cent RH. At 80 per cent RH, the coefficient of kinetic friction and its variation are higher than those at 5 and 50 per cent RH. For disks with untreated and partially bonded lubricant films (Figs 2b and c), the coefficient of kinetic friction at 50 per cent RH is a little higher than that at 5 per cent RH. The coefficient of kinetic friction at 80 per cent RH fluctuates strongly as compared with those at 5 and 50 per cent RH. For disks with a fully bonded lubricant film (Fig. 2d), however, there is a negligible effect of humidity on the kinetic friction at all humidity levels. The large variations in the coefficient of kinetic friction on the unlubricated disks and disks with untreated and partially bonded lubricant films (Figs 2a to c) suggest a strong effect of an adsorbed water film on the disk surface. The stable profiles of the kinetic friction for disks with the fully bonded lubricant film implies that the effect of the adsorbed water film on the slider surface is not significant.

On the surfaces of unlubricated disks and disks with a fully bonded lubricant film, there is no mobile fraction of a liquid film except the adsorbed water film from the humid environment. At 80 per cent RH, the unlubricated disks show some effect of the humidity on the coefficient of kinetic friction whereas the fully bonded disks do not (Figs 2a and d). One reason may be the larger contact angle

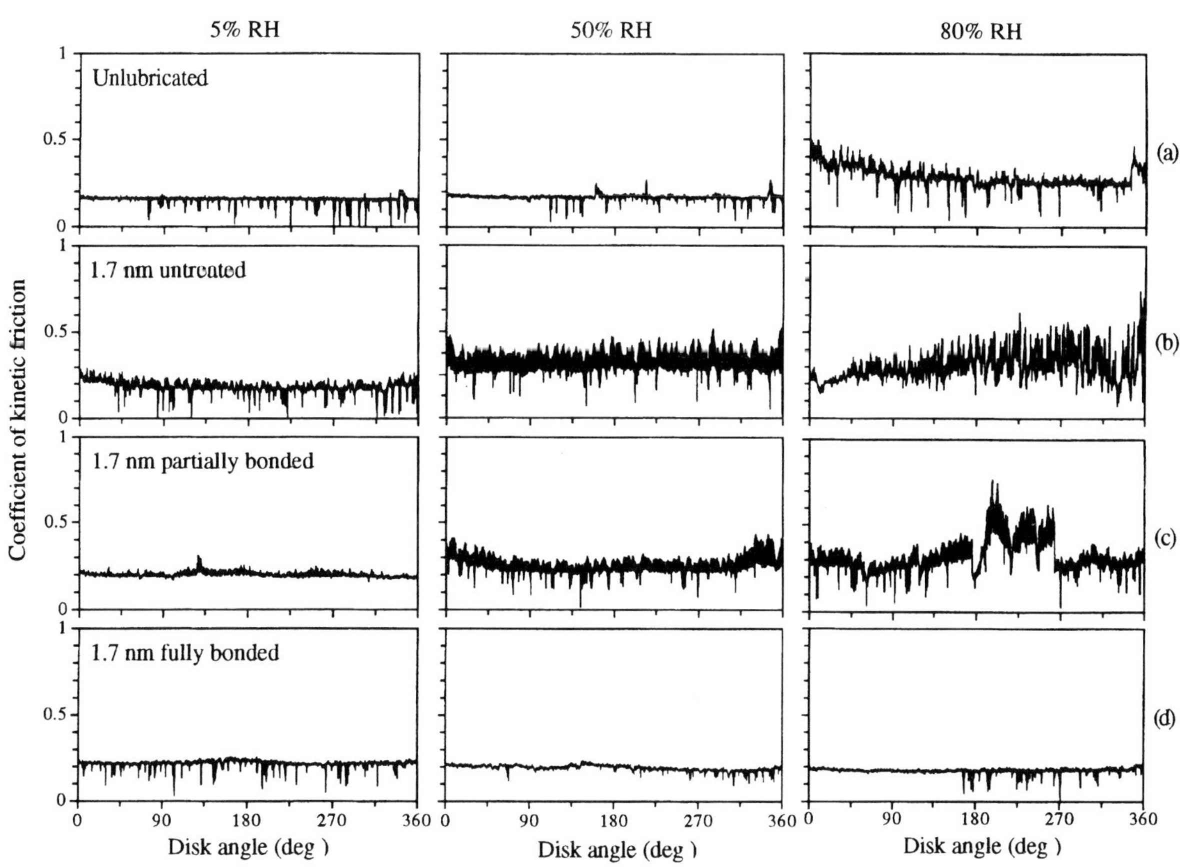

Fig. 2 Coefficient of kinetic friction over one revolution on virgin unlubricated and lubricated disks, measured at a disk speed of 0.0016 m/s and a normal load of 7 g

of water on the fully bonded disks, which is measured to be about 90° on all lubricated disks, than that on the unlubricated disks, measured to be about 68°.

3.3 Durability tests

Friction as a function of distance profiles used to obtain durability data is shown in Fig. 3 for drag tests and in Fig. 4 for CSS tests. A summary of durability in terms of disk revolutions in drag tests and CSS cycles in CSS tests as a function of relative humidity is presented in Fig. 5. In addition, the coefficient of static friction during the first cycle of CSS tests is included in Fig. 5. For the unlubricated disk, the durability increases with an increase in the relative humidity and the effect is more pronounced in CSS tests, as compared to drag tests. It is expected that water film acts as a lubricant at high humidities. For three types of lubricated disks, the effect of humidity on durability is different.

For fully bonded disks, the durability in the drag tests is insensitive to humidities and at high humidity it is equivalent to the unlubricated disk. In the CSS tests, durability decreases with an increase in the relative humidity and levels off at a lower level, which is equivalent to that of the unlubricated disk at moderate to high humidities. For partially bonded and untreated disks, durability generally decreases from low to moderate humidities and drops in all cases at high humidity. As stated earlier, in the case of partially bonded and untreated disks lubricant permeability results in lubricant displacement at high humidity, which is responsible for poor durability. The present observation of low durability at high humidities in lubricated disks is consistent with other authors (**5**, **8**). Their results are all from drag tests for untreated disks.

For partially bonded and untreated disks at 5 and 50 per cent RH in drag tests, higher durability is achieved by the partially bonded disk followed by the untreated disk. At high humidities, the durability in drag tests is comparable to that of the fully bonded and unlubricated disks. In the CSS tests, the durability of partially bonded and untreated disks is comparable at all levels of humidities. Thus, the partially bonded disk is the best in drag tests and is as good as the untreated disk in CSS tests. The existence of a mobile fraction of a lubricant film improves durability at intermediate levels of humidity. At low humidities, the presence of an immobile fraction improves durability in the case of CSS tests, while both mobile and immobile fractions are important in the case of drag tests. The mobile fraction of lubricant film allows the exposed areas to be protected and the immobile fraction can prevent the film from being disturbed by the adsorbed water molecules. Besides, the immobile fraction will not form meniscus forces in a long-term stiction test. It is clear that the degree of bonded fraction is the key to friction/stiction and durability of the disks in different environments.

A further examination of the slider surfaces after durability tests provides more information on the failure mechanisms at the head–disk interface. Figure 6 shows the index related to the amount of debris deposited on slider surfaces at leading and trailing areas against the lubricated disks after the failure point of durability tests. Figure 7 shows optical images of leading and trailing areas of the slider at 50 per cent RH after drag and CSS tests. These images provide examples of photographs used to evaluate the amount of debris deposition in Fig. 6. Based on the observations of the slider surfaces against the lubricated disk surfaces, most of the debris is deposited in the leading and trailing areas which are generally around the taper–flat intersecting area and/or the sides and back edges of the trailing end of the slider surfaces. The trailing area shows a higher probability of debris accumulation at the head–disk

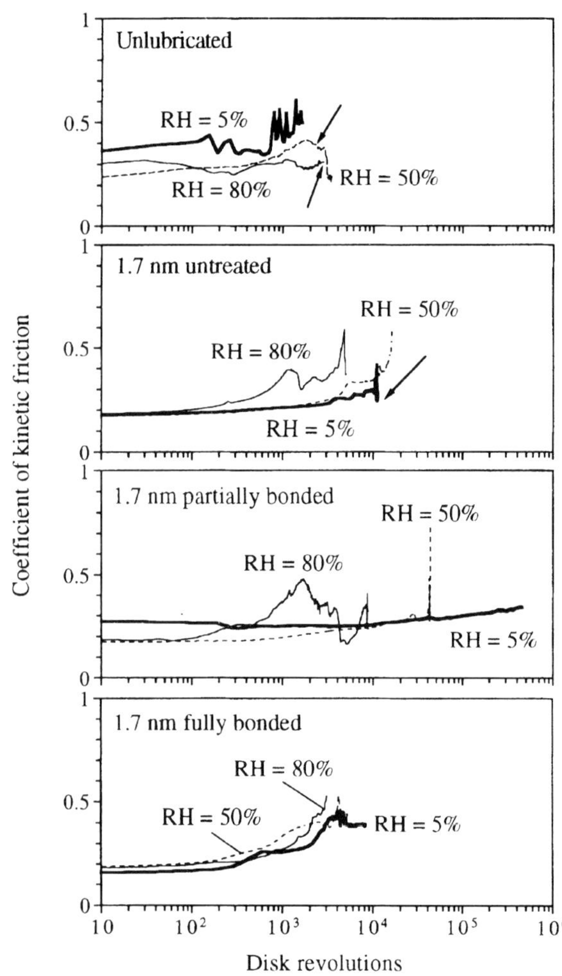

Fig. 3 Coefficient of kinetic friction as a function of the number of revolutions in drag tests, measured at a disk speed of 0.75 m/s and a normal load of 14 g. Arrows indicate that a visible wear track was formed

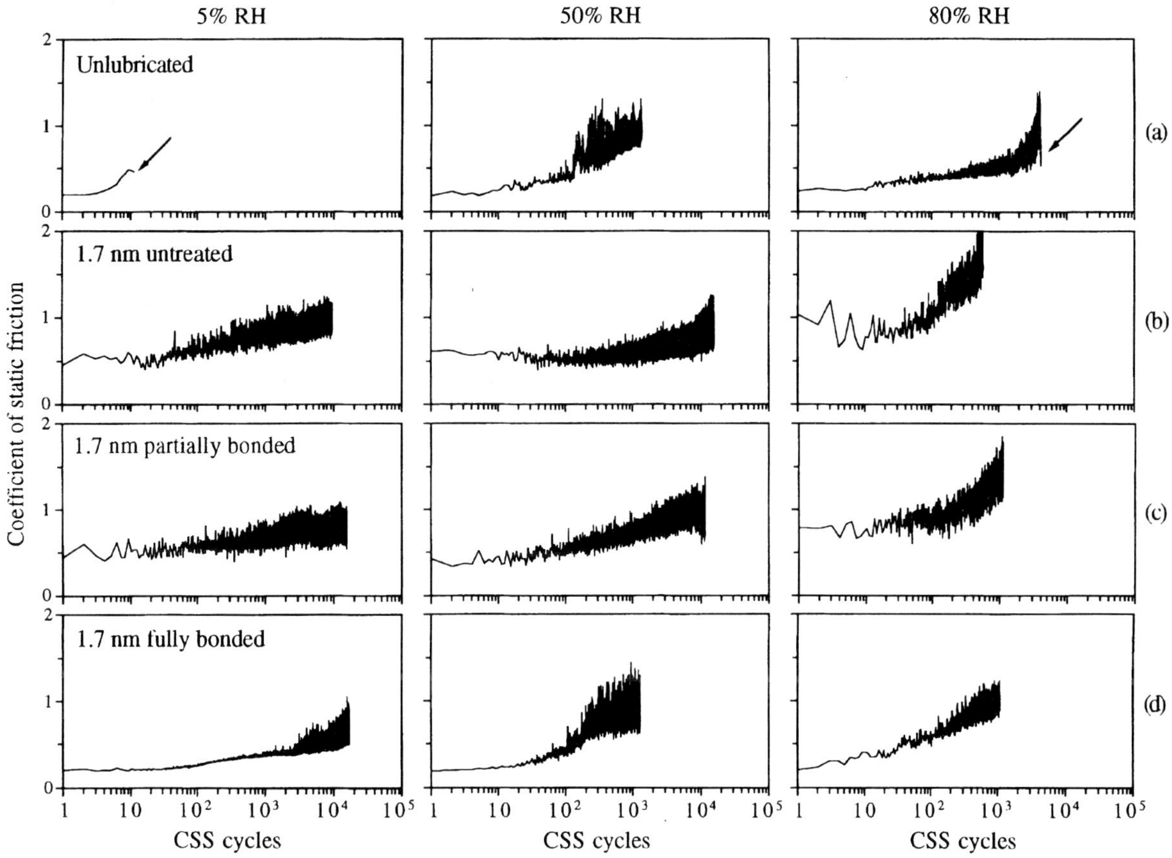

Fig. 4 Coefficient of static friction as a function of CSS cycles. Arrows indicate that a visible wear track was formed

interface as compared to the leading area. The relative sliding direction of the disk under the head slider is from the leading end to the trailing end and the trailing edge is at the minimum head flying-height position. This is responsible for more debris collected in the trailing end. There is no direct effect of the humidity on the debris deposition on slider surfaces after durability tests. The sliders against the disks with a partially bonded lubricant film show a little more debris deposition than those against the disks with untreated and fully bonded lubricant films. The deposited debris is not adherent to the slider surface. It can be removed with a dry cotton swab.

Figure 8 shows XPS spectra of regions with and without debris before and after sputtering to identify the composition of the debris. XPS spectra were obtained by collecting characteristic photoelectrons emitted after radiation of Mg–$K\alpha$ with 14 keV and 0.02 A, and with a spot size of 150 μm. The region with debris has F element which comes from the lubricant on the disk surface (Fig. 8b). After 6 min of sputtering, the region with debris does not have F element (Fig. 8c). The XPS spectra cannot identify whether the debris includes C and O elements because these also exist in the evironment. Based on optical microscopy, it is found that the debris still exists on the head surface after 6 min of sputtering. Therefore, debris on the slider surface, in addition to the lubricant, most probably consists of the disk overcoat and no slider material since there are no visible wear or scratches on the worn slider surface. In the case of the sliders against the unlubricated disks, there are visible wear scratches on the disk surface at 50 and 80 per cent RH in drag tests and at 5 and 80 per cent RH in CSS tests. If the scratches appear on the disk surface, they also appear on the slider surface. It appears that the co-existence of mobile and immobile fractions of the lubricant film is most efficient in collecting the debris, acting as cement to collect the debris on the interface, particularly the slider surface.

4 CONCLUSIONS

1. Static friction is low for unlubricated disks and disks lubricated with a fully bonded lubricant film at var-

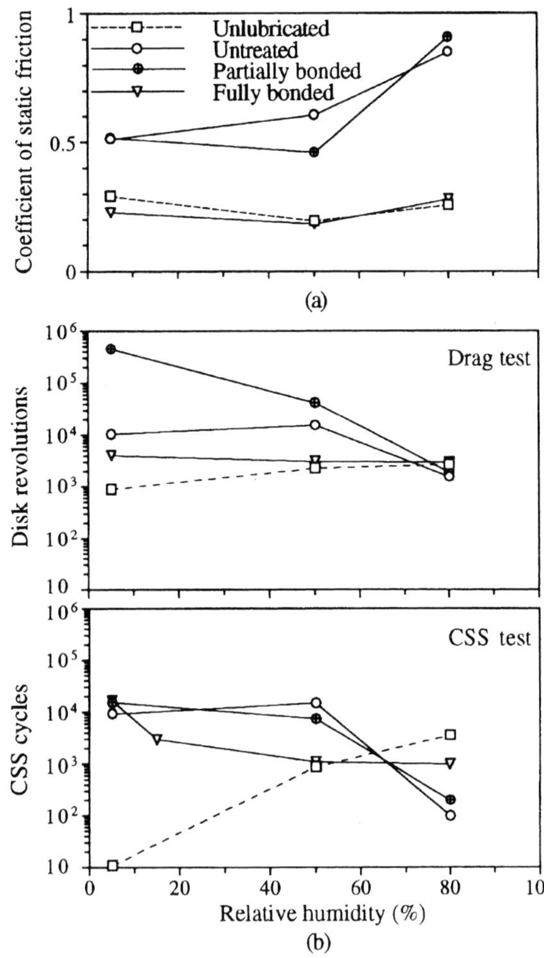

Fig. 5 (a) Coefficient of static friction and (b) durability in drag and CSS tests as a function of relative humidity

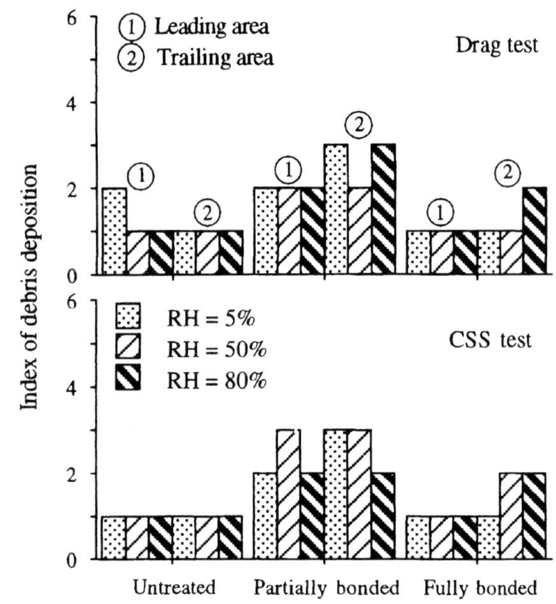

Fig. 6 Index of debris deposition on slider surfaces at leading and trailing areas after drag and CSS tests, where 0 represents a completely clean surface, 1 represents a surface with very little debris, 2 represents a surface with little debris and 3 represents a surface with some debris

ious levels of humidity whereas disks lubricated with untreated and partially bonded lubricant films have a higher static friction, which increases rapidly at 80 per cent RH.

2. A severe effect of humidity on the kinetic friction is observed at 80 per cent RH for unlubricated disks and disks with the mobile fraction of a lubricant film, whereas humidity has little effect on the kinetic friction for the disks with a fully bonded lubricant film.
3. Durability in CSS tests is more sensitive to high humidities than that in drag tests. Durability of lubricated disks is lower at high humidities for all degrees of chemically bonded lubricant films. For partially bonded and untreated disks, the existence of the mobile fraction of a lubricant film improves durability at intermediate levels of humidity. At low humidities, the presence of the immobile fraction improves durability in the case of CSS tests, while both mobile and immobile fractions are important in the case of drag tests. For unlubricated disks, durability improves with an increase in humidity.
4. Based on the observations of the slider surfaces after durability tests, there is no direct effect of humidity on debris deposition on slider surfaces after the failure point in durability tests. The slider surfaces against disks with untreated and fully bonded lubricant films are cleaner than those against disks with a partially bonded film.
5. For lubricated disks, the environmental humidity effect on the friction/stiction and durability is beneficial at low humidities and is undesirable at high humidities.

ACKNOWLEDGEMENTS

Financial support for this study was provided by the industrial membership of the Computer Microtribology and Contamination Laboratory. Lubricant samples and solvents used in this study were kindly supplied by the following sources: lubricant Z-DOL and solvent Galden HT-70 by

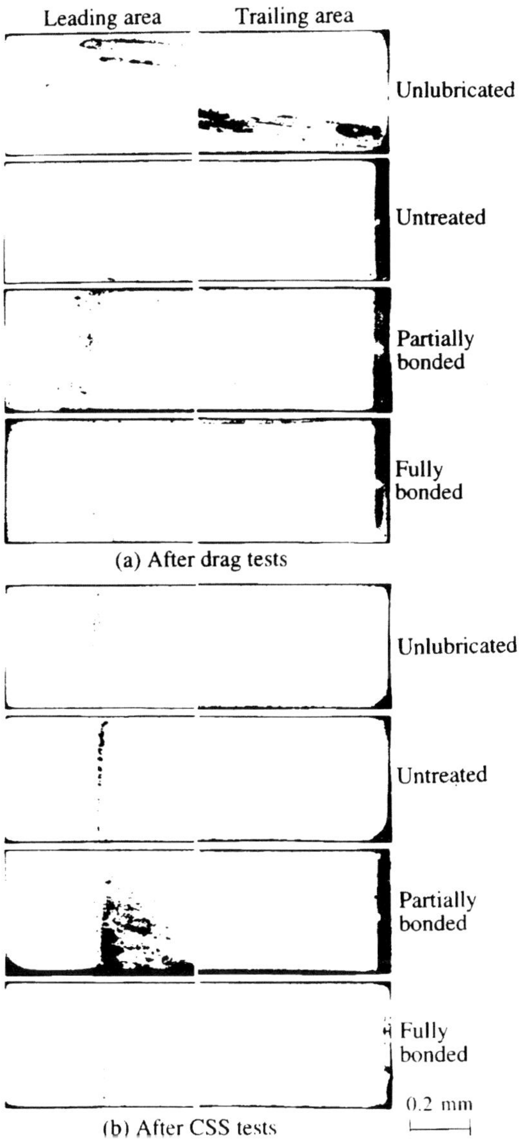

Fig. 7 Optical micrographs of slider surfaces of leading and trailing areas after (a) drag and (b) CSS tests at the relative humidity of 50 per cent

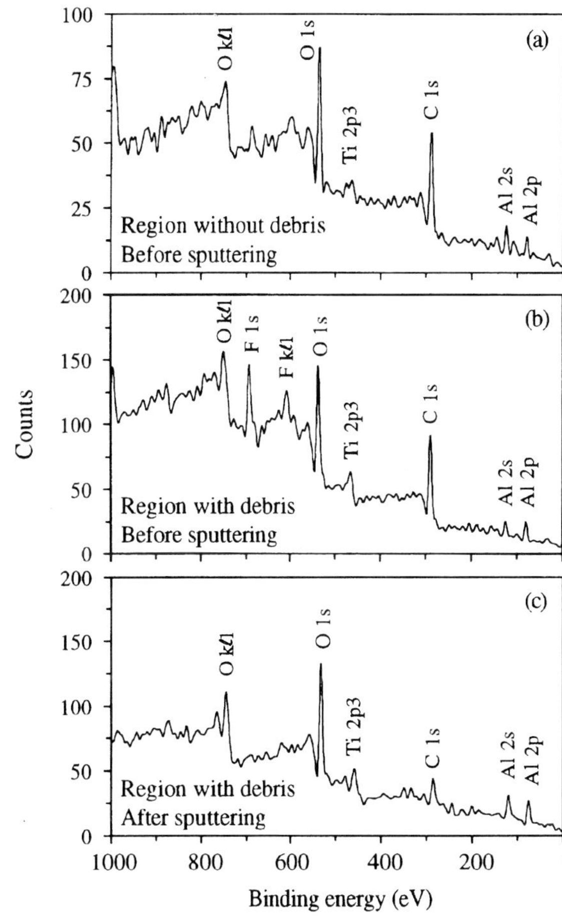

Fig. 8 XPS spectra of regions (a) without debris and before sputtering, (b) with debris and before sputtering and (c) with debris and after 6 min of sputtering, on the slider surface after the CSS test against the lubricated disk at 5 per cent relative humidity

Ausimont USA Inc., Thorofare, New Jersey, and solvent FC-72 by 3M Electronic Products Division, Austin, Texas.

REFERENCES

1 **Bhushan, B.** *Tribology and Mechanics of Magnetic Storage Devices*, 2nd edition, 1996 (Springer-Verlag, New York).

2 **Bhushan, B.** Methodology for roughness measurement and contact analysis for optimization of interface roughness. *IEEE Trans. Magn.*, 1996, **32**, 1819–1825.

3 **Liu, C. C.** and **Mee, P. B.** Stiction at the Winchester head–disk interface. *IEEE Trans. Magn.*, 1983, **19**, 1659–1661.

4 **Bhushan, B.** and **Dugger, M. T.** Liquid-mediated adhesion at the thin-film magnetic disk/slider interface. *Trans. ASME, J. Tribology*, 1990, **112**, 217–223.

5 **Li, Y.**, **Trauner, D.** and **Talke, F. E.** Effect of humidity on stiction and friction of the head/disk interface. *IEEE Trans. Magn.*, 1990, **26**, 2487–2489.

6 **Tian, H.** and **Matsudaira, T.** The role of relative humidity, surface roughness and liquid build-up on static friction behavior of the head/disk interface. *Trans. ASME, J. Tribology*, 1993, **115**, 28–35.

7 **Bhushan, B.** and **Zhao, Z.** Friction/stiction and wear studies of magnetic thin-film disks with two polar perfluoropolyether lubricants. *IEEE Trans. Magn.*, 1997, **33**, 918–925.

8 **Yang, M.** and **Talke, F. E.** Effects of gas composition, humidity and temperature on the tribology of the head/disk interface—part I: experiments. *Tribology Trans.*, 1996, **39**, 615–620.

9 **Zhao, Z.** and **Bhushan, B.** Effect of bonded-lubricant films on the tribological performance of magnetic thin-film rigid disks. *Wear*, 1996, **202**, 50–59.

Silicon carbide—a potential material for magnetic recording sliders

V Sharma and **F E Talke**
Center for Magnetic Recording Research, University of California, San Diego, California, USA

Abstract: The feasibility of using silicon carbide as a magnetic recording slider material is investigated using stiction tests with different dwell times, contact start–stop (CSS) tests and drag tests at different temperatures and humidities. The stiction and frictional properties of silicon carbide sliders are compared with standard Al_2O_3–TiC sliders and it is concluded that their tribological behaviour is superior to that of conventional Al_2O_3–TiC sliders. The results of this study show that silicon carbide is a promising material for magnetic recording sliders.

Keywords: silicon carbide, tribology

NOTATION

A_a	apparent area of contact
A_{re}	real area of contact
E_c	composite Young's modulus
p_a	apparent pressure
R_p	composite radius of the asperity peaks of the equivalent rough surface
ν	Poisson ratio
σ_p	standard deviation of peak height distribution of the equivalent rough surface

1 INTRODUCTION

Currently, the majority of disk drives use magnetic recording sliders made of alumina–titanium carbide. Due to the differences in hardness between alumina and titanium carbide, a differential wear rate exists between the two phases during lapping, resulting in a rougher surface than would be encountered if a single-phase material were used. Due to the increased interactions between the head and the disk as a result of continued reduction in the head–disk spacing, other materials such as monophase ceramics with good tribological properties are being studied as alternative magnetic recording head materials.

Among monophase ceramics, silicon carbide has a unique combination of physical, thermomechanical and tribological properties. In particular, silicon carbide has high thermal conductivity, high hardness, high elastic modulus and a low coefficient of thermal expansion. The higher thermal conductivity of silicon carbide with respect to Al_2O_3–TiC results in lower flash temperatures during head–disk asperity contacts. In addition, for elastic head–disk contact, the high elastic modulus of silicon carbide is expected to result in a smaller area of contact and lower coefficient of friction compared to conventional slider material. Furthermore, silicon carbide has generally a low coefficient of friction and other tribologically desirable properties (**1–5**).

The humidity and load-dependent frictional properties of silicon carbide pins against silicon carbide disks were studied by Takadoum *et al.* (**6**, **7**). In their tests, the coefficient of friction was found to decrease after a critical sliding distance. The critical sliding distance as well as the wear rate were observed to decrease as the load was decreased and the relative humidity (RH) level was increased from 20 to 90 per cent. The initially high value of the friction coefficient was attributed to the generation and interposition of abrasive wear debris. A subsequent decrease in the coefficient of friction was attributed to the compaction of wear particles which formed a protective film of hydrated amorphous silica. In a different study, the tribological performance of hot pressed silicon carbide, CVD silicon carbide and Al_2O_3–TiC against lubricated thin-film magnetic disks was studied via pin-on-disk tests by Koka (**8**). The coefficient of friction was observed to increase in the first 4000 cycles before levelling off to about the same value for all three materials. For both types of silicon carbides, a flat circular wear region with wear debris and smear regions was observed, whereas for Al_2O_3–TiC no such wear region was found. The presence of a flat circular wear region in silicon carbides and its

The MS was received on 5 September 1996 and was accepted for publication on 27 October 1997.

absence in Al_2O_3–TiC was attributed to the formation of an oxide layer (silicon dioxide) on the silicon carbide pins. The wear behaviour of different ceramic sliders against thin-film magnetic disk was studied by Chu *et al.* (**9**). The goal of their research was to obtain the factors that affect slider performance and only a limited number of tests were performed on silicon carbide sliders. Chu *et al.* observed a significantly smaller drag coefficient for silicon carbide sliders compared to Al_2O_3–TiC sliders. Based on their initial results, the use of silicon carbide as a potential slider material appeared promising. However, in order to be able to use silicon carbide in future magnetic recording heads, more information is needed to understand the frictional behaviour of silicon carbide sliders against thin-film magnetic recording disks.

The purpose of this research is to investigate the tribological properties of silicon carbide magnetic recording sliders using contact start–stop (CSS) tests, stiction tests and constant speed drag tests under different humidities and temperatures. Similar tests were also conducted on standard Al_2O_3–TiC sliders to establish a reference for comparison. The observed differences in the frictional behaviour of silicon carbide sliders and Al_2O_3–TiC sliders are explained on the basis of differences in their physical and mechanical properties, and it is concluded that silicon carbide is a promising material for magnetic recording sliders.

2 EXPERIMENTAL PROCEDURE

Commercially available thin-film magnetic recording disks with a centre-line surface roughness *Ra* of 5.6 nm and a 2–3 nm thick layer of PFPE lubricant were used in all the tests. Two different types of 50 per cent style silicon carbide and Al_2O_3–TiC sliders, namely two-rail and subambient pressure sliders, were used in this study. The two-rail sliders were made from hot-pressed silicon carbide, whereas the subambient sliders were made from CVD silicon carbide. The air-bearing surfaces of the sliders were manufactured using mechanical lapping with 0.25 μm oil-based diamond compound. The crown for the two-rail sliders was in the range of 5–15 nm and the crown for the subambient sliders was in the range of 20–25 nm. All air-bearing surface parameters were similar for the two-rail silicon carbide sliders and the Al_2O_3–TiC sliders and for the subambient pressure silicon carbide sliders and the Al_2O_3–TiC sliders. The two-rail sliders were used with a preload of 68.6 mN, whereas the subambient sliders were used with a preload of 34.3 mN. The centre-line surface roughness *Ra* of the silicon carbide sliders, as measured by atomic force microscopy (AFM), was less than 0.6 nm, while the average surface roughness of the Al_2O_3–TiC sliders was about 2.5 nm. Figure 1 shows typical AFM images of the air-bearing surface for (a) SiC and (b) Al_2O_3–TiC sliders.

Stiction and CSS tests were conducted under ambient conditions (23 °C, 45 per cent relative humidity) using a commercially available friction tester with a maximum axial and radial spindle run-out of less than 1.3 μm. The preload and frictional force were measured by two sets of strain gauges mounted on two perpendicular I-beams. Stiction tests with zero dwell time and 24 h dwell time were performed before CSS testing. In addition, a stiction test was performed with 24 h dwell time after CSS testing. In a typical stiction test, the maximum linear speed used was 0.005 m/s. The time-dependent friction force was measured for the first 0.8 s while the disk was accelerated to 0.005 m/s at an acceleration of 0.005 m/s^2.

CSS tests, up to 10 000 cycles, were performed on both types of two-rail and subambient sliders. During each CSS cycle the linear speed was increased from rest to 9 m/s in 3 s, was held constant at 9 m/s for 3 s and was thereafter decreased down from 9 m/s to rest in the next 3 s. A 'soak' time of 3 s was chosen between consecutive CSS cycles. For each cycle, the maximum coefficient of friction during the first 0.2 s was defined as the 'coefficient of stiction', while the maximum coefficient of friction between 0.1 and 3 s was defined as the 'maximum coefficient of friction' during take-off.

All drag tests were performed in a custom-built environmental chamber which allowed independent control of temperature and humidity. The environmental chamber consists of a double-walled steel chamber which is connected to a temperature-controlled water circulator and a humidifier (**10**). The temperature-controlled water is circulated around the chamber before entering the humidifier to stabilize the temperature of air. The dry air enters the humidifier where it is split into dry and wet lines by means of a flow controller. The wet line passes through a water bubbler which is at a higher temperature than the water circulating around the steel chamber and around the helical coils in the humidifier. The supersaturated air flows through the helical coils where the extra moisture in the air condenses due to water circulating around the coils at a lower temperature. The wet air (100% RH) mixes with the dry air (0% RH) and passes through the rest of the helical coil before it enters the chamber. By controlling the flowrate of the dry and the wet lines, any level of humidity in the chamber is achieved. Constant speed drag tests were performed, up to 20 000 disk revolutions, at a linear speed of 0.16 m/s and under (a) ambient conditions (23 °C, 45% RH), (b) hot and humid conditions (50 °C, 80% RH) and (c) cold and dry conditions (5 °C, 10% RH).

3 RESULTS

Figure 2 shows typical friction profiles for zero dwell time before CSS testing for two-rail and subambient pressure silicon carbide and Al_2O_3–TiC sliders respectively. It is

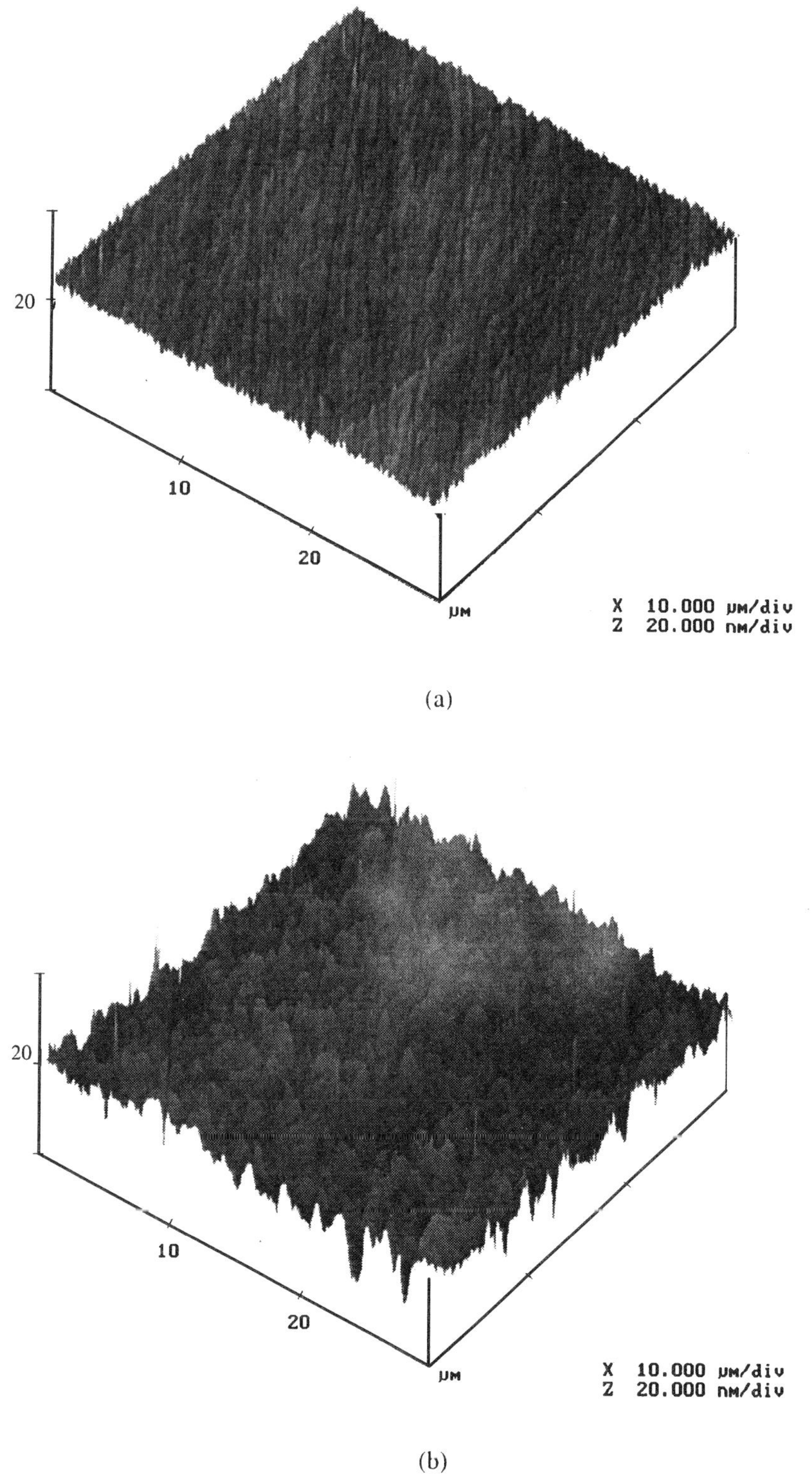

Fig. 1 Typical air-bearing surface images obtained using AFM for (a) SiC and (b) Al_2O_3-TiC sliders

observed that the maximum coefficient of friction (i.e. the coefficient of stiction) for two-rail SiC sliders is about half that of two-rail Al_2O_3-TiC sliders (0.25). For subambient sliders, the coefficient of stiction occurs between 0.1 and 0.2 s. The coefficient of stiction in the present Al_2O_3-TiC slider (0.9) is higher than the corresponding coefficient of stiction for the SiC slider (0.5).

Figure 3 shows the coefficient of friction for 24 h of dwell time before CSS testing. Comparing Figs 2 and 3, it can be seen that the coefficient of stiction for the two-rail SiC sliders remains nearly constant with dwell time, while

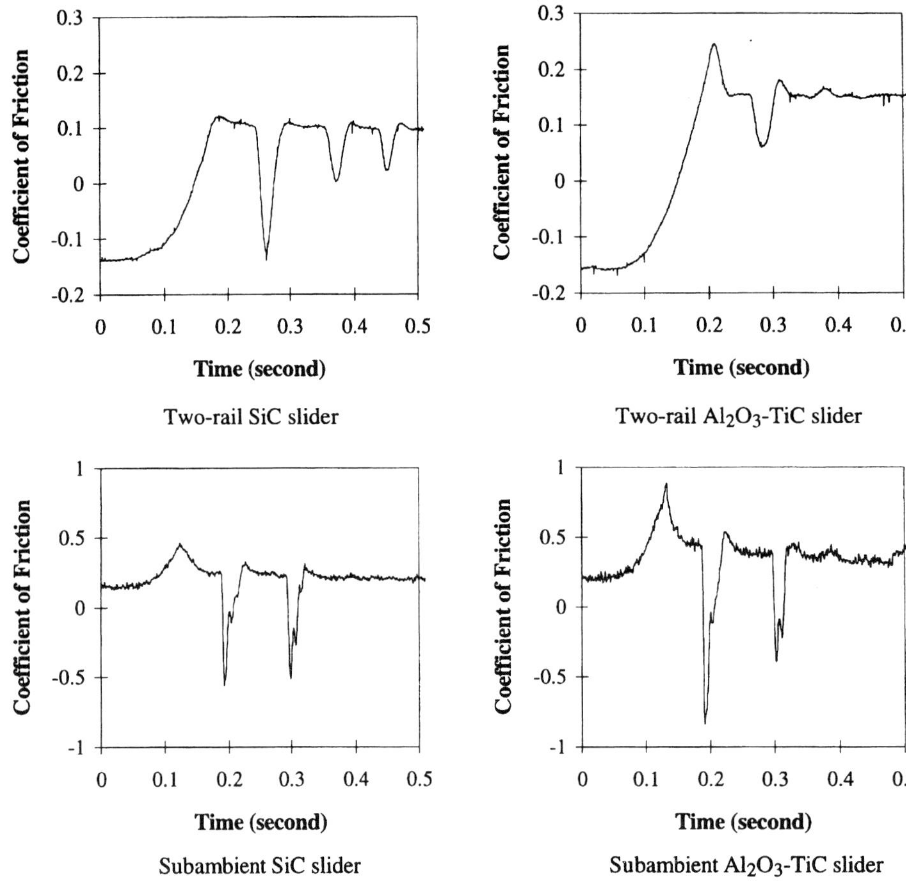

Fig. 2 Coefficient of friction as a function of time for zero dwell time before the CSS test

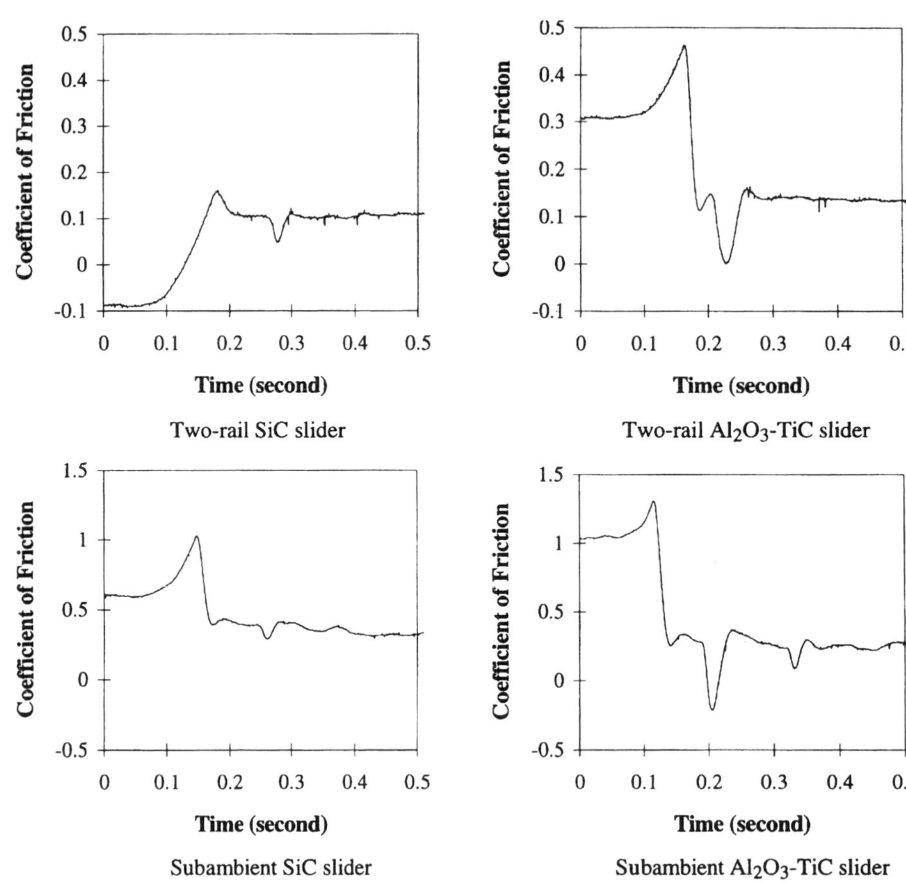

Fig. 3 Coefficient of friction as a function of time for 24 h of dwell time before the CSS test

the coefficient of stiction for the Al_2O_3–TiC sliders increases from 0.25 to 0.45 with 24 h of dwell time. The stiction behaviour of the subambient sliders was observed to be similar for Al_2O_3–TiC and SiC sliders, with the Al_2O_3–TiC sliders exhibiting a slightly larger coefficient of stiction than the SiC sliders.

The variation of the coefficient of stiction as a function of the number of CSS cycles for two-rail and subambient SiC and Al_2O_3–TiC sliders is shown in Fig. 4. From Fig. 4 it can be seen that the stiction coefficient of SiC sliders is considerably lower than the corresponding stiction coefficient of Al_2O_3–TiC sliders. While the fluctuations in the stiction coefficient for subambient SiC and Al_2O_3–TiC sliders is nearly the same, the fluctuations in the stiction coefficient for two-rail Al_2O_3–TiC sliders is considerably higher compared to two-rail SiC sliders. The coefficient of stiction for both types of SiC sliders and the subambient Al_2O_3–TiC slider increases in the first 2000 cycles before reaching a steady value. In contrast, the stiction coefficient of the two-rail Al_2O_3–TiC slider increases monotonically and exceeds 0.6 after 10 000 CSS cycles. Figure 5 shows the maximum coefficient of friction as a function of the number of CSS cycles. From Fig. 5 it can be seen that the maximum coefficient of friction for SiC sliders is lower than that of Al_2O_3–TiC sliders. For the two-rail SiC slider the maximum friction profile is the same as the stiction profile. The maximum friction coefficient for the two-rail Al_2O_3–TiC slider is higher than that of the SiC slider. For subambient SiC sliders, a slight decrease in the coefficient of maximum friction is observed before a steady state constant value of 0.25 is reached.

The coefficient of friction for 24 h of dwell time after 10 000 CSS cycles is shown in Fig. 6 for two-rail and subambient SiC and Al_2O_3–TiC sliders. It is observed that the maximum coefficient of friction for 24 h of dwell time after CSS testing increases for all sliders compared to the corresponding maximum friction coefficient for 24 h of dwell time before CSS testing. This increase is much higher for Al_2O_3–TiC sliders than for SiC sliders. For two-rail sliders the maximum coefficient of friction for SiC is observed to be 0.8, whereas the maximum coefficient of friction for Al_2O_3–TiC is about 2.9. For subambient sliders, the maximum coefficient of friction for SiC and Al_2O_3–TiC is observed to be 1.0 and 3.7 respectively.

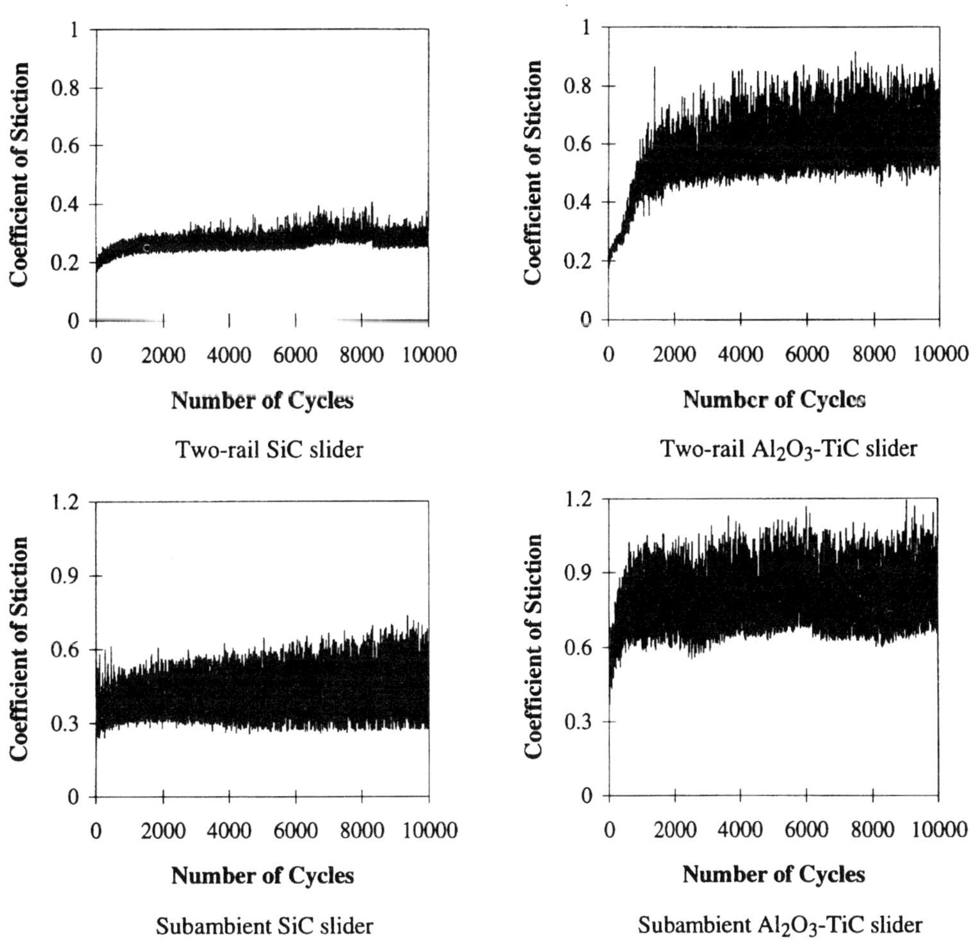

Fig. 4 Coefficient of stiction as a function of number of CSS cycles for SiC and Al_2O_3–TiC sliders

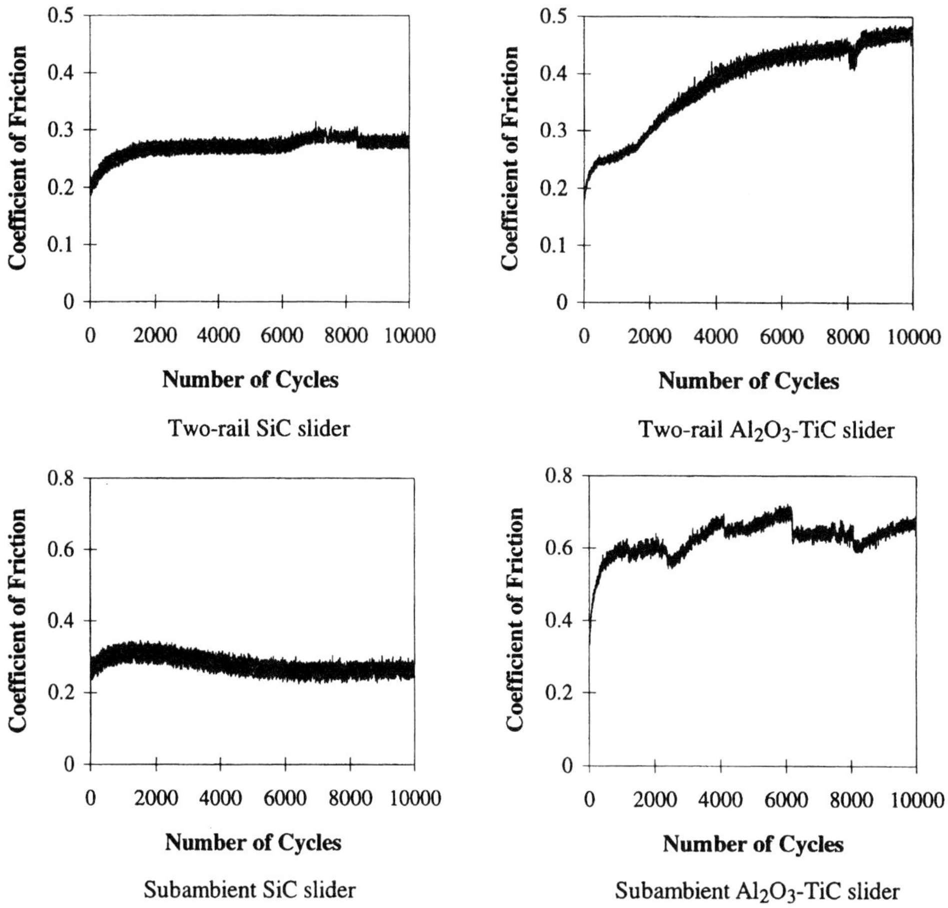

Fig. 5 Maximum coefficient of friction as a function of number of CSS cycles for SiC and Al_2O_3-TiC sliders

In addition to CSS testing, constant speed drag tests were conducted on two-rail and subambient SiC and Al_2O_3-TiC sliders, at a linear velocity of 0.16 m/s, under different temperatures and humidities. Figure 7 shows the variation of the coefficient of friction versus the number of disk revolutions for two-rail SiC and Al_2O_3-TiC sliders under ambient conditions (23 °C, 45% RH), hot and humid conditions (50 °C, 80% RH) and cold and dry conditions (5 °C, 10% RH). From Fig. 7 it can be seen that the initial coefficient of friction for both types of two-rail sliders is between 0.15 and 0.25. At 23 °C and 45% RH, the Al_2O_3-TiC slider–disk interface failed in less than 16 000 revolutions, while the coefficient of friction for the SiC slider–disk interface increased gradually from 0.25 to 0.4 in 20 000 revolutions. At 50 °C and 80% RH, a slight decrease in the coefficient of friction is observed for the SiC slider–disk interface after 5000 revolutions. For the Al_2O_3-TiC slider–disk interface the coefficient of friction increased rapidly from 0.25 to 0.55 in 1000 revolutions and the interface failed in less than 2100 revolutions. At 5 °C and 10% RH, the Al_2O_3-TiC slider–disk interface and SiC slider–disk interface failed after 800 and 4000 revolutions respectively.

Figure 8 shows the variation of the coefficient of friction versus the number of disk revolutions for subambient pressure SiC and Al_2O_3-TiC sliders for 23 °C and 45% RH, 50 °C and 80% RH, and 5 °C and 10% RH respectively. At 23 °C and 45% RH, the coefficient of friction for the SiC slider–disk interface is observed to be slightly smaller than that of the Al_2O_3-TiC slider–disk interface. When the temperature is increased to 50 °C and the humidity is increased to 80 per cent, the friction coefficient for the SiC slider–disk interface increases from an initial value of 0.1 to 0.2 in 20 000 revolutions. For the Al_2O_3-TiC slider–disk interface, however, the coefficient of friction increased from 0.1 to 0.8 in 20 000 revolutions. At 5 °C and 10% RH, the Al_2O_3-TiC slider–disk interface failed in 1200 cycles, while for the SiC slider–disk interface the friction coefficient increased from 0.1 to 0.28 in 20 000 revolutions.

4 DISCUSSION

The results of this investigation show that silicon carbide sliders have superior stiction and friction behaviour compared to conventional Al_2O_3-TiC sliders. One of

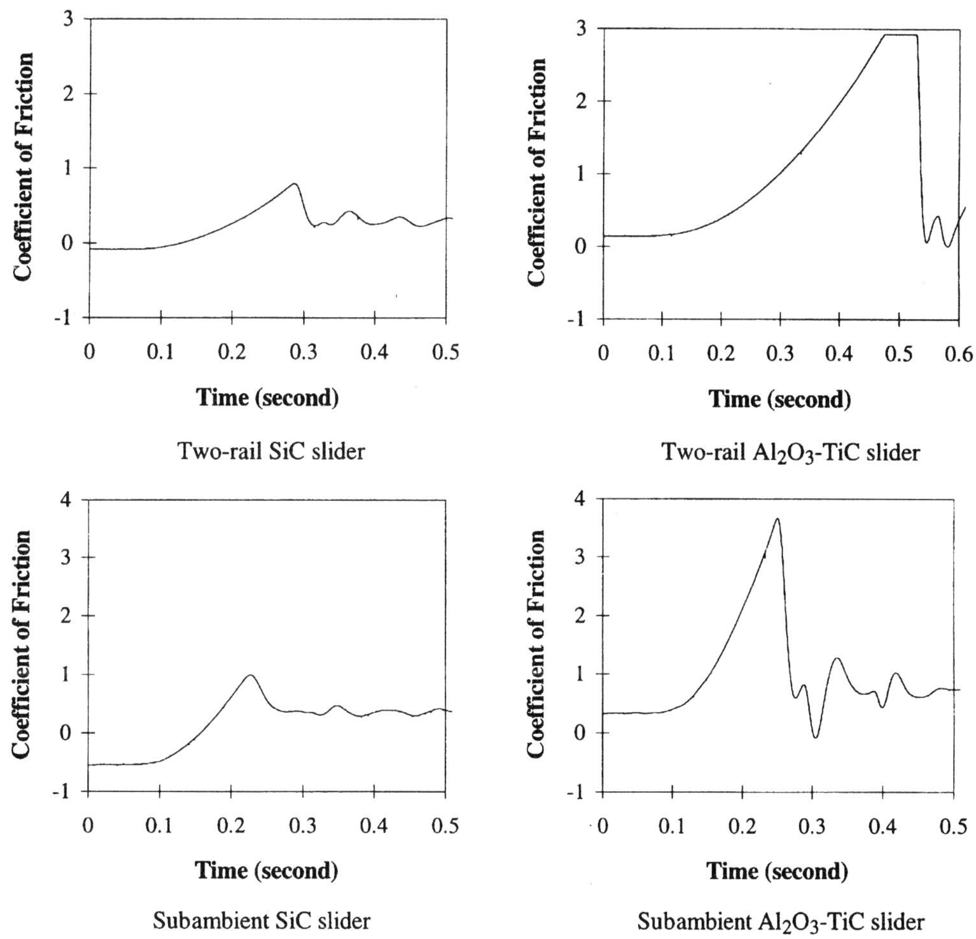

Fig. 6 Coefficient of friction as a function of time for 24 h of dwell time after the CSS test

the reasons for the superior tribological behaviour of SiC sliders seems to be related to differences in the real area of contact. That is to say, when a slider is placed on a disk, contact occurs at discrete points where asperities of the slider and the disk meet. Very high stresses can be generated at asperity contacts depending upon the normal load, the geometry of the contact and the material properties of the contacting surfaces. The deformation at the asperity contacts can either be elastic or plastic. For smooth surfaces the contact is mostly elastic, whereas for rough surfaces some amount of plastic flow is likely to occur. If the peak heights of the mating surfaces have a Gaussian distribution and the asperities are spherical near their summit with a constant radius, then the approximate real area of elastic contact (A_{re}) is given by (**11**)

$$\frac{A_{re}}{3.2 p_a A_a} \approx \frac{1}{E_c \surd(\sigma_p / R_p)} \tag{1}$$

where p_a is the apparent pressure, A_a is the apparent area of contact, R_p is the composite radius of the asperity peaks, σ_p is the standard deviation of peak height distribution of the equivalent rough surface and E_c is the composite Young's modulus. R_p, σ_p and E_c are given by the following equations:

$$\frac{1}{R_p} = \frac{1}{R_{p1}} + \frac{1}{R_{p2}} \tag{2a}$$

$$\sigma_p = \surd(\sigma_{p1}^2 + \sigma_{p2}^2) \tag{2b}$$

and

$$\frac{1}{E_c} = \frac{1 - \nu_1^2}{E_1} + \frac{1 - \nu_2^2}{E_2} \tag{2c}$$

where the subscripts 1 and 2 refer to the two rough surfaces (disk and slider in the present case) in contact and ν is the Poisson ratio of the contacting surface.

To obtain an estimate for equation (1), the average radius of asperity peaks and the standard deviation of peak heights for magnetic disks and both types of sliders is determined using an optical interferometric microscope (WYKO, ×40). Using the typical mechanical properties of Al_2O_3–TiC, SiC and rigid disk shown in Table 1, it can be seen that the real area of contact for SiC sliders on

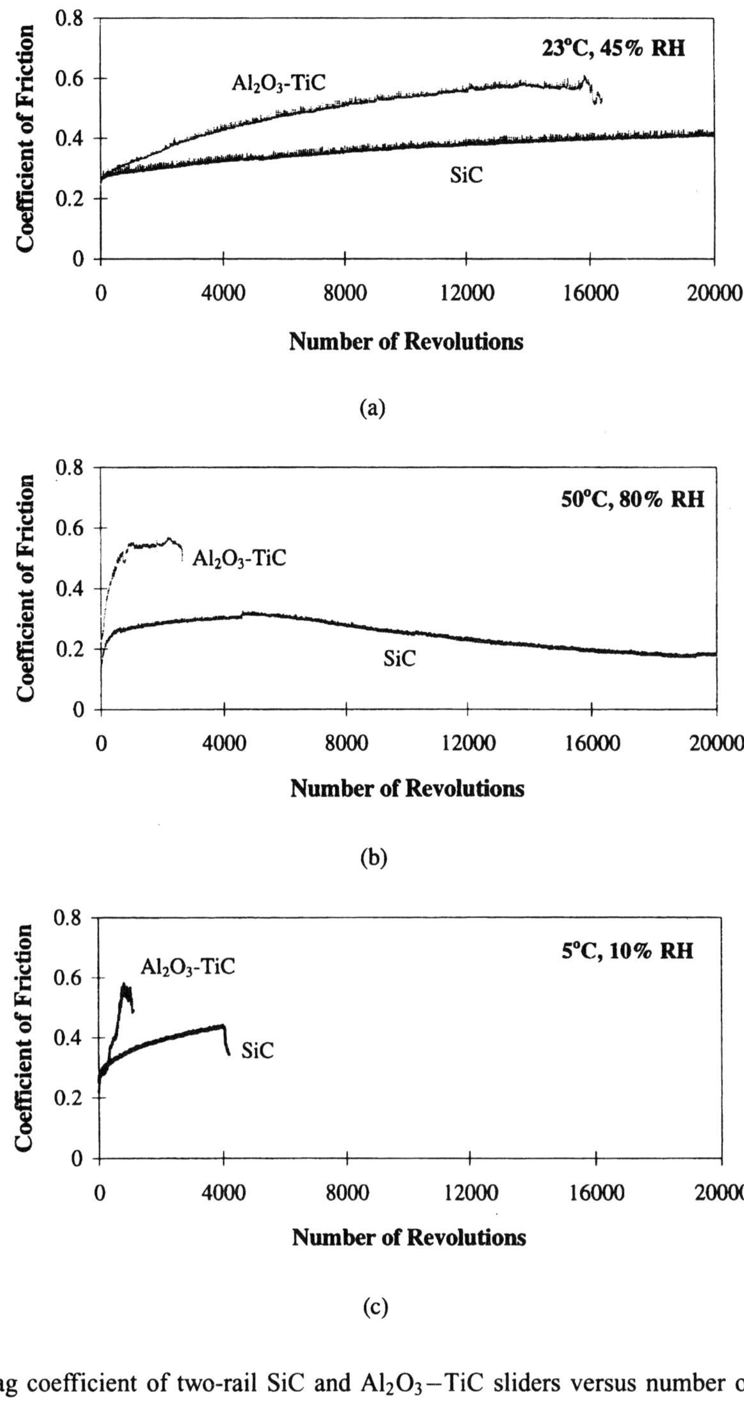

Fig. 7 Drag coefficient of two-rail SiC and Al_2O_3-TiC sliders versus number of disk revolutions for (a) 23 °C, 45% RH, (b) 50 °C, 80% RH and (c) 5 °C, 10% RH

the carbon overcoated disk is about 9 per cent less than the real area of contact for Al_2O_3-TiC sliders on the same disk. Since the observed differences in the coefficient of stiction between SiC and Al_2O_3-TiC sliders are much more than 9 per cent, it is unlikely that the differences in the stiction and friction behaviour are only due to the differences in the real area of contact.

One additional reason that may contribute to the observed superiority of SiC as compared to Al_2O_3-TiC sliders is possibly related to lubricant degradation in the presence of Al_2O_3-TiC. In a recent study (**12**), the thermal stability of perfluoropolyether lubricants was studied in the presence of alumina, and it was observed that at elevated temperatures alumina reacts with the acetal units of the perfluoropolyethers, thereby degrading the lubricant. Since high flash temperatures are created during sliding, thermal degradation of long-chain polymers in the presence of Al_2O_3-TiC is likely. Although no such degradation studies have been performed with SiC, the higher thermal conductivity of SiC compared to

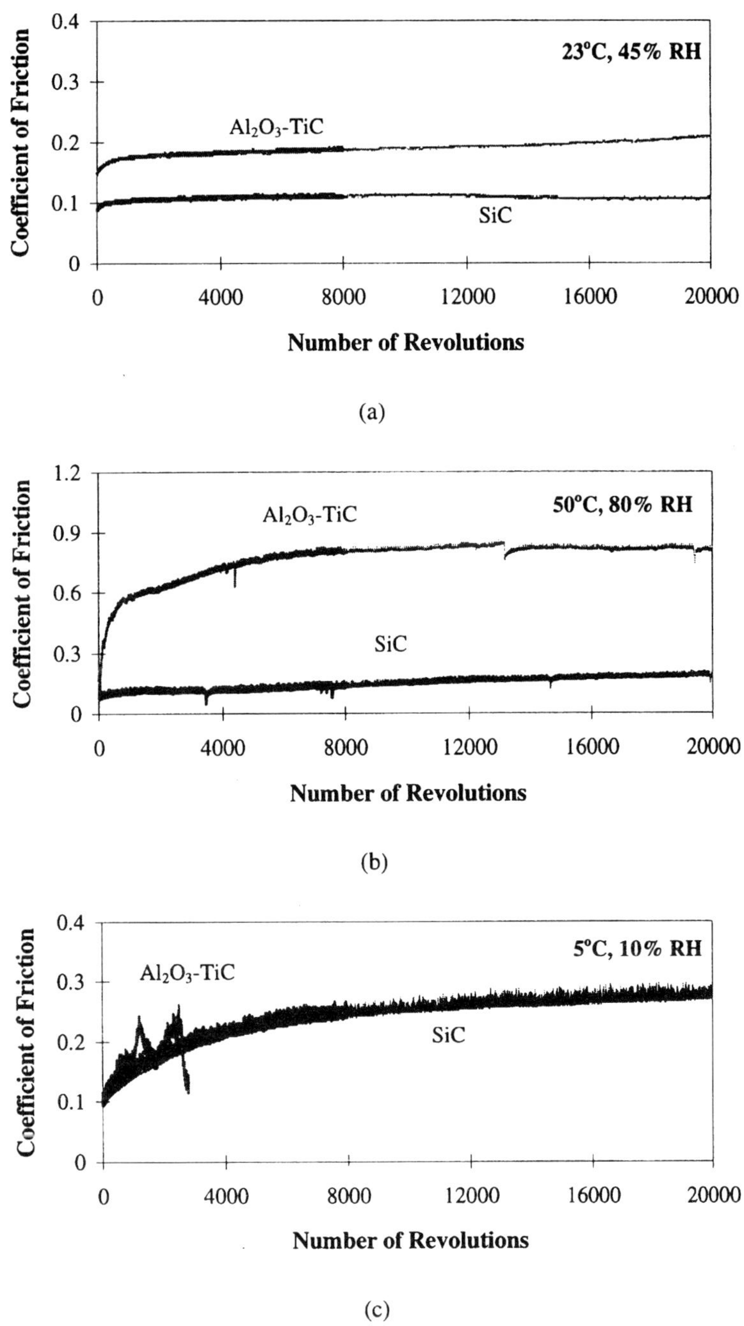

Fig. 8 Drag coefficient of subambient pressure SiC and Al_2O_3–TiC sliders versus number of disk revolutions for (a) 23 °C, 45% RH, (b) 50 °C, 80% RH and (c) 5 °C, 10% RH

Table 1 Typical properties of Al_2O_3–TiC, SiC and carbon overcoated hard disks

Material	E (GPa)	ν	Hardness (H) (GPa)	σ_p (nm)	$1/R_p$ (mm^{-1})
Al_2O_3–TiC	390	0.21	18	2.32	21.81
Silicon carbide	450	0.17	25	0.59	31.54
Rigid disk	100–130	?	5–6	5.77	32.6

Al_2O_3–TiC implies that the flash temperatures for SiC should be substantially lower, thereby reducing the tendency of the lubricant to degrade by thermal processes. It is justifiable to hypothesize that this reduced tendency to thermal degradation is one of the major reasons for the improved frictional behaviour of SiC.

The above point concerning the improvement in the thermal stability of perfluorinated polymers in the presence of SiC does not explain the reason why the stiction behaviour is improved in the presence of SiC. Other factors such as wettability, contact angle and lubricant bridge formation seem to be responsible for the improved stiction characteristics of SiC sliders. At present, those effects are being studied in more detail and it is hoped that the results of this work will be presented in the future.

5 CONCLUSIONS

For SiC and Al_2O_3-TiC sliders, the coefficient of stiction was observed to increase with dwell time before and after CSS testing. The increase in the coefficient of stiction with dwell time was much smaller for SiC sliders compared to Al_2O_3-TiC sliders. The coefficient of maximum friction and the coefficient of stiction during CSS testing for SiC sliders was observed to be much smaller than for conventional Al_2O_3-TiC sliders. The coefficient of friction during drag testing under ambient, hot and humid, and cold and dry conditions was observed to be superior for SiC sliders compared to Al_2O_3-TiC sliders. The results of this investigation indicate that SiC is a promising material for magnetic recording sliders.

ACKNOWLEDGEMENTS

The authors would like to thank Dr Don Rice of Applied Magnetics Corporation and Mike Boots and Glen Albert of Brumko Magnetics for providing test samples and for helpful discussions.

REFERENCES

1 **Alcock, J., Sorensen, O. T., Jensen, S.** and **Kjeldsteen, P.** Comparative wear mapping techniques—1. Friction and wear mapping of tungsten carbide/silicon carbide. *Wear*, 1996, **194**, 219.

2 **Blau, P. J.** Effects of surface preparation on the friction and wear behaviour of silicon nitride silicon carbide sliding pairs. *J. Mater. Sci.*, 1992, **27**, 4732.

3 **Yamamoto, Y.** and **Ura, A.** Influence of interposed wear particles on the wear and friction of silicon carbide in different dry atmospheres. *Wear*, 1992, **154**, 141.

4 **Blau, P. J.** An observation of role reversal effects in unlubricated sliding friction and wear tests of alumina and silicon carbide. *Wear*, 1991, **151**, 193.

5 **Sasaki, S.** The effects of the surrounding atmosphere on the friction and wear of alumina, zirconia, silicon carbide and silicon nitride. *Wear*, 1989, **134**, 153.

6 **Takadoum, J., Zsiga, Z., Ben Rhouma, M.** and **Roques-Carmes, C.** Correlation between friction coefficient and wear mechanism of SiC/SiC system. *J. Mater. Sci. Lett.*, 1994, **13**, 474.

7 **Takadoum, J., Zsiga, Z.** and **Roques-Carmes, C.** Wear mechanism of silicon carbide: new observations. *Wear*, 1994, **174**, 239.

8 **Koka, R.** Wear of silicon carbide in sliding contact with lubricated thin-film rigid disk. *Tribology Trans.*, 1995, **38**(2), 417.

9 **Chu, M. Y., Bhushan, B.** and **DeJonghe, L. C.** Wear behavior of ceramic sliders in sliding contact with rigid magnetic thin-film disks. *Tribology Trans.*, 1992, **35**(4), 603.

10 **Yang, M.** and **Talke, F. E.** Effect of gas composition, humidity and temperature on the tribology of the head/disk interface—Part I: experiments. *STLE Trans.* (to be published).

11 **Johnson, K. L.** *Contact Mechanics*, 1985 (Cambridge University Press).

12 **Kasai, P. H., Tang, W. T.** and **Wheeler, P.** Degradation of perfluoropolyethers catalyzed by aluminum oxide. *Appl. Surf. Sci.*, 1991, **51**, 201.

Friction and durability of ceramic slider materials in contact with lubricated thin-film rigid disks

J Xu and **B Bhushan**
Computer Microtribology and Contamination Laboratory, Department of Mechanical Engineering, The Ohio State University, Columbus, Ohio, USA

Abstract: Slider materials with improved static and kinetic friction and durability of the head–disk interface are required for minimum separation between the head slider and magnetic medium and miniaturization of sliders. Mn–Zn ferrite, calcium titanate, Al_2O_3–TiC and SiC sliders were used in drag tests, stiction tests and contact start–stop (CSS) tests. The SiC slider exhibits the best tribological performance at all humidities as compared to other sliders. Relevant mechanisms are revealed with measurement of the contact angle of perfluoropolyether (PFPE) lubricants on slider materials, lubricant transfer tests, scanning electron microscopy (SEM), X-ray photoelectron spectroscopy (XPS) and Fourier transform infra-red (FTIR) analyses on the slider surfaces before and after tests. Best friction and wear performance of SiC is attributed to lubricant transfer from the lubricated disk to the slider surface and formation of tribochemical films. A lubricant layer on the slider surface protects the slider from wear and also increases the contact angle with the water film, resulting in a low meniscus contribution leading to low stiction. The tribochemical reaction forms a gel-like hydroxide layer on the SiC slider surface which performs as a boundary lubricant, leading to a low coefficient of friction and low wear.

Keywords: friction, durability, ceramic slider, thin-film rigid disks

NOTATION

A/D	analogue to digital
CSS	contact start–stop
FTIR	Fourier transform infra-red
PFPE	perfluoropolyether
P-V	peak to valley
RH	relative humidity
SEM	scanning electron microscopy
XPS	X-ray photoelectron spectroscopy
μ_s	coefficient of static friction

1 INTRODUCTION

Ever increasing magnetic recording density requires a head slider to fly close to the disk surface which may lead to an intermittent contact condition resulting in wear on both the slider and rigid disk surface (**1**). In order to maintain or increase the durability of the head–disk interface, alternate slider materials are being investigated. Ceramic materials are used for the construction of sliders because of their high hardness, low adhesion to the contacting surface, chemically inert in a wide range of environments, low friction and wear under a wide range of sliding contact conditions (**1–4**). Al_2O_3–TiC (70–30 wt%) and polycrystalline and single-crystal Mn–Zn ferrite are the most commonly used slider materials and calcium titanate is used to some extent (**5**). Among them, Mn–Zn ferrite exhibits superior tribological performance when used against magnetic disks, followed by Al_2O_3–TiC and calcium titanate (**6–8**).

Recently, there has been growing interest in silicon carbide (SiC) as a substrate material for picosliders (35 per cent) and femtosliders (30 per cent), with magnetoresistive-type read heads. Silicon carbide has a high thermal conductivity which provides better cooling during powering of the magnetoresistive heads. Silicon carbide is also a preferable substrate for small sliders because it machines with less edge rounding, less pull-outs and high smoothness.

In addition, its high thermal conductivity leads to lower interface temperatures which generally results in less degradation of the interface (**9, 10**). There is a limited number of papers on the tribological performance of silicon carbide as a slider material. Chu *et al.* (**8**) reported that single-phase materials are superior to two-phase materials and that silicon carbide exhibits a long durability in drag tests. Based on wear tests using ceramic ball contact with lubricated thin-film disks, Koka (**11**) reported that silicon

The MS was received on 6 January 1997 and was accepted for publication on 15 July 1997.

carbide exhibits the best tribological performance, which was attributed to the formation of smears as a result of tribochemical wear of the SiC ball. Contact stresses in these tests were unrealistically high as compared to that in the head–disk interface, which renders the tests unrealistic. High wear reported in their tests is unrealistic. Based on drag tests using ceramic sliders in sliding contact with the magnetic disks, Koka *et al.* (**12**) confirmed the observations made by Chu *et al.* (**8**) that SiC material is superior to Al_2O_3–TiC material. Koinkar and Bhushan (**13**) recently reported superior microfriction, microscratch and microwear resistance of SiC as compared to other ceramic materials used in their study. No published data exists on more realistic contact start–stop (CSS) tests. Furthermore, a clear understanding of mechanisms responsible for better tribological performance of SiC does not exist.

The objective of this research is to conduct drag and CSS tests using various ceramic sliders and detailed material analyses of the interface component to elucidate the failure mechanisms, and particularly to address the question of why silicon carbide exhibits better friction and wear properties. The results of the study are the subject of this paper.

2 EXPERIMENTAL

2.1 Experimental apparatus and test procedure

Figure 1 shows the experimental apparatus used for drag, stiction and CSS tests. All tests were conducted in a clean environment with class 100 (less than 100 particles which are equal to larger than 0.3 μm in diameter per cubic feet) at an ambient temperature of 21 ± 1 °C. The disk drive is mounted in a plexiglass box which allows relative humidity (RH) to be controlled from 3 to 95 per cent. Three levels of

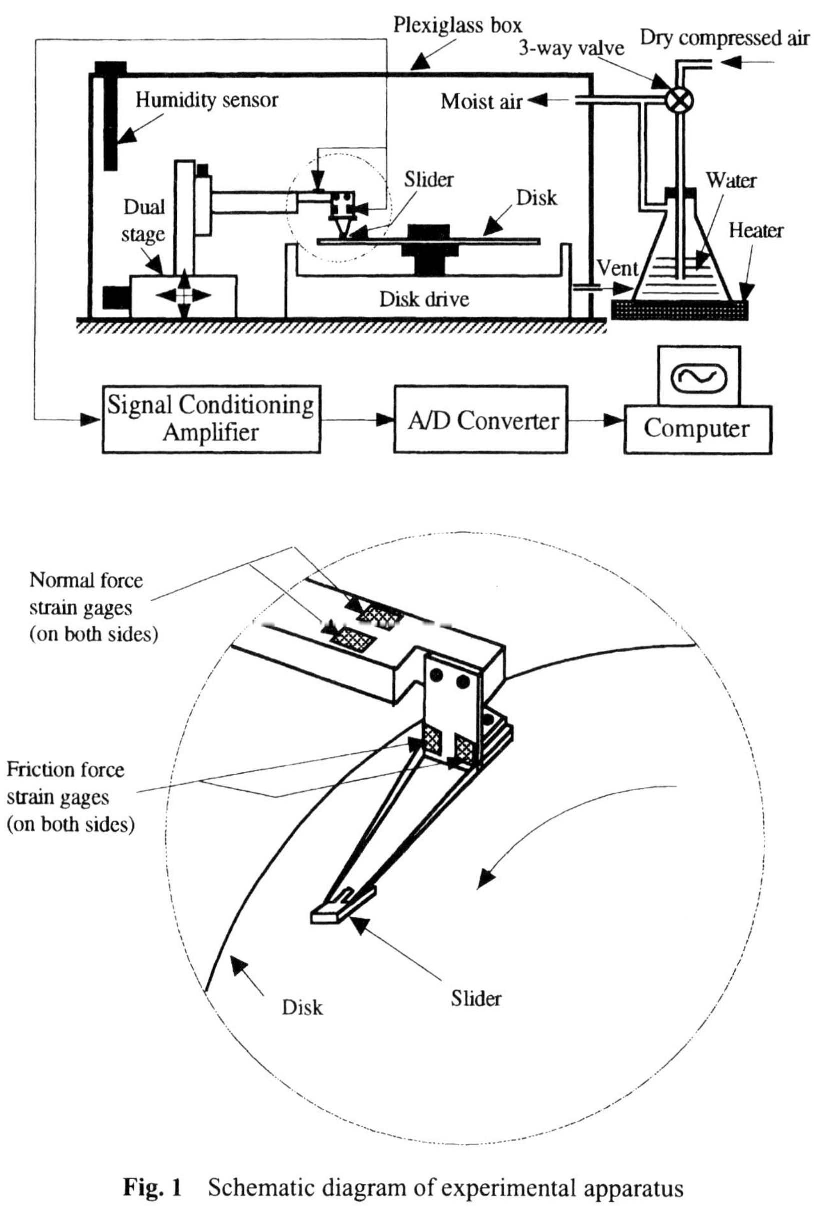

Fig. 1 Schematic diagram of experimental apparatus

relative humidity were fixed to 5, 45 and 85 per cent. A low humidity environment was obtained using bottled dry air. Lab ambient humidity was 45 ± 5% RH. High humidity was obtained by supplying humid air to the plexiglass box. The normal load and friction force were measured using strain gauges. The force signals, after suitable amplification using a signal conditioning amplifier, were fed to an A/D converter and digitized at a frequency of 25 Hz for drag tests and 3 kHz for stiction and CSS tests. Digitized data were recorded on a computer. For drag tests, data were averaged over five points and recorded. For CSS tests, all data were recorded and selected data were plotted.

2.1.1 *Drag tests*

The load selected for drag tests is 30 mN which approximately results in an apparent contact pressure of about 33 kPa. Tests were conducted at a track radius of 35.8 mm and a rotational speed of 200 r/min (0.75 m/s). A test was stopped when the head–disk interface failed. The failure criteria were as follows:

1. The coefficient of friction exceeds by a factor of 2 of the initial value.
2. A sudden increase of friction coefficient occurs.
3. A visible wear track occurs.

2.1.2 *Stiction tests*

For stiction tests, a rest time of 180 s was used. The disk was accelerated to a maximum velocity of 12 m/s in 3 s, which corresponds to an acceleration of 4 m/s^2. Five measurements of static friction force were made and an average value of these measurements was reported.

2.1.3 *CSS tests*

Periods for the acceleration to the full speed, steady rotation, deceleration and stop were selected as 10, 10, 10 and 30 s respectively. The maximum rotational speed was 3600 r/min. Tests were mostly conducted at a track radius of 35.8 mm which corresponds to a steady rotational speed of 13.5 m/s. The maximum tangential force during the start of a CSS cycle was recorded as the static friction force. The test was stopped when the head–disk interface met one of the failure criteria that was the same as that used in drag tests.

2.2 Material analyses

2.2.1 *SEM observation*

Surface morphologies of all the sliders were observed using a scanning electron microscope (SEM) (JEOL, JSM 820) with an electron beam of 10 kV. No conductive coating, with the exception of the calcium titanate slider, was used so as not to alter the surface.

2.2.2 *XPS analysis*

Surfaces of Al_2O_3–TiC and SiC sliders were analysed using X-ray photoelectron spectroscopy (XPS) (VG Scientific, ESCA Lab Mark 2) to detect if the surface was oxidized. XPS spectra were obtained by collecting characteristic photoelectrons emitted after radiation of Mg–K_α with 14 keV and 0.02 A, and with a spot size of 150 μm. The new slider surface, the surfaces after sliding and after 6 minutes of sputtering were analysed to detect any changes before and after sliding.

2.2.3 *FTIR analysis*

Surfaces of Al_2O_3–TiC and SiC sliders were analysed using Fourier transform infra-red (FTIR) to detect if the surface was hydroxidized. The FTIR spectra were obtained using an FTIR instrument (Bruker, Equinox 55) equipped with an FTIR microscope sampling accessory from the 4500 to 500 cm^{-1} frequency range. The instrument was used in the reflectance mode. Spectra were recorded using a narrow-band mercury cadmium telluride (MCT) detector and plotted as transmittance calculated from S/R, where S is the raw spectrum from the sample and R is the spectrum from the background. The spot size of the IR beam used was 25 μm in diameter with a scan number of 1000, in order to increase the sample volume to obtain a reasonably strong signal intensity.

Samples used for XPS analyses were used for FTIR analyses as well.

2.3 Contact angle measurement

The contact angle was measured using a contact angle measurement instrument (Ramè-Hart, NRL Contact Angle Goniometer 100-00). The contact angle measurements were generally reproducible to ±2°. The advancing contact angle was measured as the data. Coupons of solid materials for contact angle measurement used are Mn–Zn ferrite, Al_2O_3–TiC, calcium titanate, SiC and Al_2O_3, with the surface roughness similar to the slider surfaces and size of 10 mm × 10 mm × 2 mm. A silicon (111) wafer and unlubricated thin-film disk were measured for comparison. Liquids for the contact angle measurement are two kinds of PFPE lubricants (Z-Dol and Z-15) and water. Before measurement, the surfaces were carefully cleaned with acetone, fluorinert solvent (FC-72) and distilled water.

2.4 Lubricant transfer test

Lubricant transfer tests were conducted with ceramic specimens, the same as used in contact angle measurement, pressed on a lubricated thin-film disk at a normal load of 0.4 N (contact pressure of 4 kPa) for various periods of time. The rest time was varied from 1 to 10^5 s. An ellipsometer (**1**) was used to measure the film thickness of transferred lubricant film on the ceramic surface. The average value was obtained from the nine point locations and five repeated measurements were made at every location. The contact angle of ceramic material with water was also measured after lubricant transfer tests.

2.5 Test specimens

Mn–Zn ferrite, calcium titanate, Al_2O_3–TiC and SiC nanosliders (50 per cent) were used in this study. Physical properties of various materials can be found in Bhushan and Lowry (**14**) and Koinkar and Bhushan (**13**). Disks used in the study were coated with 20 nm thick hydrogenated diamond-like carbon (DLC) deposited by sputtering and a 2.6 nm thick perfluoropolyether (Demnum) lubricant film. The surface roughnesses of the bearing surface for all sliders and disks are given in Table 1.

3 RESULTS

3.1 Drag tests

Figure 2 shows the results of drag tests for Mn–Zn ferrite, calcium titanate, Al_2O_3–TiC and SiC sliders at relative

Table 1 Surface roughness measurements of various sliders and magnetic disk using AFM (scan size = 20 μm × 20 μm)

Materials	Standard derivation of surface heights σ (nm)	Peak-to-valley distance (nm)	Peak-to-mean distance (nm)	Autocorrelation distance (μm)
Mn–Zn ferrite slider	2.5	31.2	12.3	0.84
Calcium titanate slider	1.7	17.9	7.2	0.48
Al_2O_3–TiC slider	2.1	25.2	9.9	0.63
SiC slider	2.0	24.8	9.4	0.59
Magnetic disk	2.7	29.2	17.9	1.3

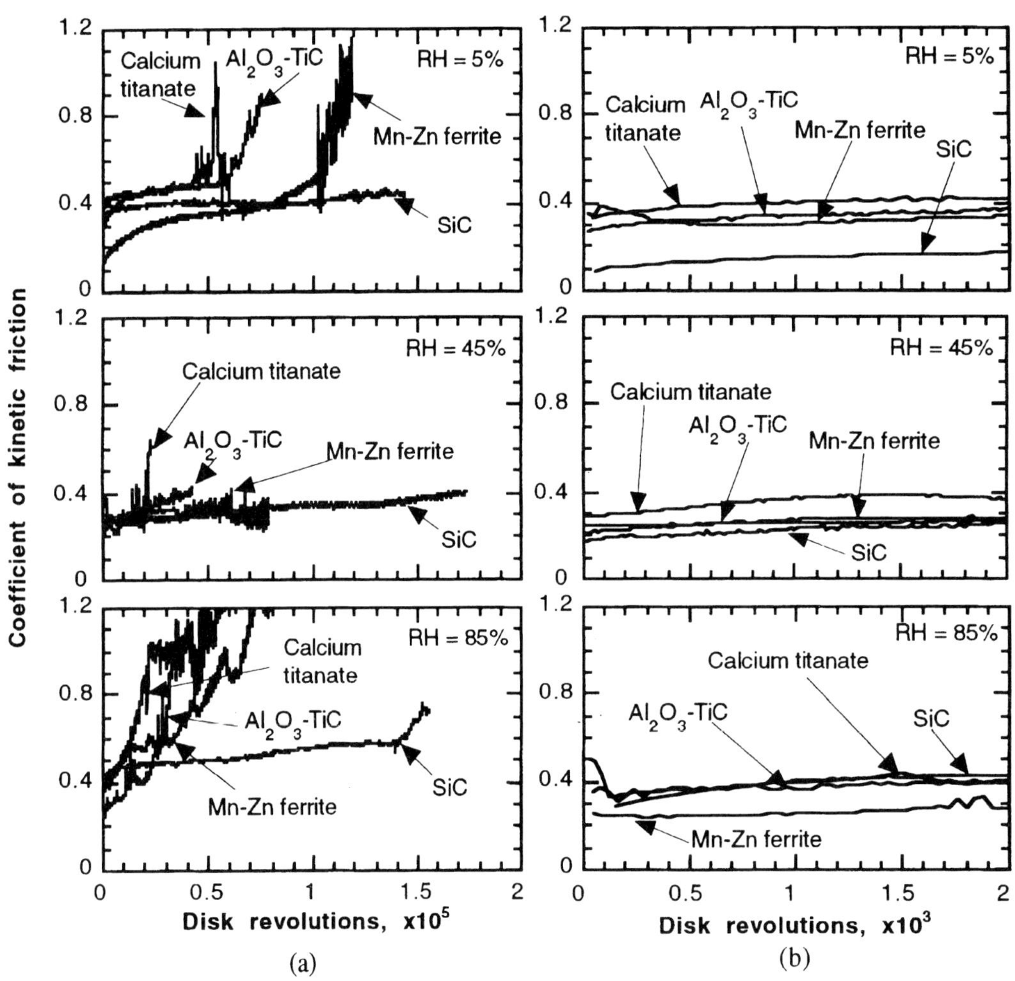

Fig. 2 Coefficient of kinetic friction as a function of disk revolutions in drag tests (a) until failure, (b) during initial 2000 disk revolutions

humidities of 5, 45 and 85 per cent. Figure 2a shows the friction data until failure occurs, while Fig. 2b shows the friction data during the first two thousand disk revolutions, where changes in the initial coefficient of kinetic friction are observed. The coefficient of kinetic friction increases with increasing disk revolutions at all humidities for all sliders. At relative humidities of 5 and 45 per cent, the coefficient of kinetic friction increases suddenly after a certain number of disk revolutions for Mn–Zn ferrite, calcium titanate and Al_2O_3–TiC sliders, while the coefficient of kinetic friction for SiC shows only a slight increase until the end of the test. The SiC slider also has the lowest coefficient of kinetic friction, followed by Mn–Zn ferrite, Al_2O_3–TiC and calcium titanate sliders during the first two thousand disk revolutions. At the relative humidity of 85 per cent, the coefficients of kinetic friction for Mn–Zn ferrite, calcium titanate and Al_2O_3–TiC sliders increase significantly with increasing disk revolutions and the durabilities are extremely limited. However, the SiC slider still maintains low friction until the disk revolution of 1.4×10^5, though the coefficient of kinetic friction during the first two thousand disk revolutions is not the lowest. Figure 3 summarizes the durability data for four kinds of sliders at various levels of relative humidity. It can be seen that SiC has the longest durability and maintains good durability even at higher relative humidities. Followed by SiC, the long durability is observed with Mn–Zn ferrite and Al_2O_3–TiC. The durability of calcium titanate is the shortest. The latter three sliders have decreased durability with increasing relative humidity and the durability is very short at the high relative humidity (85 per cent).

3.2 Stiction tests

Figure 4 shows typical friction profiles of Al_2O_3–TiC and SiC sliders. The coefficient of static friction (μ_s) is obtained by taking the maximum value of the friction force. Figure 5a presents static friction data for various sliders tested at the relative humidity of 45 per cent. The initial coefficient of kinetic friction is also presented in the figure for comparison. It can be seen that the SiC slider has the

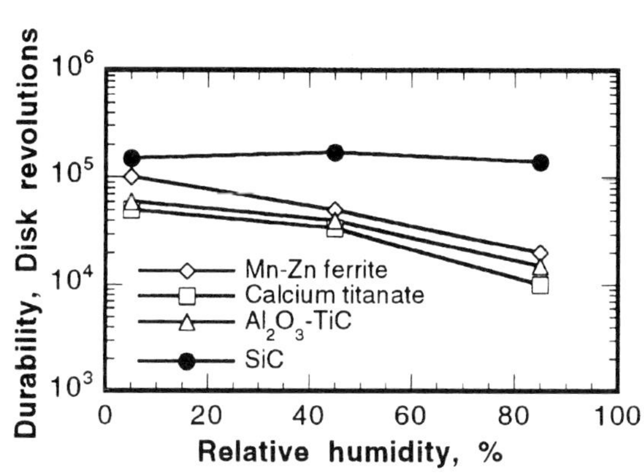

Fig. 3 Durability of various sliders at various levels of relative humidity in drag tests

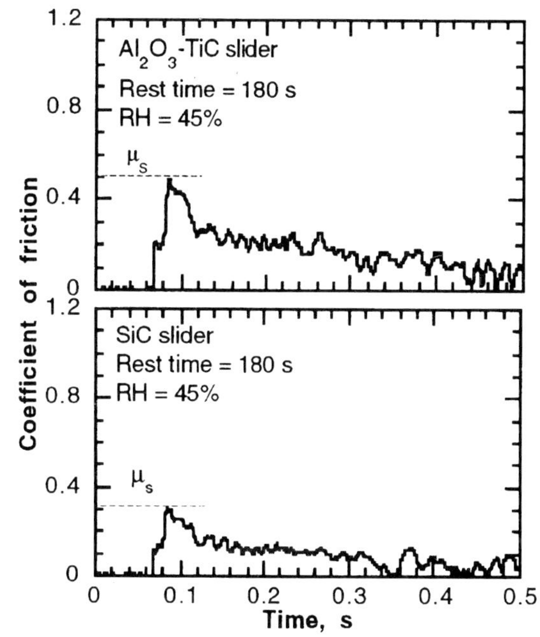

Fig. 4 Representative friction profiles for static friction measurements

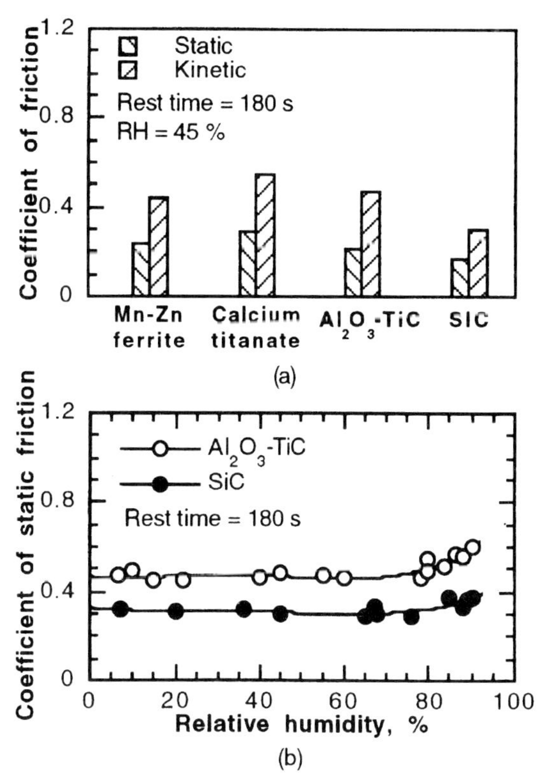

Fig. 5 (a) Typical data of coefficient of static and kinetic friction for various sliders. (b) Coefficient of static friction of two sliders as a function of relative humidity

lowest coefficient of static friction followed by Mn–Zn ferrite, Al_2O_3–TiC and calcium titanate sliders. Figure 5b shows the changes in the coefficient of static friction of Al_2O_3–TiC and SiC sliders with increasing relative humidity. It can be seen that the coefficient of static friction for the SiC slider is lower than that for the Al_2O_3–TiC slider at all humidities.

Figure 6 shows the friction profiles of Al_2O_3–TiC and SiC sliders during the initial five CSS cycles. It is interesting to note that the coefficient of static friction for the SiC slider at the first cycle is 0.74 and decreases in the following cycles, and then remains at a steady value which is about half of the first cycle from the third cycle onwards. However, the coefficient of static friction for the Al_2O_3–TiC slider shows only a little change with increasing CSS cycles. The significant decrease in the coefficient of static friction with increasing cycles at the beginning implies that something changes at the head–disk interface when an SiC slider is used. This phenomenon will be discussed later in combination with other results.

3.3 CSS tests

Figure 7a shows the results of CSS tests for Mn–Zn ferrite, calcium titanate, Al_2O_3–TiC and SiC sliders at a relative humidity of 45 per cent and for Al_2O_3–TiC and SiC sliders at relative humidities of 5 and 85 per cent as well. Durability data are summarized in Fig. 7b.

At the relative humidity of 45 per cent, the coefficient of static friction increases with increasing CSS cycles for Mn–Zn ferrite, Al_2O_3–TiC and calcium titanate sliders.

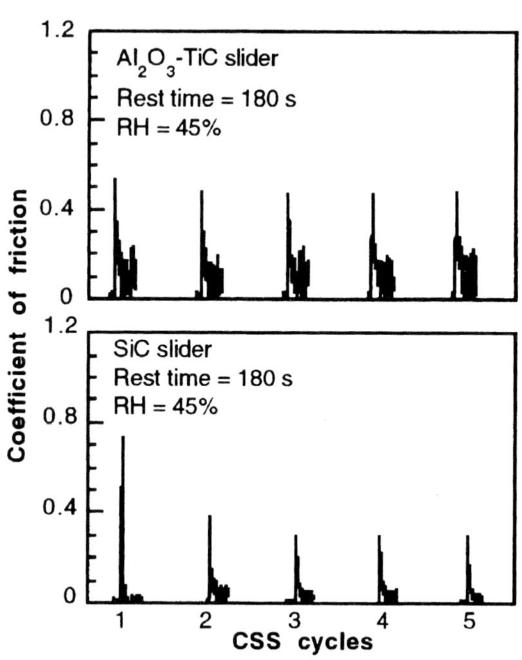

Fig. 6 Changes in coefficient of friction profiles with an increase in CSS cycles during the initial five cycles

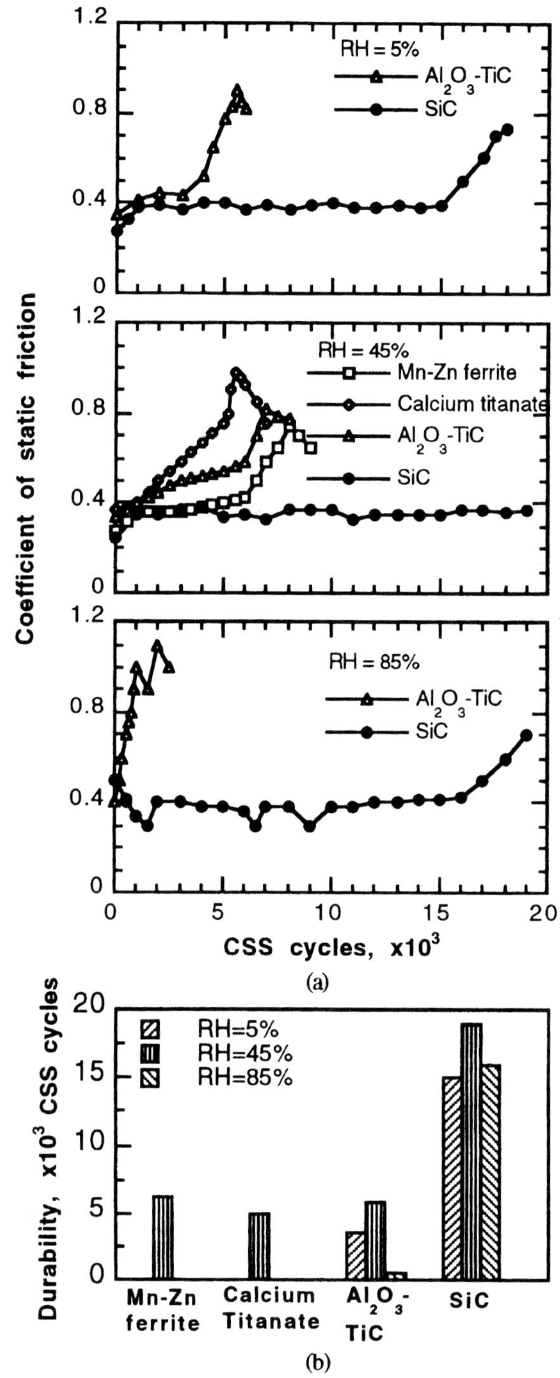

Fig. 7 (a) Coefficient of static friction as a function of CSS cycles in CSS tests at three levels of relative humidities. (b) Summary of durability of various slider materials in different humidity levels

The coefficient of friction increases up to a maximum and then drops at a certain number of CSS cycles at which the disk fails or reaches its lifetime, which is the durability of the interface. The durability for Mn–Zn ferrite, Al_2O_3–TiC and calcium titanate is 7.0×10^3, 6.0×10^3 and 5.0×10^3 CSS cycles respectively. The increase in the coefficient of static friction of the SiC slider is significantly less than that of the others. It still does not reach its lifetime after about 1.9×10^4 CSS cycles.

At the relative humidity of 5 per cent, the durability for the Al_2O_3–TiC slider is about 4.0×10^3 while that for SiC is about 1.5×10^4 CSS cycles. Compared with the durability at the relative humidity of 45 per cent, the durability for Al_2O_3–TiC and SiC sliders at the relative humidity of 5 per cent is shorter. This suggests that the existence of a reasonable amount of water vapour is better for head–disk interface operation.

At the relative humidity of 85 per cent, the coefficient of static friction for the Al_2O_3–TiC slider increases significantly from the beginning of the test. The durability is only about 1.0×10^3 CSS cycles. The coefficient of static friction for the SiC slider decreases from 0.5 to 0.4 in the first 2.0×10^3 CSS cycles and then remains approximately steady around the value of 0.4, finally increasing significantly after 1.6×10^4 CSS cycles.

3.4 SEM observations

Figure 8a shows the SEM micrographs of the Al_2O_3–TiC slider surface after 1.0×10^4 disk revolutions in the drag test. Lubricant stain was observed on the bearing surface, which probably was formed from the transfer of lubricant on to the lubricated disk. That is an important observation which may closely relate to the tribological performance of the head–disk interface since keeping the interface lubricated results in prevention of an increase in friction. The lubricant stain was not observed on Mn–Zn ferrite and SiC sliders. That implies that either no lubricant was transferred from the lubricated disk to the slider or the transferred lubricant is homogeneously distributed on the slider bearing surface which makes it indistinguishable in SEM. This question will be addressed with later contact angle measurements and lubricant transfer tests. In addition, although a lubricant stain on the bearing surface of calcium titanate slider was observed, a high-quality SEM micrograph could not be obtained because of electrical charging of the sample surface during the SEM observation.

Figure 8b shows the SEM micrographs of Mn–Zn ferrite, Al_2O_3–TiC, calcium titanate and SiC slider surfaces after failure in drag tests. It can be seen that the debris has accumulated on the bearing surfaces, especially around the leading edge. The debris is believed to be responsible for the final failure of the head–disk interface by resulting in either a large increase in coefficient of friction or a visible wear track.

Figure 9 shows the SEM micrograph of Al_2O_3–TiC and SiC slider surfaces after failure in CSS tests. Both Al_2O_3–TiC and SiC slider surfaces are covered with debris which is similar to that observed after drag tests. This means that the final stages for both drag and CSS tests are always accompanied with accumulated debris, especially around the leading edge, which is believed to be responsible for the large increase in the coefficient of friction and making a wear track.

3.5 Contact angle measurements

In order to study wettability of various slider materials, contact angle measurements were made. Figure 10 shows the results of contact angle measurements of typical lubricants and water against slider materials. The lubricants are Z-Dol and Z-15, which are at present commonly used in the hard disk industry. The contact angle data for lubricants on silicon, unlubricated disk and alumina are also given in the figure for comparison. It can be seen that the contact angles of Z-Dol, Z-15 and water against SiC are 7, 13 and 44° respectively, which are comparable to the unlubricated disk, while against Mn–Zn ferrite they are 15, 22 and 82°, against Al_2O_3–TiC they are 14, 27 and 77° and against calcium titanate they are 17, 32 and 84° respectively. The contact angles of Z-Dol, Z-15 and water on SiC are significantly smaller than those against Mn–Zn ferrite, Al_2O_3–TiC and calcium titanate. These data indicate that lubricant will wet the SiC and minimize subsequent wetting by water. This may closely relate to low stiction and good durability of the SiC head slider against the disk surface.

3.6 Lubricant transfer tests

Figure 11a shows the transferred film thickness of lubricant on the slider materials as a function of rest time after static contact with the disk surface. It can be seen that the film thickness on the SiC increases with time and approaches a value of about 4 nm after a rest time of 10^5 s. This suggests that the lubricant can be easily transferred from the lubricated disk to the bearing surface of the SiC slider. The lubricant film can also be measured for the other materials when the rest time is over 10^4 s. Lubricant transfer is most in Mn–Zn ferrite followed by Al_2O_3, Al_2O_3–TiC and calcium titanate. Since the film thickness on Al_2O_3–TiC and calcium titanate materials varies significantly in terms of measured location, this implies that the transferred lubricant is not uniformly distributed on the surface. This is in good agreement with the SEM observation of the slider bearing surface (see Fig. 8) which shows a lubricant stain of local lubricant transfer for the Al_2O_3–TiC slider. Combined with the results of Fig. 11a, it can be seen that no stain found on the SiC slider surface with the SEM observation probably indicates a uniform lubricant transfer.

The contact angle of water against tested slider materials was also measured after the lubricant transfer test. Contact angle data as a function of rest time are shown in Fig. 11b. The contact angle for SiC increases significantly with increasing rest time initially and then approaches a value of 65°. This suggests that the SiC surface changes from being more hydrophilic to more hydrophobic when the surface is covered with transferred lubricants. It is believed to be the reason why the coefficient of static friction for SiC drops dramatically after initial one or two CSS cycles (see Fig. 6), since a hydrophobic surface is supposed to have less stiction (**1**). The contact angles for Mn–Zn ferrite, Al_2O_3–TiC, calcium titanate and Al_2O_3 do not change when the

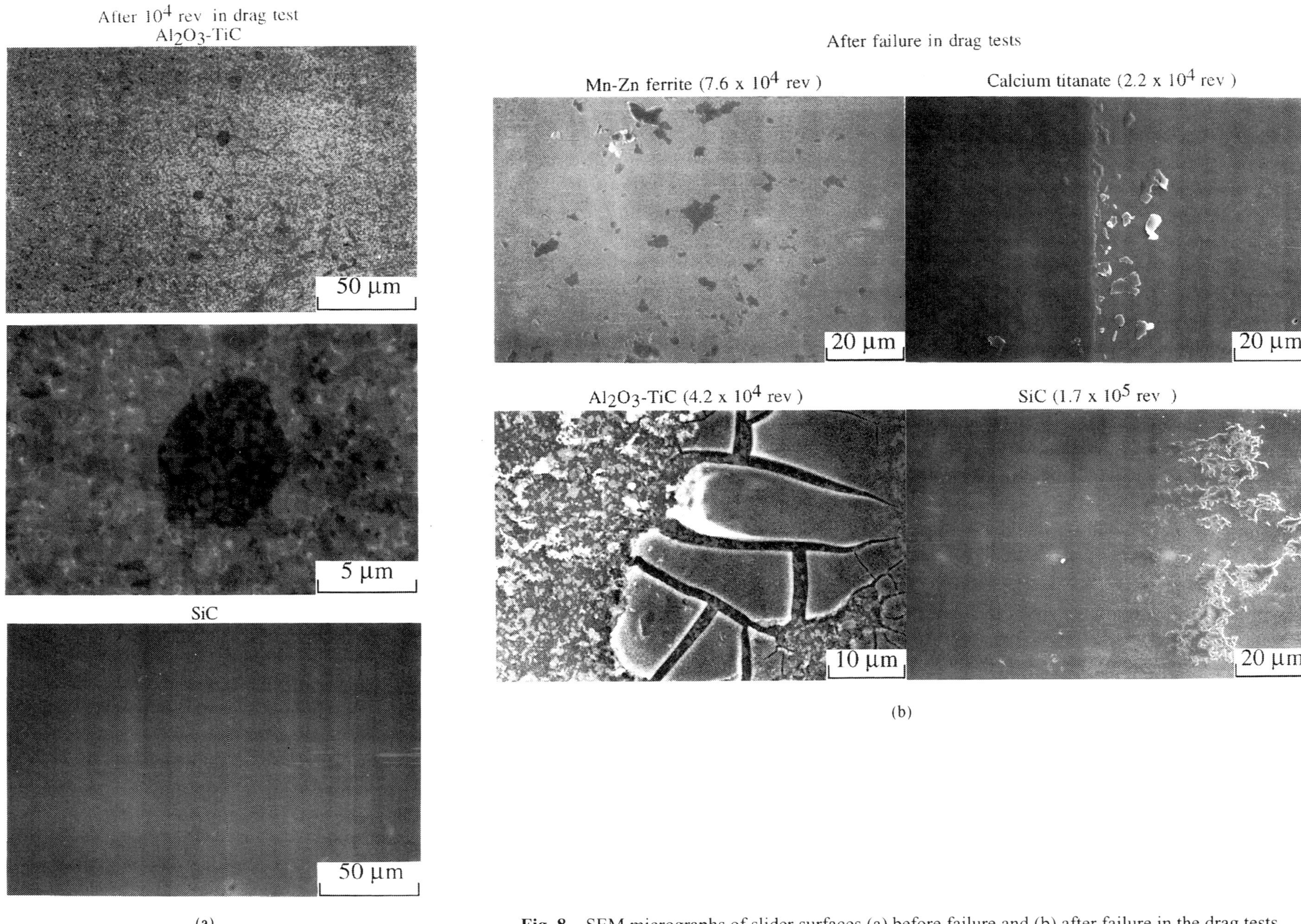

Fig. 8 SEM micrographs of slider surfaces (a) before failure and (b) after failure in the drag tests

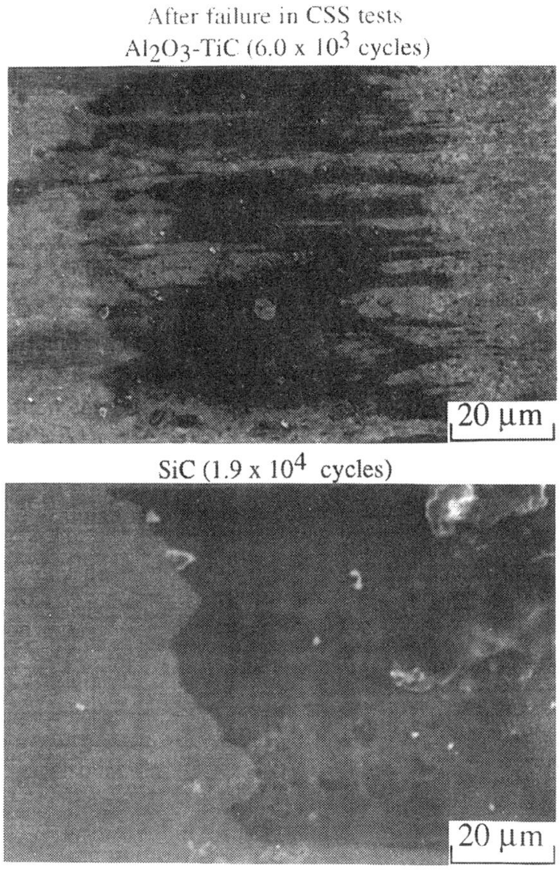

Fig. 9 SEM micrographs of two slider surfaces after CSS tests

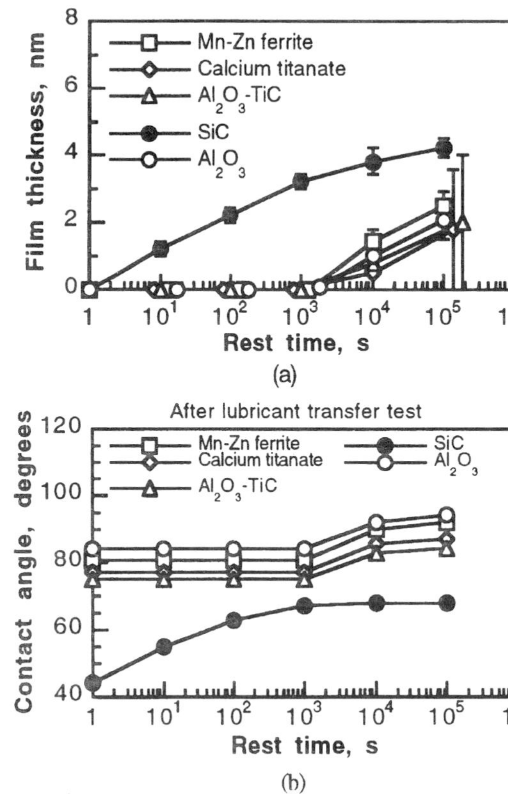

Fig. 11 (a) Thickness of transferred lubricant film and (b) the change of contact angle with water as a function of rest time of various material specimens after contact with a lubricated disk

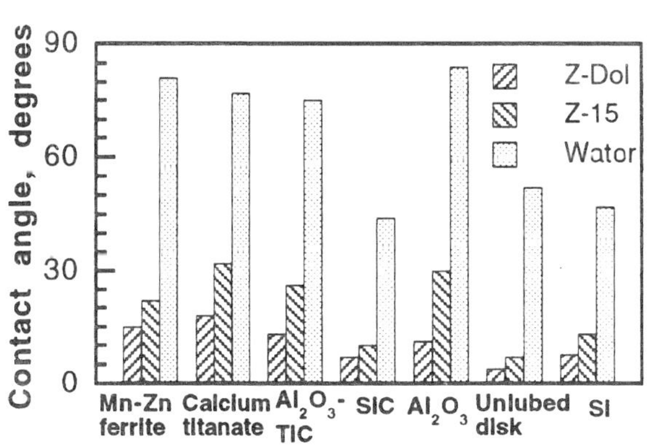

Fig. 10 Contact angle measurements for two lubricants and water on various slider materials, magnetic disk and silicon (for reference).

rest time is less than 10^3 s and increase by about 10° when the rest time is over 10^4 s. Compared with the results of transferred lubricant thickness, it can be concluded that the lubricant transfer is responsible for the increase in contact angle of water against slider materials with increasing rest time. The reason why SiC has less contact angle with water compared with other slider materials may be attributed to the oxidized surface which is hydrophilic.

3.7 XPS analyses

Figure 12 shows the results of XPS analyses for SiC and Al_2O_3–TiC sliders after the drag test at relative humidities of 5, 45 and 85 per cent. Figure 12a shows the results based on Si $2p^3$ electron and O 1s electron analyses. When the surface is oxidized, the binding energy of Si $2p^3$ electrons shows that there is growth in the SiO_2 peak (**15**). A deconvolution technique was used to estimate the relative amount of silicon dioxide and silicon carbide.

It can be seen from the Si $2p^3$ electron analyses that there is some amount of silicon dioxide on the slider virgin surface. The results of the slider surface after sliding at relative humidities of 5, 45 and 85 per cent show that the surfaces are more oxidized after sliding. The higher the relative humidity the more is the oxidation. It can be concluded that the tribochemical reaction occurs on the slider surface during sliding, which leads to oxidation of SiC on the surface, and higher humidity increases the rate of oxidation. After surfaces were analysed, they were sputtered with argon ion for 6 min and then analysed again. Only one typical result is shown in the figure since all

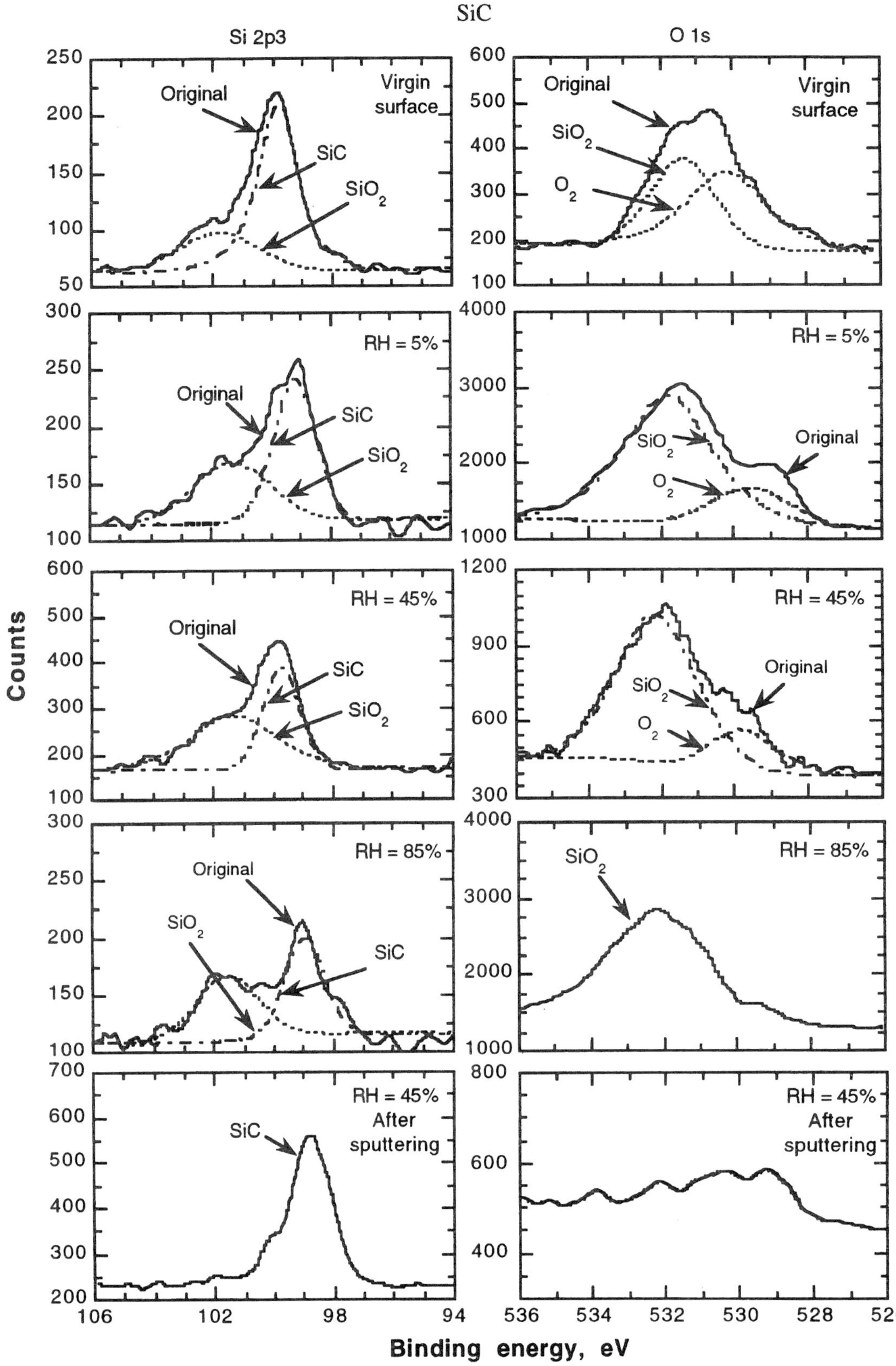
SiC
Si 2p3
O 1s
Original
Virgin surface
SiC
SiO2
O2
RH = 5%
RH = 45%
RH = 85%
RH = 45% After sputtering
Counts
Binding energy, eV

(a)

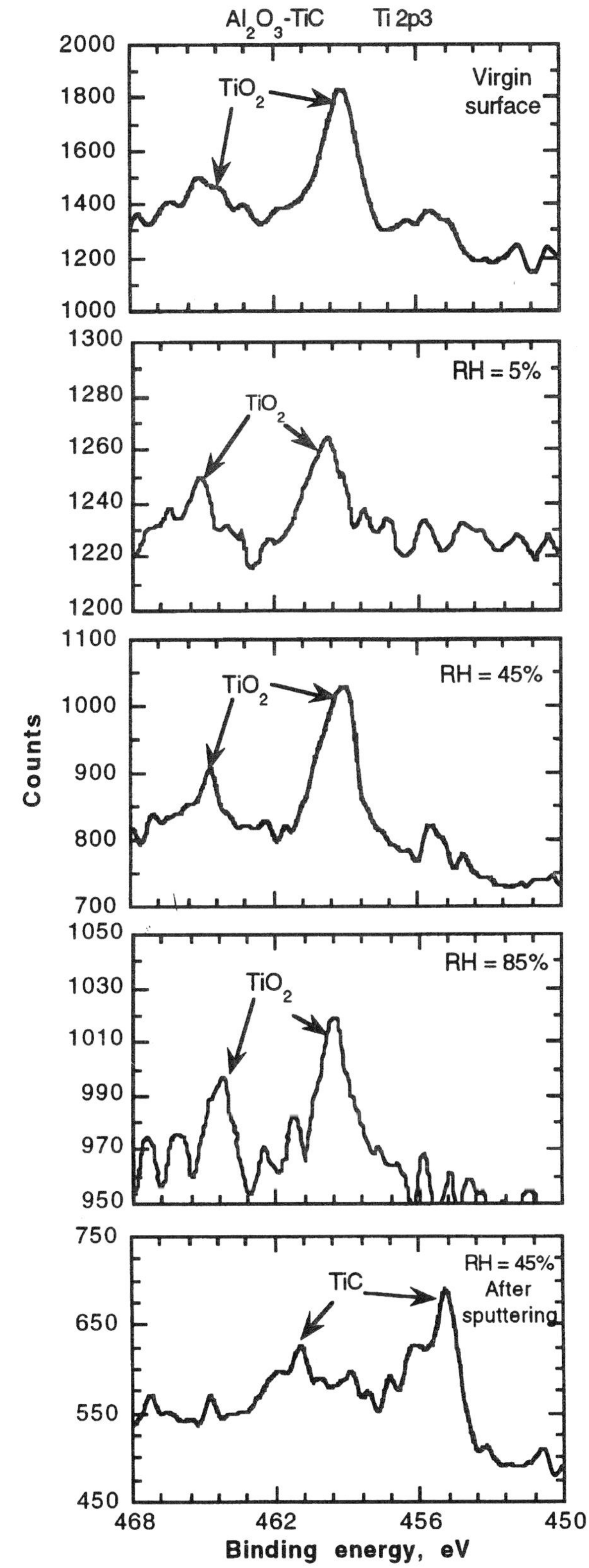

(b)

Fig. 12 XPS analyses of slider surfaces before and after drag tests for (a) SiC and (b) Al_2O_3–TiC sliders

analyses after sputtering show similar results. It can be seen that only Si $2p^3$ electrons for SiC appear on the spectrum, which suggests that the thickness of oxide is less than 2 nm since 6 min of sputtering approximately leads to a 2 nm loss of surface material.

From the O 1s electron analyses, it can be seen that there are comparable amounts of silicon dioxide and adsorbed oxygen on the virgin surface. After sliding, the relative amount of silicon dioxide increases while that of adsorbed oxygen decreases. The relative amount of silicon dioxide increases with increasing relative humidity used in the sliding tests. At high humidity (RH = 85 per cent), almost all of O 1s electrons come from the silicon dioxide rather than adsorbed oxygen. That suggests that there is a larger amount of silicon dioxide formed on the slider surface and less adsorbed oxygen at the head–disk interface at a higher relative humidity. Since neither silicon dioxide nor oxygen was detected after sputtering, this confirms that the thickness of silicon oxide is less than 2 nm.

Figure 12b shows the XPS data for the Al_2O_3–TiC slider based on Ti $2p^3$ electrons. It can be seen that the surface is covered with titanium dioxide rather than TiC for the virgin surface. No change was found after sliding at all humidity levels. The thickness of titanium dioxide is less than 2 nm since there is no more titanium dioxide detected after 6 min of sputtering.

3.8 FTIR analyses

Figure 13 shows the FTIR analysis results of Al_2O_3–TiC and SiC sliders after drag tests at relative humidities of 45 and 85 per cent. The FTIR spectra for the virgin surface and after drag tests at the relative humidity of 5 per cent are not shown in the figure since they appear similar to that after drag tests at the relative humidity of 45 per cent. It can be seen from the figure that there are adsorbed IR signals for the SiC slider tested at the relative humidity of 85 per cent at wave numbers from 3600 to 3900 cm^{-1}, which indicates the existence of Si—OH stretches (**16**). Therefore, it can be concluded that the surface consists of hydroxide, $Si(OH)_x$, rather than oxide when the slider is slid at high humidity. Since there are not any adsorbed signals on the IR spectra for the Al_2O_3–TiC slider tested, even at the relative humidity of 85 per cent, it may be concluded that there is no or extremely small amounts of hydroxide on the Al_2O_3–TiC slider surface.

4 DISCUSSION

All drag tests, stiction tests and CSS tests show that the SiC slider has the lowest coefficient of static and kinetic friction and the longest durability as compared with Mn–Zn ferrite, Al_2O_3–TiC and calcium titanate sliders (see Figs 2, 3, 5 and 7). The difference in tribological performances is more significant at high relative humidity since the SiC slider still maintains long durability while the other sliders fail quickly at the relative humidity of 85 per cent. These observations suggest that there must be some unique properties in the SiC material. Based on the present investigation, it is believed that the following mechanisms are responsible for the low coefficient of friction and long durability of the SiC slider.

One of the properties of SiC is the ease of lubricant transfer. The contact angles of PFPE lubricants on SiC are significantly smaller than those on other slider materials and are comparable to the thin-film disk (see Fig. 10). Consequently, SiC can be easily wetted and adsorbed by lubricants since a low contact angle means an easy wetting of the lubricant on the solid materials. Figure 11 confirms that lubricants can easily be transferred from the lubricated disk during the static contact. It is believed that sliding facilitates lubricant transfer more easily than static contact. The local lubricant stain found on the Al_2O_3–TiC slider surface after sliding with SEM is also direct evidence of lubricant transfer (see Fig. 8); however, the film is patchy. Based on the fact that the lubricant layer on the thin-film disk is composed of adsorbed layer and mobile layer, it is reasonable to imagine that the transferred lubricant is mainly from the mobile layer on the disk (**1**). Figure 14 shows schematically the transfer process of lubricant from a lubricated disk to a slider. Transferred lubricant is adsorbed on the surface of SiC slider and forms a protective layer. That reduces the risk of direct solid contact, reduces wear and provides longer durability. Transferred lubricant layer on the SiC slider also changes the surface from hydrophilic to hydrophobic, which reduces the stiction force especially at high relative humidity (see Figs 6 and 10). However, transferred lubricant may increase the tendency to collect debris, which is undesirable.

The other property of SiC is related to tribochemical reaction. XPS analyses show that the silicon dioxide is formed on the surface of the SiC slider during sliding and FTIR analyses further show that the oxide is hydrated. Similar results have been reported by several investigators and it is concluded that silicon-based non-oxide ceramics, such as silicon nitride and silicon carbide, easily react with oxygen derived from oxygen in the air or from water vapour (**17, 18**). Oxide ceramics will react with water whether it is present as liquid or as vapour, sliding in humid air. In the case of non-oxide ceramics, oxidation can be followed by hydration. For example, for SiC, the following reactions can take place in the humid environment:

$$SiC + 2H_2O \rightarrow SiO_2 + CH_4$$

and

$$SiO_2 + 2H_2O \rightarrow Si(OH)_4$$

These tribological effects are responsible for the influence of atmospheric composition and sliding on friction which is commonly observed in sliding (**3, 18–23**). These hydroxides are gel-like silica which work as boundary lubricants and lead to a low friction coefficient (**18, 21, 24**). They also protect the surface from wear and avoid the formation of wear debris. It is obvious that the hydroxides formed on the SiC slider surface benefit the head–disk interface and

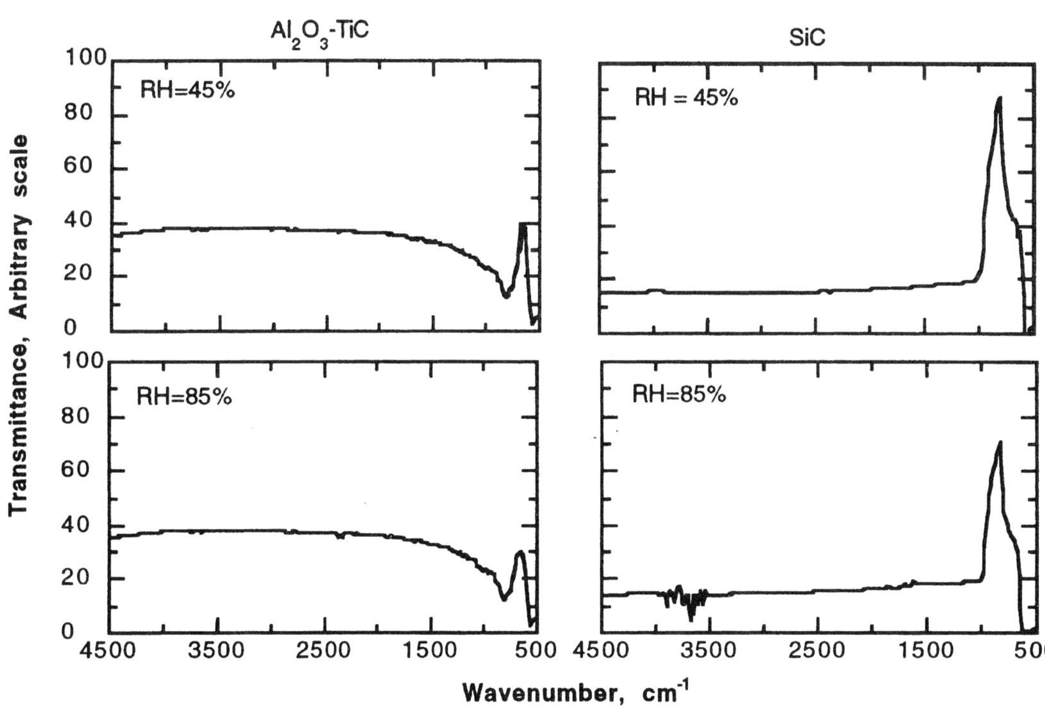

Fig. 13 FTIR analyses of slider surfaces before and after drag tests for (a) Al_2O_3–TiC and (b) SiC sliders

result in a low coefficient of friction and provide a long durability. Since high humidity makes easy formation of hydroxides on the SiC surface, it is believed to be an important reason why the SiC carbide slider still has a good durability even at high relative humidity while the other slider materials fail quickly.

Al_2O_3 is known to be catalytic to the degradation of the perfluoropolyether lubricants (**25**). Since tribological performances of not only Al_2O_3–TiC but Mn–Zn ferrite and calcium titanate as well are poor as compared to SiC, it is believed that the catalytic activity of sliders is not responsible for poor tribological performance.

5 CONCLUSIONS

The SiC slider is found to have the best stiction, kinetic friction and durability among the sliders tested in drag and CSS tests. The reasons for low kinetic friction and good durability can be attributed to the lubricant transfer, which leads to the formation of a lubricant layer on the slider surface, and tribochemical reaction, which leads to the formation of a gel-like hydroxide layer. Both of them protect the head–disk interface from direct solid contact. The reasons for low stiction may be that SiC wetted by lubricant exhibits a high contact angle, which minimizes water condensation and consequently meniscus contribution. The following are detailed conclusions:

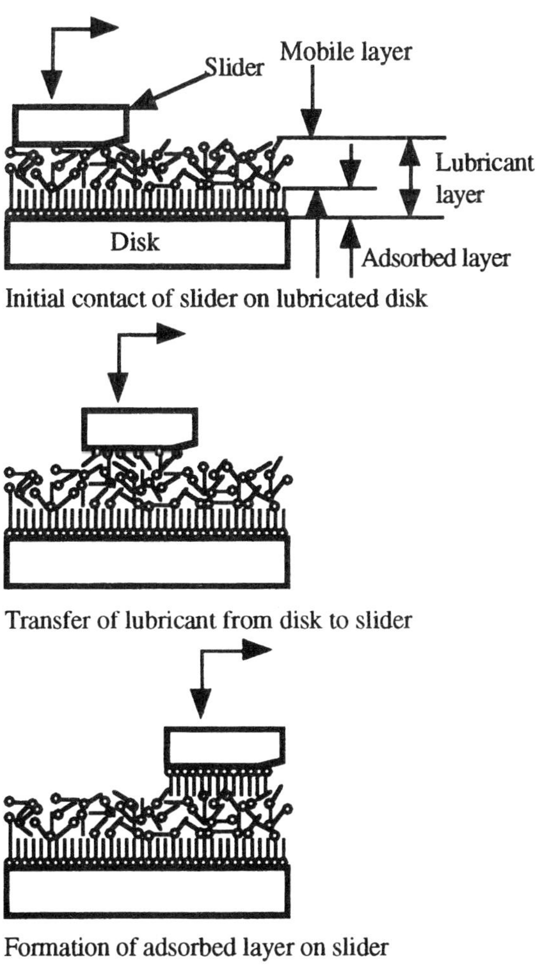

Fig. 14 Schematic model for lubricant transfer from a lubricated disk to a slider

1. The SiC slider has the lowest coefficient of static and kinetic friction and the longest durability, followed by Mn–Zn ferrite, Al_2O_3–TiC and calcium titanate sliders.
2. High relative humidity (85 per cent) reduces the durability of the Al_2O_3–TiC slider. The high stiction force induced by adsorbed water in the head–disk interface is responsible for the shorter durability. However, the SiC slider exhibits a long durability at high humidity. Hydroxides formed on the SiC surface result in a lower coefficient of friction which leads to a long durability.
3. Local lubricant stain was found on Al_2O_3–TiC and calcium titanate slider surfaces after sliding. However, it was not found on Mn–Zn ferrite and SiC slider surfaces.
4. PFPE lubricants (Z-Dol and Z-15) have the lowest contact angles against SiC compared with those against Mn–Zn ferrite, Al_2O_3–TiC and calcium titanate. This indicates that SiC is easily wetted and transferred by the lubricants.
5. The thickness of transferred lubricant on the SiC surface reaches 4 nm in static contact tests. Mn–Zn ferrite shows a thinner thickness of transferred lubricant layer compared with SiC, while Al_2O_3–TiC and calcium titanate only have locally transferred lubricant layers. After lubricant transfer, the SiC surface becomes more hydrophobic which is desirable for low stiction.
6. Oxide was found on the virgin surface of SiC and Al_2O_3 TiC sliders. The amount of oxide on the SiC slider surface increases after sliding and high relative humidity leads to a larger amount of oxide formation. No changes were found on the Al_2O_3–TiC slider surface before and after sliding. The oxide thicknesses for both SiC and Al_2O_3–TiC are less than 2 nm.
7. Hydroxides were found on the SiC surface after sliding at high relative humidity (RH = 85 per cent) while no hydroxide was found on the Al_2O_3–TiC surface.

ACKNOWLEDGEMENTS

The ceramic materials for the construction of sliders and specimens used in this study were supplied by Coors Ceramics Components (AD995 Al_2O_3), 3M Ceramic Materials Department (3M 310 Al_2O_3–TiC), Philips Components (3H7 polycrystalline Mn–Zn ferrite and

861-D calcium titanate), the Norton Company Electronics (ČRB093 SiC). The authors thank Mr Bob Wood of Brumko Magnetics Corp., Elkhorn, Nebraska, for preparation of sliders and Mr Matt Dugas of Advanced Research Corp., Minneapolis, Minnesota, for preparation of ceramic specimens. The authors also thank Mr John Mitchell and Mr Joel Haynes for technical assistance in SEM and XPS analyses respectively. Financial support for this project was provided by the industrial membership of the Computer Microtribology and Contamination Laboratory.

REFERENCES

1 **Bhushan, B.** *Tribology and Mechanics of Magnetic Storage Devices*, 2nd edition, 1996 (Springer-Verlag, New York).

2 **Buckley, D. H.** and **Miyoshi, K.** Friction and wear of ceramics. *Wear*, 1984, **100**, 333–353.

3 **Ishigaki, H., Kawaguchi, I., Iwasa, M.** and **Toibana, Y.** Friction and wear of hot pressed silicon nitride and other ceramics. *Trans. ASME, J. Tribology*, 1986, **108**, 514–520.

4 **Jahanmir, S.** (Ed.) *Friction and Wear of Ceramics*, 1994 (Marcel Dekker, New York).

5 **Bhushan, B.** Magnetic slider/rigid disk substrate materials and disk texturing techniques—status and future outlook. *Adv. Inf. Stor. System*, 1993, **5**, 175–209.

6 **Chandrasekar, S.** and **Bhushan, B.** Friction and wear of ceramics for magnetic recording applications—Part I: a review. *Trans. ASME, J. Tribology*, 1990, **112**, 1–16.

7 **Chandrasekar, S.** and **Bhushan, B.** Friction and wear of ceramics for magnetic recording applications—Part II: friction measurements. *Trans. ASME, J. Tribology*, 1991, **113**, 313–317.

8 **Chu, M. Y., Bhushan, B.** and **DeJonghe, L. C.** Wear behavior of ceramic sliders in sliding contact with rigid magnetic thin-film disks. *Tribology Trans.*, 1992, **35**, 603–610.

9 **Bair, S., Green, I.** and **Bhushan, B.** Measurement of asperity temperatures of a read/write head slider bearing in hard magnetic recording disks. *Trans. ASME, J. Tribology*, 1991, **113**, 547–554.

10 **Bhushan, B.** Magnetic head–media interface temperatures—Part 3: application to rigid disks. *Trans. ASME, J. Tribology*, 1992, **114**, 420–430.

11 **Koka, R.** Wear of silicon carbide in sliding contact with lubricated thin-film rigid disk. *Tribology Trans.*, 1994, **38**, 417–423.

12 **Koka, R., Holtz, A.** and **Armatis, F.** Tribology of different ceramic sliders on lubricated thin-film disks. *Tribology Trans.*, 1997 (in press).

13 **Koinkar, V. N.** and **Bhushan, B.** Microtribological studies of Al_2O_3, Al_2O_3–TiC, polycrystalline and single-crystal Mn–Zn ferrite and SiC head slider materials. *Wear*, 1996, **202**, 110–122.

14 **Bhushan, B.** and **Lowry, J. A.** Friction and wear studies of various head materials and magnetic tapes in a linear mode accelerated test using a new nano-scratch technique. *Wear*, 1995, **190**, 1–15.

15 **Klauber, C.** X-ray photoelectron and Auger electron spectroscopy. *Microbeam Analysis*, 1995, **4**, 341–368.

16 **Herres, W.** *HRGC-FTIR: Capillary Gas Chromatography—Fourier Transform Infrared Spectroscopy*, 1987 (Dr A. H. Verlag, Heidelberg, Germany).

17 **Dong, X.** and **Jahanmir, S.** Wear transition diagrams for silicon carbide. *Tribology Int.*, 1993, **28**, 559–572.

18 **Xu, J., Kato, K.** and **Hirayama, T.** The transition of wear mode during the running-in process of silicon nitride sliding in water. *Wear*, 1997, **205**, 55–63.

19 **Tomizawa, H.** and **Fischer, T. E.** Friction and wear of silicon nitride and silicon carbide in water: hydrodynamic lubrication at low sliding speed obtained by tribochemical wear. *ASLE Trans.*, 1987, **30**, 41–46.

20 **Fisher, T. E.** Tribochemistry. *Ann. Rev. Mater. Sci.*, 1988, **18**, 303–323.

21 **Mizuhara, K.** and **Hsu, S. M.** Tribochemical reaction of oxygen and water on silicon surface. In *Wear Particles: From the Cradle to the Grave* (Eds D. Dowson, C. M. Taylor, T. H. C. Childs, M. Godet and G. Dalmaz), 1992, pp. 323–328 (Elsevier Science Publishers BV, Amsterdam).

22 **Mizuhara, K.** and **Hsu, S. M.** Tribochemistry of silicon-based ceramics under aqueous solution. In *Proceedings of Eurotrib '93*, Budapest, 30 August–2 September 1993, Vol. 3, pp. 52–57.

23 **Bhushan, B.** and **Venkatesan, S.** Friction and wear studies of silicon in sliding contact with thin-film magnetic rigid disks. *J. Mater. Res.*, 1993, **8**, 1611–1628.

24 **Sasaki, S.** The effects of the surrounding atmosphere on the friction and wear of alumina, zirconia, silicon carbide and silicon nitride. *Wear*, 1989, **134**, 185–200.

25 **Bhushan, B.** Magnetic recording surfaces. In *Characterization of Tribological Materials* (Ed. W. A. Glaeser), 1993, pp. 116–133 (Butterworths–Heinemann, Boston, Massachusetts).

Head–tape interface technology

Friction, wear and magnetic performance of metal evaporated and particulate magnetic tapes

S T Patton and **B Bhushan**
Computer Microtribology and Contamination Laboratory, Department of Mechanical Engineering, The Ohio State University, Columbus, Ohio, USA

Abstract: Metal evaporated (ME), metal particle (MP) and barium ferrite (BaFeO) magnetic tapes are leading candidates for ultra-high-density magnetic tape recording applications. Using a commercial video cassette recorder as a magnetometer and the Wallace equation, changes in the r.m.s. head output level were correlated to changes in head-to-tape spacing as tape wear occurred during play/rewind cycling tests. Interface stability and recording performance at a 0.6 μm recording wavelength were measured to bit level resolution using a drop-out counter. Some pause mode testing was done for comparison with streaming mode experiments. Methodologies to measure head and tape wear were developed and applied to worn specimens. Development of the experimental apparatus with nanometer vertical and submicrosecond temporal resolutions has enabled unprecedented understanding of the interplay of friction, wear and surface topography in a sliding contact. Waviness of ME tape and high roughness of BaFeO tape led to poor high-density recording performance as compared to the excellent performance of the flat MP tape. Interface stability generally improved as the tapes were used and worn smooth by the rotary heads, and head-to-tape spacing was reduced by about 10 nm over 1000 play/rewind cycles for the tapes. ME tape showed the least durability of all of the tapes, and damage areas initiated at high points or bumps on the tape surface were connected by lateral cracks (driven by longitudinal tension) across the tape width at tape failure after about 800 play/rewind cycles. MP tape performance improved gradually through 1000 play/rewind cycles. High asperities on the virgin tapes were severed off the tapes during the record pass, and those that remained on the tape surface increased friction force and tape wear by three-body abrasion early on in play/rewind cycling tests. Lower friction and virtually no wear were observed later in cycling tests when loose wear debris were no longer on the tape surface and the wear mechanism was adhesive. The wear coefficient in the streaming mode was larger than that in the pause mode due to abrasive particles in the contact interface in the streaming mode. MP and BaFeO tapes caused head stains on the metal core and glass surfaces of composite metal-in-gap heads. Anisotropic high tape stiffness led to asymmetrical head wear and head contours with large radii of curvature.

Keywords: friction, wear, magnetic tape and head, spacing loss, tribological and magnetic performance

NOTATION

d	head-to-tape spacing
d_1	head-to-tape spacing for MP tape during the first play/rewind cycle
d_s, d_p	head-to-tape spacing in streaming and pause modes respectively
dB	head output decibel level
F	friction force
F_s, F_p	friction force in streaming and pause modes respectively
H	hardness
k	wear coefficient
k_s, k_p	wear coefficient in streaming and pause modes respectively
l	data track length
L	length of the chord which has end-points at either end of the tape contact region of the head surface
R	radius of curvature
T	tape tension
T_s, T_p	tape tension in streaming and pause modes respectively
v_h	head velocity
V	wear volume
V_s, V_p	wear volume in streaming and pause modes respectively
w	data track width

The MS was received on 17 September 1996 and was accepted for publication on 8 August 1997.

W normal load
x sliding distance
x_s, x_p sliding distance in streaming and pause modes respectively

α fractional contact area
α_s, α_p fractional contact area in streaming and pause modes respectively
λ recording wavelength
μ coefficient of friction
μ_s, μ_p coefficient of friction in streaming and pause modes respectively

1 INTRODUCTION

In order to further increase recording densities, new magnetic tapes are being used and continue to be developed for high-density magnetic recording applications (**1–3**). Metal evaporated (ME), metal particle (MP) and barium ferrite (BaFeO) tapes are leading candidates for consumer digital audio, video and data processing applications. Higher tape smoothness, thinner magnetic layers and larger head-to-tape relative velocities place new demands on the durability of the head-to-tape interface in ultra-high-density recording applications. High surface energy, reactivity and ductility of the metal particles and films may affect the durability of the interface (**2**). ME tapes introduce additional tribological concerns such as corrosion resistance, crack formation, adhesion of coating to substrate and chemical stability of topical lubricant, some of which are not major concerns with particulate tapes (MP and BaFeO).

There is still debate as to which of the emerging tape technologies is best suited for use in the next generation of ultra-high-density tape recorders. The thin continuous reactively evaporated magnetic coating of ME tape has among its advantages: lower demagnetization fields, high packing density of magnetic material, good overwrite, high remanent magnetization, narrow isolated pulse width and a high carrier-to-noise ratio at short recording wavelengths (**4**). Some disadvantages of ME tape include: asymmetric isolated pulse shape, susceptibility to corrosion, questionable durability, production complexity and high cost.

Recent advances in MP and BaFeO tape technologies have reduced advantages presented by ME tape (**5–10**). Innovations of reduced and more uniform particle size and a new double coating technology which makes possible ultra-thin particulate magnetic coatings by utilizing a non-magnetic underlayer of fine particulates have led to a dramatically improved short-wavelength performance for MP and BaFeO tapes (**5**, **7**). Double coating technology produces thin ultra-smooth coatings for both MP and BaFeO tapes, and thus reduces spacing loss, demagnetization fields and isolated pulse width. BaFeO tape has the disadvantage of low remanent magnetization, but low medium noise results in high carrier-to-noise ratios at short recording wavelengths. Generally, ME and MP tapes are best suited for applications where system noise is dominant and BaFeO tape is best suited for systems where medium noise is dominant and low carrier amplitude is not detrimental to system performance.

The pause mode performance of the tapes is well understood because several studies have investigated the performance of the tapes in pause mode (**11–16**). However, little is known about their streaming (play) mode performance, which is the normal mode of operation and is pertinent to data processing applications. Wear mechanisms and key performance parameters may be different in streaming and pause modes due to the fact that there is tape motion in the streaming mode. Tape wear generates debris which may cause head clogging and signal loss during streaming mode operation (**17**). Performance of the tapes must not degrade when they are subjected to play/rewind cycling, if they are to be used in applications in which no loss of information is tolerable over the lifetime of the tape. Previous durability studies of ME and MP tapes subjected to play/rewind cycling showed that MP tape is more durable than ME tape (**18**, **19**). Wear mechanisms of ME, MP and BaFeO tapes, which affect the magnetic performance of the tapes, must be better understood to further improve their reliability and performance. In a previous study, changes in friction force, r.m.s. head output and signal drop-out frequency were measured in cycling tests with ME and MP tapes to determine the performance of the tapes, but detailed analyses of the interface components and elucidation of critical tribological mechanisms is needed to gain further understanding (**19**).

Metal core recession (recession of the metal core from the ferrite air bearing surface) and head stains (tenaciously adherent films which are not easily wiped away with a cloth and solvent) increase the distance between a tape and the read/write gap of a head transducer which results in signal loss. Previous studies showed that metal core recession depended on the relative wear rates of various head materials and the kind of tape being used in the tape drive (**20**, **21**). Experiments with an 8-mm camcorder which uses amorphous CoNbZr metal in gap (MIG) heads showed that metal core recession reached a stable value in about 15 h of streaming mode operation and that ME tape caused larger metal core recession than MP and BaFeO tapes (**22**). However, these studies used a non-contact optical profiler (NOP) to measure metal core recession, and the NOP cannot correctly determine level differences between dissimilar materials. A recently developed methodology which uses atomic force microscope (AFM) imaging to correctly measure level differences between dissimilar materials can be used to measure metal core recession (**14**, **23**, **24**). This method can be used to determine metal core recession caused by ME, MP and BaFeO tapes run against crystalline Sendust MIG heads. Head stains have a propensity to develop under low humidity and low tension conditions and are thought to protect against head wear or to be more prevalent when head wear is low (**25–27**). Severity of head staining for MIG heads depends on the metal core material, the region of the composite head and the kind of tape (**21**, **27**, **28**). Generally, head stains may be composed of both organic and inorganic species and the film may be spotty to continuous in its coverage (**21**, **25**–**28**). Experiments that utilized an optical microscope and an 8 mm camcorder which uses amorphous CoNbZr MIG heads showed that MP and BaFeO tapes caused head stains primarily on the metal core and glass surfaces, whereas ME tape did not cause head stains (**27**). However, the relative propensity of the tapes to stain Sendust MIG heads is not known and a more sophisticated technique of stain characterization using AFM imaging needs to be developed to further understand stain formation.

Users of video cassette recorders (VCRs), camcorders, R-DAT and other rotary head recorders may routinely use different kinds of magnetic tapes in their recorders. Past experiments with ME, MP and BaFeO tapes showed that the pause mode lifetime of a tape increased as a head ran-in with the same kind of tape (**11**, **15**). Interchanging ME and particulate tapes (MP and BaFeO) in a tape recorder caused excessive head and tape wear, while a tape formed its preferred head shape in pause mode experiments (**14**). It is well known that for increased tape thickness or stiffness, there is a tendency for heads to be contoured with larger radii of curvature as compared to contours formed by thinner tapes (**29**, **30**). Previous studies showed that MP and BaFeO tapes formed similar head contours which had smaller radii of curvature than the respective contours formed by ME tape (**14**, **15**, **22**). It has been suggested that anisotropy of mechanical properties in the machine direction (MD) and transverse direction (TD) primarily determine head contour or worn head shape (**29**, **31**). It is important to determine how mechanical properties of the tapes influence head contour, if the problem of tape conformity to the heads upon interchanging tapes in a drive is to be solved by tailoring mechanical properties of the various tapes to be used in the drive.

The objective of this investigation was to study the friction and wear mechanisms of the tapes during streaming mode operation and their interplay with the magnetic performance of the tapes. The authors present a new complementary set of measurement techniques and analyses to assess tape performance. Results of this study are the subject of this paper.

2 EXPERIMENTAL

2.1 Apparatus and test procedure

A commercial Hi-8 VCR (Sony EVC-100) with four heads was used as a tape transport and magnetometer. In this transport, the linear tape speed is 14.3 mm/s and the speed of the rotary heads is 3.8 m/s with inlet and outlet tensions of about 0.10 and 0.15 N respectively. The given tape tension, a head width of 60 μm, a measured head longitudinal radius of curvature of 7.8 mm and a wrap angle of about 0.1 rad (contact length of about 0.8 mm) produced a normal load of about 15 mN and an average contact pressure at the contact region of the head and tape of about 300 kPa. The VCR was instrumented to measure friction force between the rotary heads and tape, r.m.s head output and signal drop-outs down to submicrosecond duration (**15**, **16**, **19**). The experimental apparatus is shown schematically in Fig. 1a. It consists of an Hi-8 VCR, buffers/amplifiers, r.m.s. signal converters, a drop-out counter, an analogue-to-digital (A/D) converter and computer. The VCR consists of a loading/unloading mechanism, a rotary upper drum, a stationary lower drum, electronics and a tape transport. The tape transport consists of various guide pins, a roller guide, a pinch roller, a capstan and a tension regulator. A sine wave pattern was recorded on to a virgin tape at a wavelength of 0.6 μm using VCR electronics. A 0.6 μm recording wavelength was used to attain high sensitivity of reproduced head output to changes in head-to-tape spacing.

Friction force was measured by monitoring the voltage across the motor used for drum rotation; friction force was obtained using a previously developed calibration procedure (**15**). The calibration constant was 54 mV/mN and resolution of the friction force measurement was 100 μN. Friction force of about 7 mN was measured in the streaming mode and corresponds to a coefficient of friction of about 0.4. Head output was measured in an r.m.s. voltage format using an r.m.s. signal converter. Playback signal from the video heads and friction signal from the motor used for drum rotation were processed through buffer/amplification stages, and the r.m.s. voltage of each signal was measured using wide-bandwidth (about 10 MHz) r.m.s. signal converters. Head output data are presented in a voltage decibel format, and the reference 0 dB voltage is usually the initial playback r.m.s. voltage for a given tape. Thus, 0 dB for a given tape should be interpreted as an arbitrary voltage level. Motor voltage and r.m.s. head output were sampled at a rate of 4 Hz over the duration of 7 min streaming and pause mode experiments using an A/D converter, and data were stored in a personal computer.

Signal drop-outs were measured using a 40 MHz bandwidth single-channel drop-out counter (System 20 VTT multilevel drop-out tester by Doradus Corp., Minneapolis, Minnesota). Twenty classes of drop-outs defined by ranges of drop-out depth and duration can be measured simultaneously by the tester. Drop-out depth and duration are selectable in the ranges of 1–22 dB in 0.5 dB increments and 0.5–300 μs in 0.5 μs increments respectively. In these experiments, the various ranges of drop-out depths and duration selected are: 4–6, 6–9, 9–12 and >12 dB; 0.5–3, 3–10, 10–20, 20–50 and >50 μs.

To study durability of the tapes, streaming mode experiments in which the tapes were subjected to play/rewind cycling were conducted in a class 10 000 laboratory at a temperature of 22 ± 1 °C and 45 ± 5% RH. VCR heads were run-in for two days with the same kind of

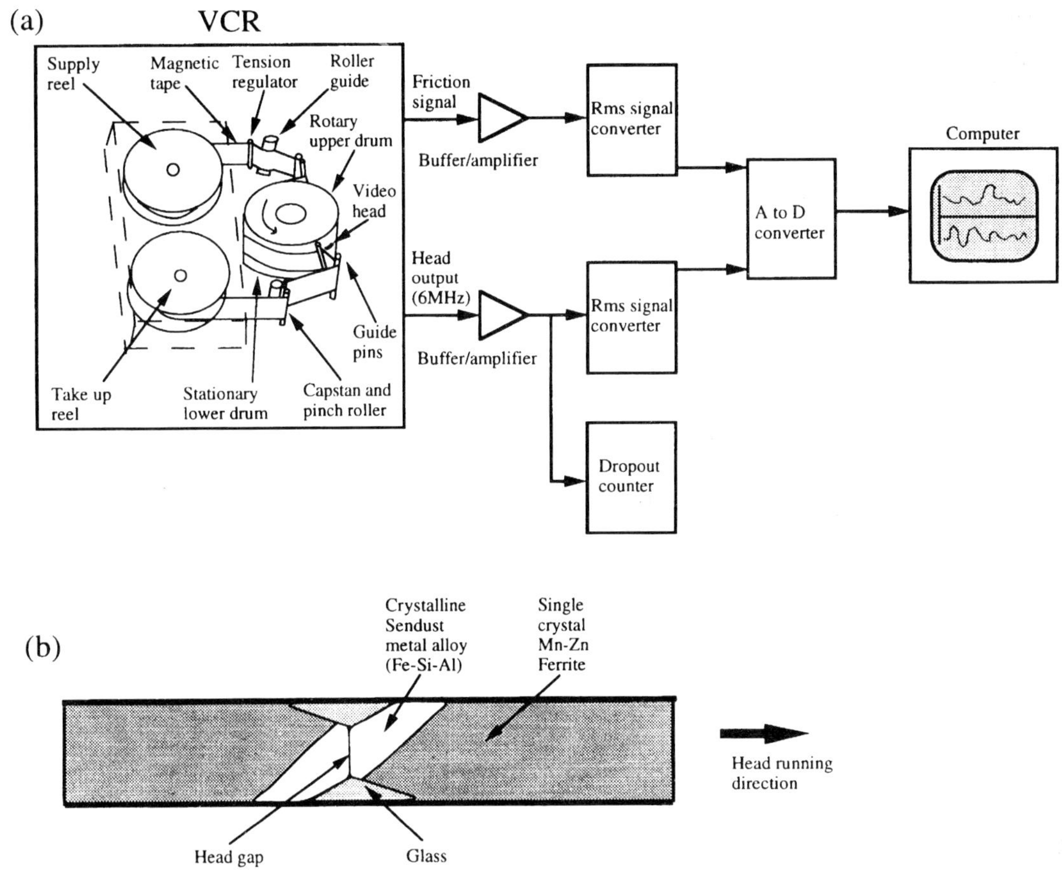

Fig. 1 Schematics of (a) commercial Hi-8 VCR used as a tape transport and magnetometer and associated instrumentation and (b) MIG read/write video head

tape to be used in an experiment that allowed head shape and metal core recession to saturate (**14**, **22**). A virgin cassette was recorded for 10 min (about 9 m length of tape) in the VCR. It was then played back and rewound (which constitutes one play/rewind cycle) repeatedly. Friction force, r.m.s. head output and signal drop-outs were measured on the same section of tape during selected play/rewind cycles, and head and tape samples were observed periodically with an optical microscope. To ensure that changes in friction force, head output and signal drop-outs resulted from tape wear as opposed to changes in head performance, reference measurements were made using a virgin cassette of the same kind of tape at the end of a cycling experiment. It was determined that head performance did not change over the course of cycling experiments. At selected numbers of play/rewind cycles, surface topography of a tape specimen was measured using an NOP. The scan size for NOP measurements was 250×250 μm with a lateral resolution of about 1 μm. Changes in head output level were correlated with changes in head-to-tape spacing using the following Wallace equation:

$$\Delta \, \mathrm{dB} = -54.6 \Delta d/\lambda \tag{1}$$

where Δ dB is the incremental change in head output decibel level, Δd is the incremental change in head-to-tape spacing and λ is the recording wavelength. At a recording wavelength of 0.6 μm, a 1 dB increase in head output during playback corresponds to about 10 nm reduction in head-to-tape spacing based on equation (1). With a power signal-to-noise ratio of about 40 dB and a measurement system resolution of 0.1 dB at peak power, the experimental measurement system can resolve a 1 nm change in head-to-tape spacing and therefore can resolve nanometre changes in tape topography due to tape wear. Signal drop-outs were measured to determine interface stability and recording performance to bit level resolution. An experiment was stopped when the number of play/rewind cycles was 1000 or when a tape was worn to a failure condition, whichever occurred first. The criterion for tape failure was based on both magnetic and tribological considerations, and a tape was deemed to have failed when a large or sudden degradation in both magnetic and tribological performance occurred during an experiment. Pause mode experiments were conducted with ME and MP tapes for comparison with streaming mode results.

To study the effects of tape stiffness or anisotropy of mechanical properties of a tape on head wear, MD and TD stiffnesses of the tapes were measured and saturated or steady state head contours formed by the tapes were measured by AFM imaging. Bending stiffness measurements of the tapes and selected tape substrates were made with a tape loop stiffness tester, which generated a force–displacement curve from which the bending stiffness was calculated for the tapes and substrates (**32**). Tape specimens were cut in 8 mm × 8 mm squares and sandwiched between two glass slides to make loops in either the MD or TD. A linear stage with a micrometre scale was used to press a tape segment against a microbalance which measured the contact force between the tape and microbalance. The AFM was chosen over non-contact optical and stylus profilers because the AFM can measure true level differences between dissimilar materials (optical interference techniques introduce phase shifts due to different indexes of refraction for unlike materials, which causes measurement errors) and produces three-dimensional images, unlike the stylus profiler which gives only two-dimensional line scans (**14**, **24**). The AFM also has better lateral resolution (about 200 nm at a 50 × 50 μm scan size) than the other instruments. Head contour and pole tip recession have been measured by AFM imaging in previous studies (**14**, **23**, **24**). To measure the head contour for each kind of tape, the heads were first run against an abrasive lapping tape (Sony V8-25CLH) for 20 s which formed the standard initial head contour. The lapping tape contains Fe_2O_3 particles with a mean particle size of about 400 nm and is produced in the same manner as particulate magnetic tapes but with larger surface roughness. To form the saturated head contour for each kind of tape, the heads were run against a given kind of tape continuously for 3 days in the ambient laboratory condition. The contour of an MIG head was then measured by AFM imaging. These experiments were also used to study metal core recession and stain formation on the various materials that comprise the composite MIG head. AFM imaging and an optical microscope were used to characterize stain formation on the heads, and metal core recession information was extracted from AFM images.

2.2 Head and tape specimens

Two read/write MIG heads placed diametrically opposite on the drum have a core material of single-crystal Mn–Zn ferrite with crystalline Sendust metal alloy (Fe–Si–Al) in the gap (Fig. 1b). MIG heads are needed to increase magnetic field strength to write on to high-coercivity tapes used in the Hi-8 system. Two Mn–Zn ferrite heads without metal in the gap are also present on the rotary drum; one is used for automatic track finding and the other as a flying erase head.

Commercially available Hi-8 ME, MP and BaFeO tapes were used in the experiments. A detailed description of the compositions and structures of the tapes is given in a separate article (**15**). The ME tape used in these experiments has a single-layer magnetic coating of $Co_{80}Ni_{20}$. Some analyses that were performed on ME and MP tapes were not performed on BaFeO tape. Tables 1, 3 and 4 to be presented later give roughness parameters for ME and MP virgin tape samples as measured by an NOP, and data for virgin BaFeO tape were published in a separate paper (**15**). Although the r.m.s. surface height of ME tape is comparable to that of MP and BaFeO tapes, surface slope, surface curvature, summit slope and number of summits per square millimetre (summit density) are smaller than those of MP and BaFeO tapes. The ME tape appears to have fewer asperities which are not as sharp as those on MP and BaFeO tapes. The BaFeO tape has an r.m.s. surface height almost twice that of ME and MP tapes, and generally has higher roughness data than the other tapes. It has been suggested that difficulty in dispersing BaFeO particles during the tape coating process causes high roughness of BaFeO tape (**7**). High roughness of BaFeO tape may inhibit its record and playback performance at the short recording wavelength (0.6 μm) used in this study.

Figure 2 shows three-dimensional profiles and two-dimensional line scans from an NOP of virgin ME, MP and BaFeO tape surfaces. Surface asperities found on MP and BaFeO tapes with the NOP are not present on the ME tape surface. The 1 μm lateral resolution of the NOP does not allow it to resolve nanoasperities present on the ME tape surface (**15**). Waviness (of about 60 nm peak-to-valley and about 250 μm wavelength) of the ME tape surface is visible in the NOP profile and line scan, and bumps on the surface may induce localized wear and inhibit good recording performance at high linear recording densities due to spacing loss. Surface asperities on MP and BaFeO tapes appear to be similar in shape and size, but the BaFeO tape recording surface has more waviness than that of MP tape. Therefore, the MP tape recording surface appears to have superior morphology for ultra-high-density recording as compared to ME and BaFeO tapes.

It was suggested that buckling of the magnetic layer away from the substrate may be the cause of bumps on the ME tape surface (**15**). However, further investigation into this phenomenon suggests that thermal damage to the substrate during deposition of the magnetic coating causes waviness of the ME tape surface. Figure 3 shows three-dimensional profiles from an NOP of the ME substrate front side, ME tape magnetic coating side and ME tape front side with magnetic coating removed by dipping the tape in a 10 per cent (by volume) HCl solution for one minute. The ME substrate front side has no bumps or waviness prior to deposition of the magnetic coating, and as discussed previously in reference to Fig. 2, bumps are present on the magnetic coating side of ME tape. By removing the magnetic coating with HCl solution, the post-deposition condition of the substrate was exposed for NOP analysis. The ME tape front side with the magnetic coating removed (substrate front side post-deposition) has bumps and waviness, whereas the ME tape back side was found to have no bumps and waviness. Thus, localized thermal

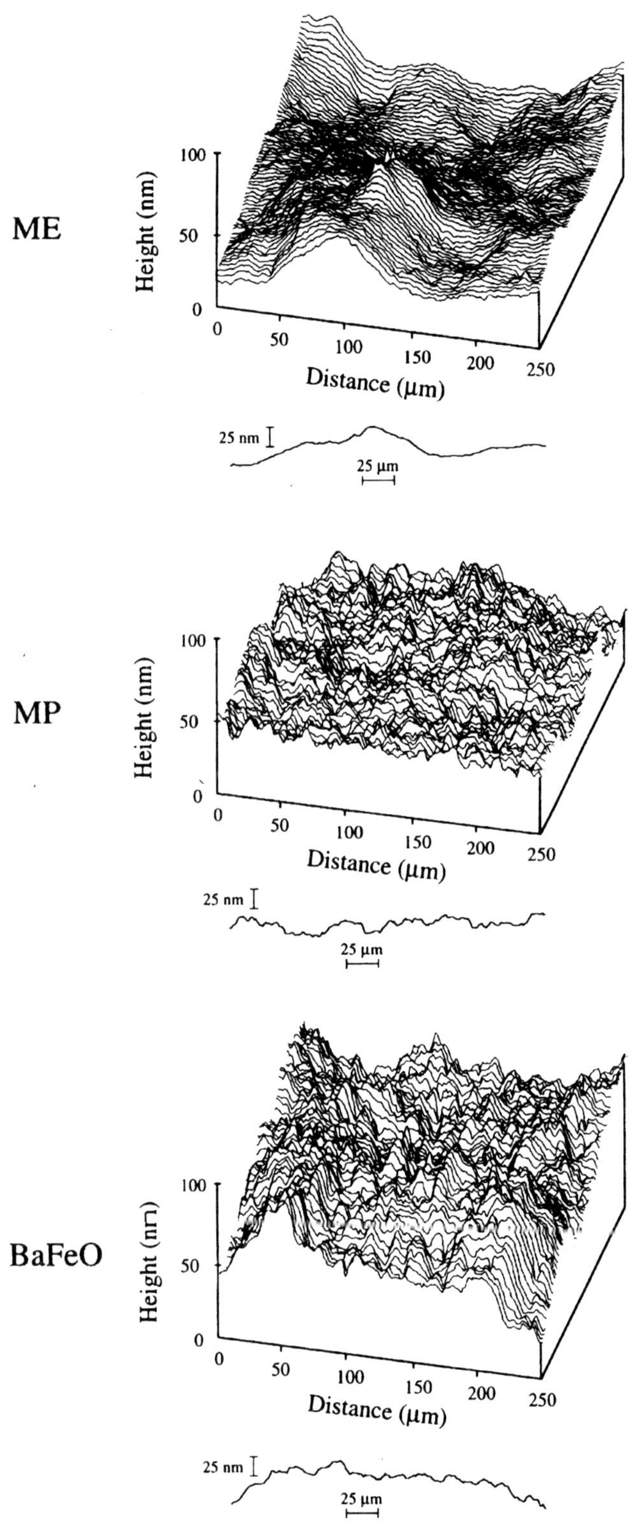

Fig. 2 Three-dimensional profiles and two-dimensional line scans from an NOP of virgin ME, MP and BaFeO tape surfaces

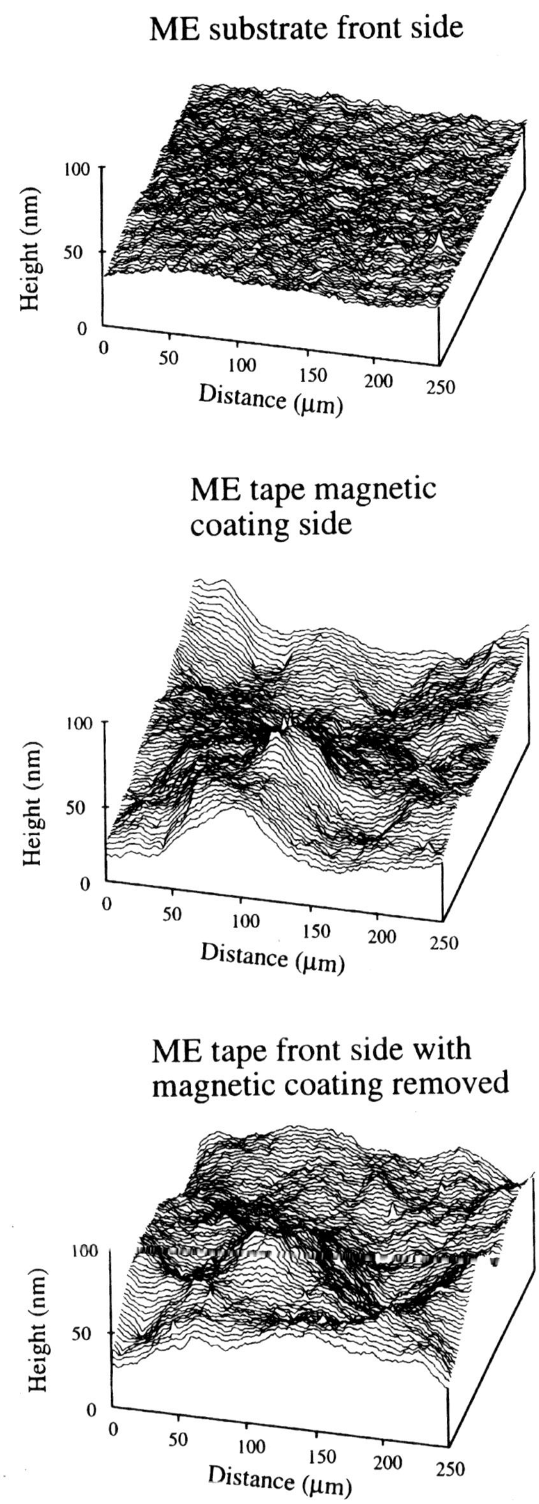

Fig. 3 Three-dimensional profiles from an NOP of ME substrate front side, ME tape magnetic coating side and ME tape front side with magnetic coating removed from the tape

effects during metal film deposition distort the substrate front side. Table 1 gives roughness parameters for the substrate used to construct ME tape, ME tape and ME tape front side with magnetic coating removed from the tape. The substrate front side roughness parameters show that the substrate is much smoother than both the magnetic coating side of the tape and the tape front side with the magnetic coating removed, which both have similar

Table 1 Roughness data for the front side and back side of ME substrate and tape as measured with an NOP*

Roughness parameter	Substrate front side	Tape magnetic coating side	Tape front side with magnetic coating removed	Substrate back side	Tape back side
R.m.s. surface height (nm)	2.6	8.3	8.3	11	14
R.m.s. profile slope x (mrad)	0.58	0.83	0.78	4.2	11
R.m.s. profile slope y (mrad)	0.42	0.67	0.64	2.9	7.0
R.m.s. surface slope (mrad)	0.71	1.1	1.0	5.1	13
R.m.s. profile curvature x (1/mm)	0.76	1.0	0.95	4.8	16
R.m.s. profile curvature y (1/mm)	0.53	0.68	0.68	2.8	8.2
R.m.s. surface curvature (1/mm)	0.48	0.62	0.60	2.9	9.0
Summit height (nm)†					
Mean	3.8/5.7	6.1/18	6.2/15	22/120	14/58
Standard deviation	3.1/0.63	7.4/2.3	6.5/2.0	21/15	13/6.8
Summit |slope| (mrad)†					
Mean	0.10/0.032	0.29/0.33	0.16/0.16	2.0/7.6	7.9/16
Standard deviation	0.10/0.014	0.34/0.33	0.15/0.16	2.8/2.7	3.4/4.1
Summit curvature (1/mm)†					
Mean	1.8/1.9	2.5/3.5	2.4/2.8	7.3/40	9.0/23
Standard deviation	0.49/0.32	1.1/1.2	1.2/1.6	6.4/12	4.5/7.1
Summit-to-valley distance (nm)	25	70	71	230	165
Summit-to-mean distance (nm)	14	31	30	140	75
Number of summits per square millimetre†	150/26	310/53	230/110	6700/53	49 000/800
Profile zero crossings x (1/mm)	33	36	35	88	160
Profile zero crossings y (1/mm)	21	15	17	51	80
Average autocorrelation distance (μm)	33	33	34	11	16

*Scan size for NOP is 250 × 250 μm.
†Mean and r.m.s. values are given for all summits (first numbers) and summits with top 25% summit height (second numbers). A threshold of 1.0 nm was used in defining summits on the surfaces.

roughness statistics. Thus, deposition of the magnetic coating increases roughness of the tape over that of the substrate. The post-deposition substrate has higher roughness than the substrate front side, which is an artefact of the substrate deformation that occurred during deposition of the magnetic coating. The back side of the substrate has higher roughness statistics than the front side, and application of the back coating by the tape manufacturer makes the tape back side rougher than the substrate back side. A detailed description of ME tape construction is given in a separate article (**15**).

Stiffnesses in the MD and TD directions and thicknesses for the tapes and selected substrates are given in Table 2 (**32**). The TD stiffness of ME tape is larger than that of the other tapes by about a factor of two. The BaFeO tape has smaller stiffnesses in both the MD and TD directions than the other tapes. The stiffer anisotropic ME substrate has a higher TD stiffness, whereas the anisotropic MP substrate is tensilized PET with a higher MD stiffness.

3 RESULTS AND DISCUSSION

3.1 Play/rewind cycling results for the various tapes

Figure 4 shows friction force and head output during streaming mode testing in an ambient laboratory condition with ME and MP tapes. Streaming mode data in Fig. 4 were taken during the first play/rewind cycle. Modulation of the friction force with an amplitude and period of about 1 mN and 5 s respectively resulted from oscillatory movement of the tape by the tension regulator of the VCR, and head output was consequently modulated in phase with the friction force due to changes in head-to-tape spacing (high tension gives high output). This kind of measurement determined the friction force and head output level at a given number of play/rewind cycles during cycling experiments. The friction force obtained with MP tape is larger than that of ME tape. The friction force and head output for ME and MP tapes maintained the same offset values

Table 2 Stiffness and thickness data for magnetic tapes and selected substrates

Tape or substrate	ME	MP	BaFeO	ME substrate	MP substrate
MD stiffness (μPa m^3)	0.55	0.62	0.38	0.21	0.16
TD stiffness (μPa m^3)	0.78	0.40	0.29	0.38	0.08
Thickness (μm)	10.8	10.8	11.0	9.8	8.7

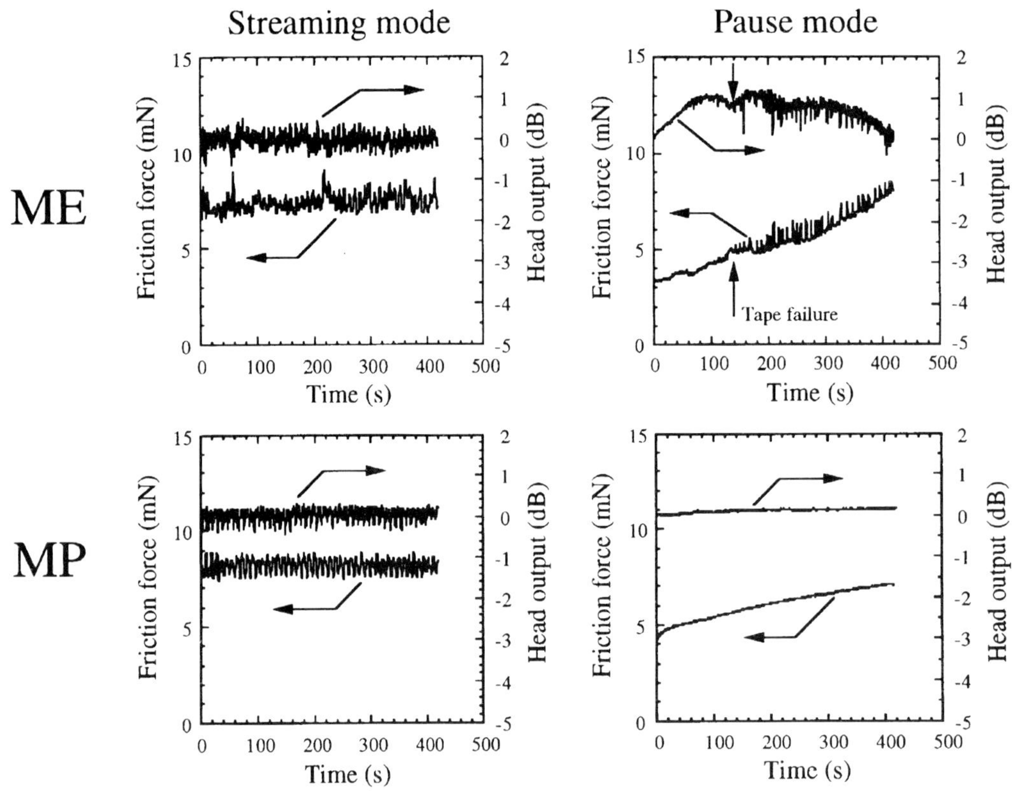

Fig. 4 Friction force and head output during streaming mode and pause mode testing with ME and MP tapes (note that 0 dB is the initial playback r.m.s. voltage for each tape and vertical arrows in the block for ME tape in the pause mode indicate tape failure)

throughout the 7 min streaming mode experiments. This is expected since the tape condition is the same (first play/rewind cycle) throughout each 7 min streaming mode experiment.

Figure 4 shows friction force and head output during pause mode testing in an ambient laboratory condition with ME and MP tapes. In the pause mode, there is no tape motion and the rotary heads pass over the same track on the tape over and over again. The friction force in the pause mode was less than in the streaming mode due to smaller tape tension, and friction force was not modulated by the tension regulator in the pause mode. The ME tape lacked stability in friction force and head output relative to that exhibited by MP tape. The head output for ME tape showed a rapid increase to about +1 dB, implying a decrease in head-to-tape spacing of about 10 nm for a recording wavelength of 0.6 μm based on equation (1). Since lubricant thickness is about 5–10 nm, this implies that essentially all of the lubricant was removed from areas of the ME tape surface that were contacting the head surfaces. After lubricant depletion from these areas, interaction of the rotary heads with the magnetic coating prevented the rapid increase in head output observed during lubricant removal and generated solid debris which, at this environmental condition, caused signal instability due to spacing loss upon traversing the tape contact region of the head surface. The friction force increased as lubricant was removed, and further increased and became unstable after lubricant depletion with friction spikes temporally correlated to signal drop-outs. The presence of instability in both head output and friction force in the pause mode was defined to be the failure condition, which is indicated by vertical arrows in Fig. 4 for ME tape. For MP tape, both friction force and head output increased as the experiment progressed due to mild burnishing of the tape surface.

Figure 5 shows the initial drop-out frequency in various classes of drop-outs for ME, MP and BaFeO tapes. The tapes were in a drop-out condition less than 10 per cent of the time, which illustrates the intermittent nature of the wear events that caused the drop-outs. It was determined in another study that wear debris of a given size caused drop-outs of varying depth and duration and that wear debris particles with diameters from 60 to 120 nm caused drop-out classes measured in this study (**33**). The ME tape generally had a larger drop-out frequency than MP tape for a given class of drop-outs. In the 4–6 and 6–9 dB classes of drop-outs, the drop-out frequencies of BaFeO tape were between that of ME and MP tapes. In the 9–12 and >12 dB classes of drop-outs, the drop-out frequencies of BaFeO tape were generally larger than those of both ME and MP tapes. For a given drop-out depth, drop-out frequencies for MP and BaFeO tapes always decreased with increased drop-out duration, whereas for ME tape the drop-out frequency was largest for a drop-out duration of 3–

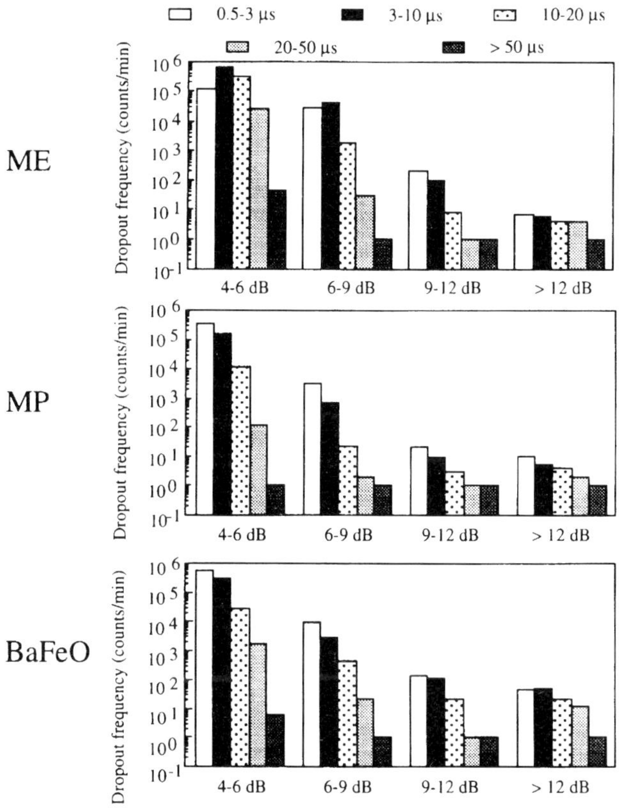

Fig. 5 Drop-out frequency in various classes of drop-outs for ME, MP and BaFeO tapes during the first play/rewind cycle

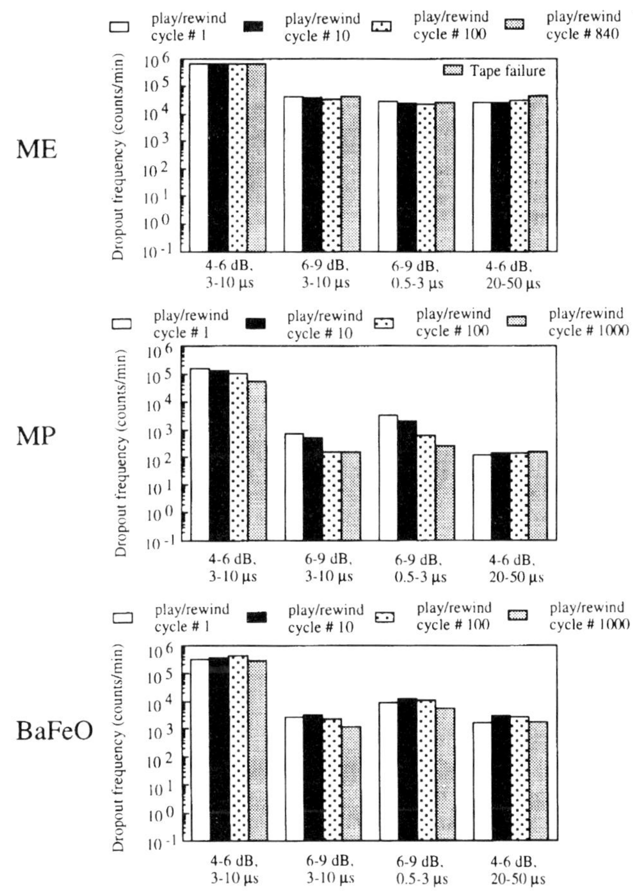

Fig. 6 Drop-out frequency in various classes of drop-outs for ME, MP and BaFeO tapes at various numbers of play/rewind cycles

10 μs for 4–6 and 6–9 dB drop-out depths. The operative drop-out mechanism may be different for ME tape as compared to MP and BaFeO tapes. For a given drop-out duration, the drop-out frequency generally decreased with increased drop-out depth for all of the tapes. Considering spacing loss in both the recording and playback of the signal at a 0.6 μm recording wavelength, the 4–6 dB and 20–50 μs class of drop-outs corresponds to the head moving a distance of about 100–200 μm (at 3.8 m/s) and a head-to-tape spacing of about 20–40 nm, which corresponds well with the waviness of the ME tape surface. The drop-out frequency for ME tape reduced by three orders of magnitude in going from the 4–6 dB and 20–50 μs drop-out class to the 4–6 dB and >50 μs drop-out class, which suggests that surface features associated with a distance not exceeding about 200 μm on the tape surface, or the waviness of the ME tape surface, caused a high drop-out frequency in the 4–6 dB and 20–50 μs drop-out class. The drop-out frequencies for MP and BaFeO tapes in the 4–6 dB and 20–50 μs class of drop-outs were more than two and one orders of magnitude less than that of ME tape respectively.

Figure 6 shows the effect of subjecting the tapes to play/rewind cycling on drop-out frequency in various classes of drop-outs. In the classes of drop-outs shown in Fig. 6, ME tape had a higher drop-out frequency which generally decreased less with tape wear than that observed with MP tape. However, the 4–6 dB and 20–50 μs drop-out frequency increased with tape wear for both ME and MP tapes, which was typical of drop-out classes of larger time duration. Performance of MP tape improved with use, and the improvement must have resulted from tape wear. The relative insensitivity of ME tape performance to tape wear suggests that the major cause of drop-outs for ME tape may be waviness of the tape surface, as opposed to wear debris of the tape. Rougher BaFeO tape performance generally degraded over the first 100 play/rewind cycles, and then improved from 100 to 1000 play/rewind cycles to a performance level only slightly better than it had during the first play/rewind cycle. Tape failure occurred for ME tape after 840 play/rewind cycles, and there was an increase in drop-out frequencies at the failure condition. No failure was observed for MP and BaFeO tapes through 1000 play/rewind cycles. The MP tape had lower drop-out frequencies, which decreased more with tape wear than that observed with BaFeO tape. Thus, MP tape had a better drop-out performance throughout the cycling experiments than both ME and BaFeO tapes.

Figure 7 shows the effect of subjecting the tapes to play/rewind cycling on friction force and head output. Although the initial behaviour of head output differed for different tapes, head output for all of the tapes eventually increased to +1 dB or higher as the number of play/rewind cycles approached 1000 and the tapes were worn smooth. This indicates that head-to-tape spacing was reduced by 10 nm or more during cycling tests. For ME and MP tapes, there was an initial decrease in head output over the first few cycles. Debris on the tape surface increased head-to-tape spacing over the first few cycles. Head-to-tape spacing increased more in the initial stages of wear and decreased more with further play/rewind cycling for MP tape, which was worn more readily than ME tape. In the case of the rougher BaFeO tape, there was no initial decrease in head output, but rather a sawtooth shaped pattern over about the first 300 play/rewind cycles preceded the eventual increase in head output to +1 dB at 1000 play/rewind cycles. Thus, a rougher tape can delay the eventual reduction in head-to-tape spacing.

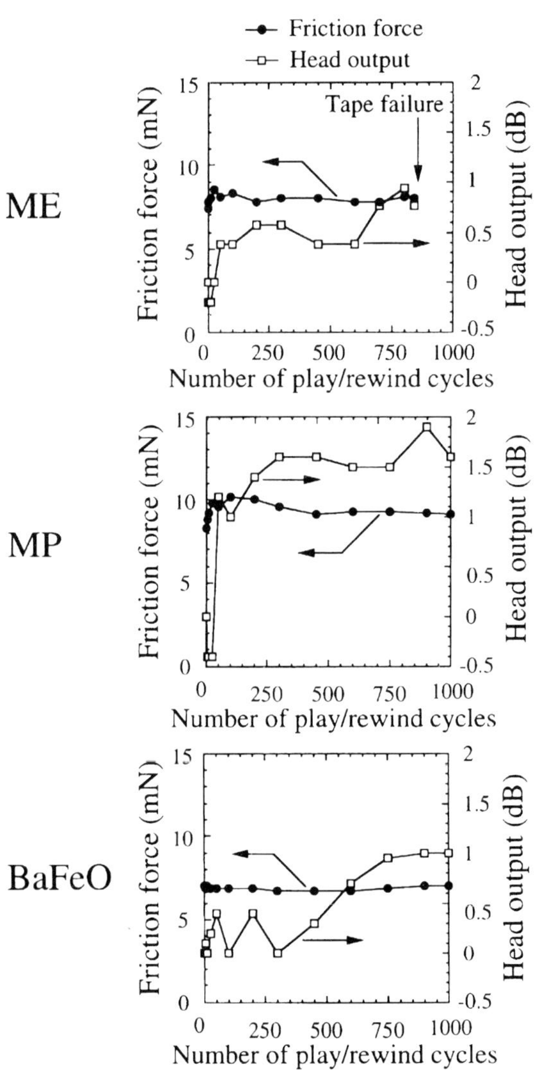

Fig. 7 Progression of friction force and head output during play/rewind cycling with ME, MP and BaFeO tapes (note that 0 dB is the playback r.m.s. voltage during the first play/rewind cycle for each tape)

ME and MP tapes showed a rapid increase in friction over the first few play/rewind cycles, whereas BaFeO tape maintained a constant friction force throughout the experiment, as shown in Fig. 7. Increased head-to-tape spacing for ME and MP tapes concurrent with the rapid increase in friction suggests that debris present on the tape surface caused increased friction force. Wear debris particles have been associated with high friction due to increased plowing in a previous study (**34**). Higher friction force was only partially maintained after debris was removed from the tapes at higher numbers of play/rewind cycles. The relatively constant friction force of rougher BaFeO tape, and in particular the lack of an initial increase in friction over the first few passes, suggests that the nature of the head-to-tape contact did not change considerably due to debris generation. Waviness of BaFeO tape may have allowed debris to settle into the tape away from higher contacting asperities.

Figure 8 shows the effect of subjecting the tapes to play/rewind cycling on head output and drop-out frequency. For ME tape, the 6–9 dB and 3–10 μs drop-outs decreased mildly as the tape was worn, which is typical of ME tape behaviour. However, the 4–6 dB and 20–50 μs drop-outs started to increase at about 100 play/rewind cycles when significant reduction (~5 nm) of head-to-tape spacing had occurred and cracks were first observed on the tape surface. A sharp increase in head output preceded tape failure, and this was followed by an abrupt decrease in head output and an abrupt increase in drop-out frequencies at tape failure. This indicates that the head-to-tape spacing decreased suddenly just prior to tape failure.

In the case of MP tape, drop-out frequencies generally decreased as the tape was worn smooth. From about 300 play/rewind cycles to the completion of the experiment, drop-out frequencies and head output shown in Fig. 8 for MP tape have reached saturated values. The saturated value of head output corresponds to about a 15 nm reduction in head-to-tape spacing and the reduction in spacing produced greater stability of the head-to-tape interface based on decreased drop-out frequencies.

In the case of BaFeO tape, attainment of drop-out frequencies lower than their initial values was delayed as compared to MP tape until the beginning of a 10 nm reduction in head-to-tape spacing at about 300 play/rewind cycles. The sawtooth-shaped head output pattern over the first 300 play/rewind cycles is out of phase with an inverted sawtooth shaped 4–6 dB and 10–20 μs drop-out class pattern, as shown in Fig. 8. These out-of-phase sawtooth patterns suggest a cyclic wear process of debris generation (decreased output and increased drop-out frequency) and debris removal (increased output and decreased drop-out frequency) over the first 300 play/re-wind cycles for rougher BaFeO tape. From about 300

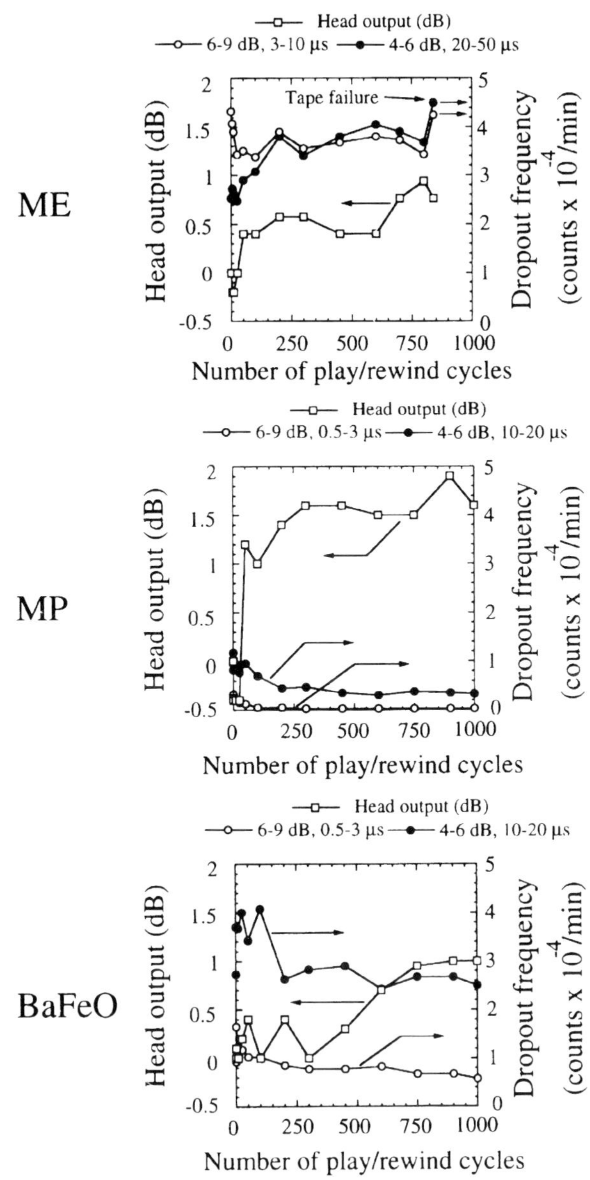

Fig. 8 Progression of head output and drop-out frequency during play/rewind cycling with ME, MP and BaFeO tapes (note that 0 dB is the playback r.m.s. voltage during the first play/rewind cycle for each tape)

play/rewind cycles to the completion of the experiment, drop-out frequencies became more stable and steadily decreased as head-to-tape spacing became smaller.

Optical micrographs in Fig. 9 of the head surfaces after running against ME and MP tapes for various numbers of play/rewind cycles show that loose wear debris was transferred to the leading edges from the tape and collected on the trailing edges of the tape contact regions of the head surfaces. In a previous study using the same tapes in the pause mode, chemical analyses showed that loose wear debris consisted of tape binder for particulate tapes and topical lubricant and magnetic coating for ME tape (**15**). Prior to running for 7 min at each condition, the heads were wiped clean of any loose debris with a lint-free cloth soaked in isopropyl alcohol. Thus, the amount of loose wear debris transferred to the heads at each condition gives an insight into the amount of tape wear and the amount of loose wear debris present on the tape surface. The amount of loose wear debris collected on the leading and trailing edges of the tape contact regions of the head surfaces was largest during the recording pass, which suggests that much debris was generated in the head-to-tape interface in the first tape pass. For both ME and MP tapes during play/rewind cycling, the amount of loose wear debris transferred to the head surfaces generally decreased from the record pass through the end of the cycling experiment. However, there was a tendency for loose wear debris to agglomerate into larger flake-like particles at larger numbers of play/rewind cycles (particularly for MP tape), which is indicative of an adhesive wear mechanism; this may explain why drop-out classes of a larger time duration increased with tape wear, as shown in Fig. 6. Aside from the record pass in which there was an excessive transfer of loose wear debris to the head surfaces from MP tape, ME and MP tapes transferred similar amounts of loose wear debris to the head surfaces. In the pause mode for each kind of tape, there was more loose wear debris transferred to the head surfaces than in the streaming mode except for the case of the record pass for the tapes. For ME tape in the pause mode, lubricant was transferred to the heads and collected in pools of liquid on the trailing edges of the tape contact regions, which was not observed in the streaming mode. The fact that much loose wear debris transferred to the head surfaces from MP tape in the pause mode is surprising since head output data in Fig. 4 shows that the reduction in head-to-tape spacing was only 1 or 2 nm during the experiment. Pause mode debris for MP tape is flake-like, which again is indicative of an adhesive wear mechanism.

Optical micrographs in Fig. 10 of the tape surfaces show the progression of wear for ME and MP tapes after various numbers of play/rewind cycles and after 7 min of pause mode testing. For both ME and MP tapes during play/rewind cycling, the amount of tape wear increased from the record pass through the end of the cycling experiment as head-to-tape spacing was reduced in each case. Tape wear initiated at high localized contact locations and is first visible after 10 play/rewind cycles as damage areas for ME tape and impressions or 'footprints' of rolling particles and high aspect ratio elliptical-shaped damage areas for MP tape. The larger and sharper asperities of MP tape, when severed from the tape surface and located in the contact interface, rolled in the interface as rigid bodies due to their irregular shape and caused impact wear (**35**). The sliding and rolling action of loose wear debris on the tape surface on and near the damage points, and the fact that the initial damage occurred at the highest summits or load-bearing asperities, led to growth in the size of the damaged areas with further play/rewind cycling as head-to-tape

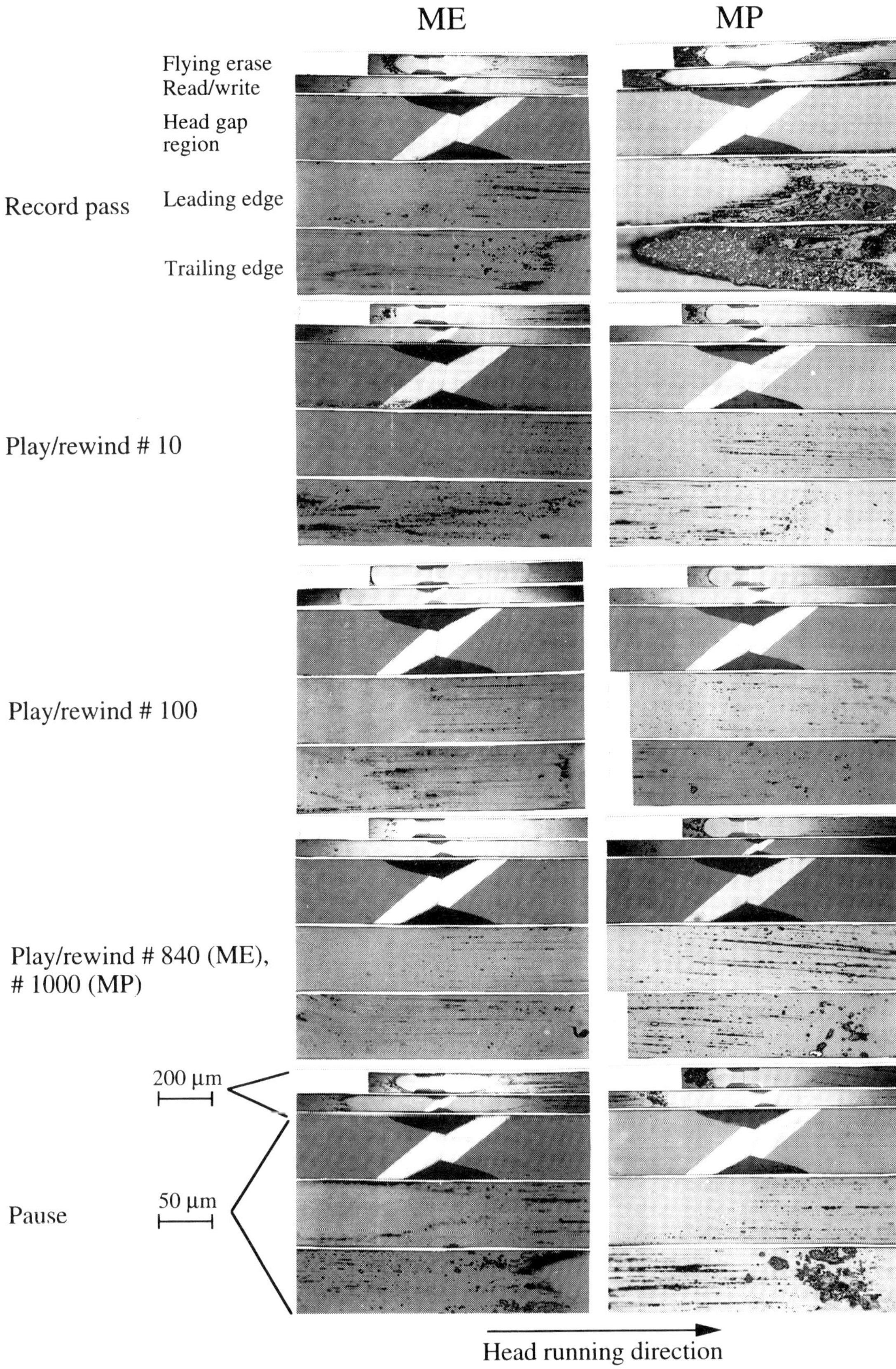

Fig. 9 Optical micrographs (from top to bottom at each condition) of the flying erase ferrite head, an MIG read/write head and magnified views of the head gap region, the leading edge and the trailing edge for MIG read/write heads all run against ME and MP tapes in streaming and pause modes for 7 min under various conditions. Note that the head was wiped clean prior to the 7 min running at each condition

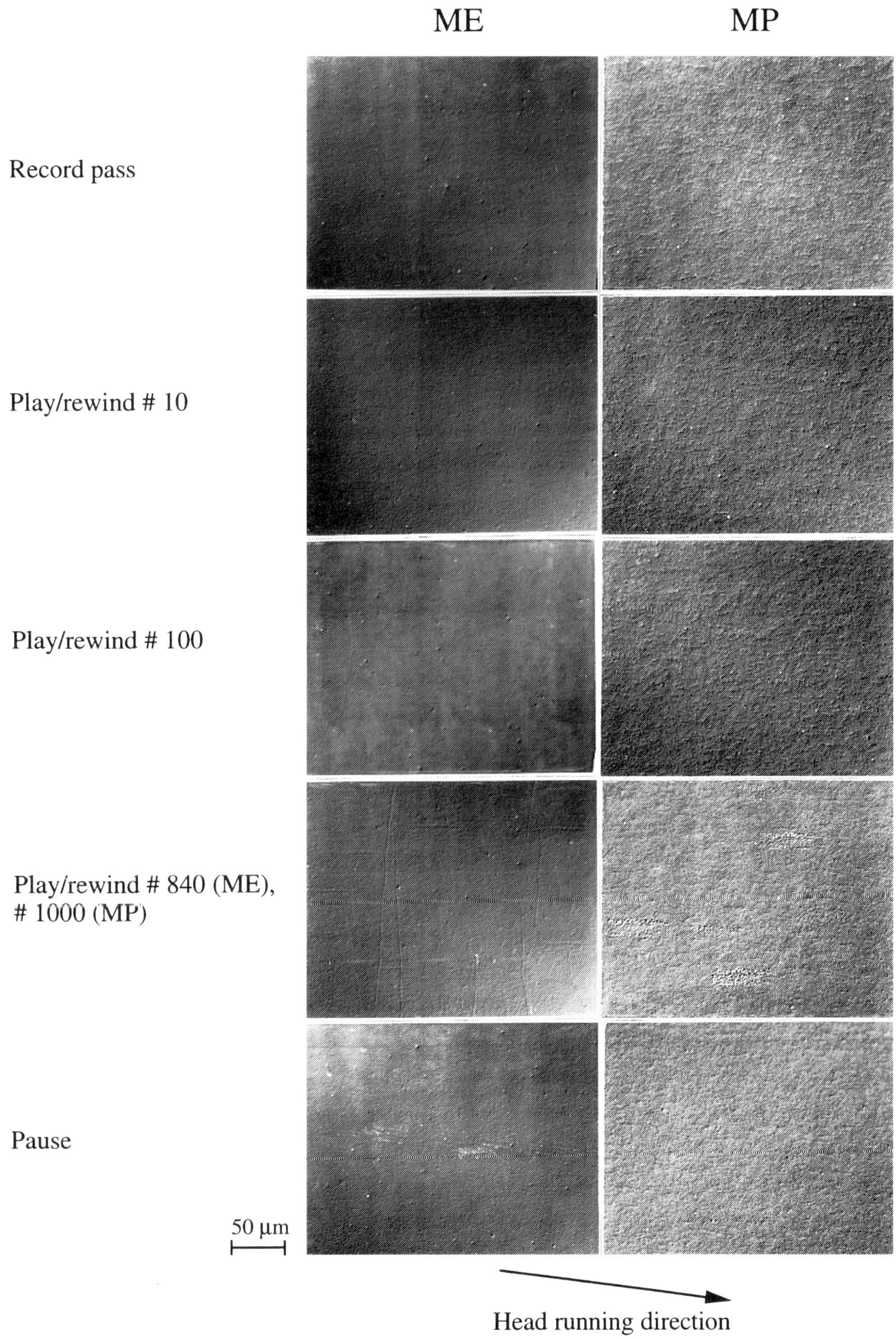

Fig. 10 Optical micrographs showing progression of wear of the ME and MP tape surfaces after the record pass, after various numbers of play/rewind cycles and after 7 min of pause mode testing

spacing became smaller. In an earlier study, it was estimated that a loose abrasive particle may spend 90 per cent of its time rolling and 10 per cent of its time sliding or abrading the surfaces (**36**). It should be noted that the percentage of time that a particle rolls in the interface should depend on the size and shape of the particle, and that a particle may wear the surfaces by impacting them even while rolling. Thus, in going from 10 to 100 and 100 to 1000 play/rewind cycles, existing damage areas grew in size and new damage areas were created as the tape was worn smooth, which allowed further reduction in head-to-tape spacing which caused additional damage. As the ellliptical wear scars became larger at higher numbers of play/rewind cycles, it is evident in Fig. 10 that the semi-major axes of the ellipses are along the head running direction.

Waviness played a role in the wear mechanism for ME tape, whereas the relative flatness of MP tape led to different wear behaviour. For ME tape, damage caused by the rotary heads occurred at a smaller number of discrete locations per unit area than that present on the MP tape surface. After 100 play/rewind cycles, damage was present at two discrete locations for ME tape, whereas damage was present at about 10 discrete locations for MP tape, as shown in Fig. 10. After about 1000 play/rewind cycles, damage to the ME tape surface was more localized than that present on the MP tape surface. The tape surface between localized damage areas after 840 play/rewind cycles for ME tape appeared to be identical with the virgin surface of ME tape (with the exception of the lateral cracks), which suggests that these regions had little or no interaction with the rotary heads. The damage to MP tape was more widespread after 1000 play/rewind cycles, and most of the surface looked as though it had interaction with the rotary heads due to the presence of wear scars.

On the ME tape surface damage initiated at high points or bumps, which were caused by thermal damage to the substrate during deposition of the magnetic coating. The initial and highest localized damage regions were about 250 μm apart, which correlates well with waviness of ME tape. A large number of cracks formed and connected localized damage areas at tape failure after 840 play/rewind cycles. A relatively small number of lateral cracks were first observed on the ME tape surface after 100 play/rewind cycles, as shown in Fig. 10. At the tape failure condition after 840 play/rewind cycles, one end of a lateral crack was located generally on a localized damage area and the other end on an adjacent (across the tape width) localized damage area. The cracks often propagated through a defect in the magnetic coating and in some cases terminated at a defect site. The lateral nature of all of the cracks suggests that the crack driving force was probably longitudinal tension maintained in the tape by the tape transport, since the most common mode of cracking or the tensile mode is the opening or separation of material by the maximum tensile stress, which acts in a plane normal to the applied force (**37, 38**).

Tables 3 and 4 give roughness data for ME and MP tapes used in play/rewind cycling experiments as measured with an NOP. The NOP was selected for tape wear measurements due to the intermittent nature of wear on the tape surfaces. The 250 × 250 μm scan size of the NOP is large

Table 3 Roughness data for ME tape used in play/rewind cycling experiments as measured with an NOP*

Roughness parameter	Virgin	After record pass	After 10 play/rewind	After 100 play/rewind	After 840 play/rewind
R.m.s. surface height (nm)	8.3	8.0	10	9.8	14
R.m.s. profile slope x (mrad)	0.83	0.69	0.72	0.82	1.1
R.m.s. profile slope y (mrad)	0.67	0.59	0.61	0.62	0.76
R.m.s. surface slope (mrad)	1.1	0.91	0.94	1.0	1.3
R.m.s. profile curvature x (1/mm)	1.0	0.83	0.81	0.86	1.3
R.m.s. profile curvature y (1/mm)	0.68	0.56	0.55	0.59	0.77
R.m.s. surface curvature (1/mm)	0.62	0.51	0.50	0.53	0.78
Summit height (nm)†					
Mean	6.1/18	7.5/‡	9.9/‡	5.4/15	3.5/25
Standard deviation	7.4/2.3	6.5/‡	4.2/‡	9.1/2.7	11/5.1
Summit slope (mrad)†					
Mean	0.29/0.33	0.13/‡	0.20/‡	0.15/0.14	0.24/0.40
Standard deviation	0.34/0.33	0.10/‡	0.19/‡	0.10/0.05	0.23/0.27
Summit curvature (1/mm)†					
Mean	2.5/3.5	2.4/‡	2.5/‡	2.5/2.5	2.2/2.7
Standard deviation	1.1/1.2	0.91/‡	0.75/‡	1.0/0.70	0.93/1.4
Summit-to-valley distance (nm)	70	56	79	81	96
Summit-to-mean distance (nm)	31	38	30	32	44
Number of summits per square millimetre†	305/53	106/16	71/16	96/27	520/52
Profile zero crossings x (1/mm)	36	16	17	23	25
Profile zero crossings y (1/mm)	15	22	15	15	12
Average autocorrelation distance (μm)	33	31	31	30	32

*Scan size for NOP is 250 × 250 μm.
†Mean and r.m.s. values are given for all summits (first numbers) and summits with top 25% summit height (second numbers). A threshold of 1.0 nm was used in defining summits on the surfaces.
‡Not enough summits in the top 25% of summit heights for partial statistics to be calculated.

Table 4 Roughness data for MP tape used in play/rewind cycling experiments as measured with an NOP*

Roughness parameter	Virgin	After record pass	After 10 play/rewind	After 100 play/rewind	After 1000 play/rewind
R.m.s. surface height (nm)	8.3	9.4	9.4	8.5	8.5
R.m.s. profile slope x (mrad)	1.8	1.5	1.9	1.9	1.5
R.m.s. profile slope y (mrad)	1.6	1.4	1.6	1.6	1.4
R.m.s. surface slope (mrad)	2.4	2.0	2.5	2.5	2.1
R.m.s. profile curvature x (1/mm)	2.2	1.5	2.3	2.4	1.7
R.m.s. profile curvature y (1/mm)	1.5	1.1	1.5	1.5	1.2
R.m.s. surface curvature (1/mm)	1.4	0.96	1.4	1.4	1.1
Summit height (nm)†					
Mean	4.7/16	6.7/16	5.6/24	5.9/20	4.4/16
Standard deviation	7.9/2.9	8.0/2.2	9.7/3.6	8.2/2.9	9.1/2.7
Summit |slope| (mrad)†					
Mean	0.91/1.3	0.55/0.67	0.96/1.1	1.1/1.3	0.82/1.2
Standard deviation	0.54/0.84	0.38/0.35	0.47/0.28	0.64/0.50	0.66/0.79
Summit curvature (1/mm)†					
Mean	1.9/2.1	1.7/1.8	1.9/2.1	1.9/2.0	1.8/1.9
Standard deviation	0.58/0.97	0.44/0.44	0.43/0.41	0.47/0.46	0.49/0.39
Summit-to-valley distance (nm)	70	71	79	69	70
Summit-to-mean distance (nm)	32	35	36	30	29
Number of summits per square millimetre†	1200/240	210/60	1600/160	1200/150	340/50
Profile zero crossings x (1/mm)	48	34	48	46	53
Profile zero crossings y (1/mm)	40	39	45	38	58
Average autocorrelation distance (μm)	27	28	29	29	26

*Scan size for NOP is 250 × 250 μm.
†Mean and r.m.s. values are given for all summits (first numbers) and summits with top 25% summit height (second numbers). A threshold of 1.0 nm was used in defining summits on the surfaces.

enough to ensure that roughness statistics do not critically depend on the location of a measurement with respect to a given damage area on the tape surface while still giving a micrometre of lateral resolution. The ease and simplicity of use of the NOP can expedite getting adequate sample statistics which requires taking numerous measurements on each tape sample. Data given in Tables 3 and 4 are the average of measurements made at 10 different locations on each tape surface. At each measurement location, 12 measurements were taken and averaged to arrive at each roughness parameter.

Roughness measurements were made on virgin tapes and on tapes at various stages of the cycling experiments. Roughness measurements were not made on tape samples used in pause mode experiments because the narrow (less than 50 μm wide) wear track and mild intermittent nature of the wear were not conducive to getting any meaningful statistics for the roughness parameters. The roughness of the tapes changed as the tapes were used and consequently worn. The evolution of roughness parameters was different for ME and MP tapes, which suggests that wear mechanisms were different for the tapes. Generally, roughness parameters for summits changed by higher percentages than those of the entire surface because localized wear occurred on high load-bearing summits.

Both ME and MP tapes were worn during the record pass. The data suggest that the highest summits on each tape surface were removed during the record pass. Surface slope, surface curvature, summit slope and summit density all decreased for the tapes in going from virgin to after the record pass. Thus, during the record pass many high asperities were removed from the tape and, as shown in Fig. 9, much loose wear debris was transferred to the heads during the record pass. Many severed asperities were likely to remain on the tape surface because it is unlikely that they all transferred to the head surfaces.

For ME tape during the cycling experiment, roughness parameters in Table 3 remained about the same from the record pass through 100 play/rewind cycles. The highly intermittent and localized nature of the wear of ME tape over the first 100 play/rewind cycles, as seen in Fig. 10, is responsible for the unchanging roughness parameters through the first 100 play/rewind cycles. Surface height, surface slope, surface curvature, summit slope, summit-to-valley distance and summit density increased sharply after 840 play/rewind cycles and tape failure. The sharp increases in roughness parameters at tape failure are a measure of the catastrophic crack formation which connected heavily worn bumps on the ME tape surface and grossly deformed the surface.

For MP tape during the cycling experiment, surface slope, surface curvature, summit slope and summit density all evolved in a similar manner. After 10 and 100 play/rewind cycles, roughness parameters were larger than after the record pass. Loose wear debris located near localized high summits accelerated the wear process by rolling and sliding in the interface, and loose wear debris on the tape surface caused high roughness parameters after 10 and 100 play/rewind cycles. Evidence that the tape was pushed away from the heads by loose wear debris on the tape surface during the first 25 play/rewind cycles is given in Figs 7 and 8. The fact that roughness data indicate that loose wear debris was present on the tape surface after 100 play/rewind cycles does not contradict the fact that head output had increased to +1 dB, since considerable wear of high points on the tape had occurred which allowed reduction in head-to-tape spacing. Loose wear debris present on the MP tape surface from 10 to 100 play/rewind cycles caused the initial rapid increase in friction force shown in Fig. 7. After 1000 play/rewind cycles, roughness parameters decreased to about the same values that they had after the record pass. Thus, loose wear debris was removed from the tape surface and high summits were worn smooth between 100 and 1000 play/rewind cycles, which allowed a reduction in head-to-tape spacing and the tape attained great running stability during streaming mode operation. Head output data in Figs 7 and 8 for MP tape reached a saturated value after about 300 play/rewind cycles, which suggests that virtually no tape wear occurred from 300 play/rewind cycles to the end of the experiment when little loose wear debris was present on the tape surface. The wear mechanism for the tape was abrasive over the first 300 play/rewind cycles due to loose wear debris on the tape surface and was adhesive thereafter, when little loose wear debris was present on the tape surface. It is known that wear coefficients for abrasive wear are about 100 times larger than those for adhesive wear (**39**).

Figure 11 shows the summit density for virgin and worn tapes as measured with an NOP and presented in Tables 3 and 4. Localized wear occurred for ME tape, which made roughness parameters somewhat insensitive to wear, but it is still evident that high summits were removed during the record pass and the summit density generally increased thereafter as cracks formed on the tape surface. The summit density for MP tape was sensitive to tape wear. High summits were removed during the record pass and summit density was high after 10 and 100 play/rewind cycles when an abrasive wear mechanism was operative. However, after 1000 play/rewind cycles the summit density was again low and the wear mechanism was adhesive. Limited data available for BaFeO tape are consistent with those of MP tape.

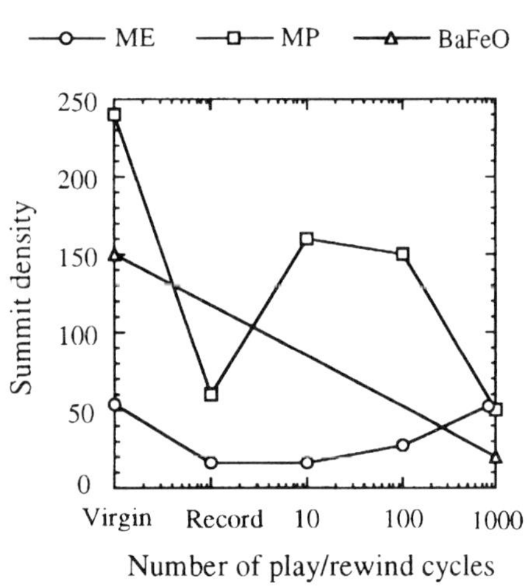

Fig. 11 Summit density (for summits with the top 25 per cent summit height) for virgin and worn tapes as measured with an NOP (note that the first data point is for the virgin tape and the second data point is for after the record pass). For BaFeO tape, data were taken only for the virgin tape and the tape ran for 1000 play/rewind cycles

Figure 12 shows a schematic illustration of the evolution of tape wear and the head-to-tape contact condition during play/rewind cycling with MP tape. High asperities on the virgin tape were severed off during the record pass and many remained on the tape surface as loose wear debris. During the first play/rewind cycle, the recording head slides across the tape surface at velocity v_h and d_1 is the distance between the recording head and the back side of the tape. Head-to-tape spacing during the tenth play/rewind cycle increased by 5 nm (head output dropped 0.5 dB) due to the presence of more loose wear debris generated by the sliding action of the original asperities severed from the virgin tape, and friction was high due to increased plowing in the interface. The contact condition during the hundredth play/rewind cycle is similar to that of the tenth play/rewind cycle except for the fact that head-to-tape spacing decreased by 15 nm due to abrasive wear. Finally, head-to-tape spacing decreased by another 5 nm and the amount of loose wear debris was low again at 1000 play/rewind cycles. At this point, the wear mechanism is adhesive and any remaining blunt and flake-like loose wear debris is unable to abrade the tape surface.

Wear mechanisms of the tapes appear to be different in the pause mode as opposed to the streaming mode. Since there is no tape motion in the pause mode, the rotary heads pass over the same track on the tape over and over again, and consequently tape damage is confined to a rectangular area along the head running direction with a width necessarily less than that of the video heads. This is evident in Fig. 10 for ME tape which was worn more in the pause mode than MP tape. Waviness played a role in the wear mechanism of ME tape in the pause mode. The optical micrograph in Fig. 10 shows that damage is more severe at two discrete locations (high points) along the rubbing track. High contrast of the head rubbing track with respect to the unworn region is probably a result of lubricant being depleted by transfer to the video heads which exposed the bare metallic magnetic coating. This contrast is not observed with ME tape in the streaming mode, where there is no evidence of lubricant depletion. This interpretation is consistent with data shown in Figs 4 and 9 of this paper. The wear scar on MP tape after 7 min of pause mode operation is barely discernible in the optical micrograph in Fig. 10, and this is consistent with head output data in Fig. 4, which suggests that head-to-tape spacing was reduced by only about 1 or 2 nm during 7 min of pause mode operation. The relatively mild wear of MP tape in the pause mode as compared to the streaming mode is of interest. It is known that small changes in operating conditions of an

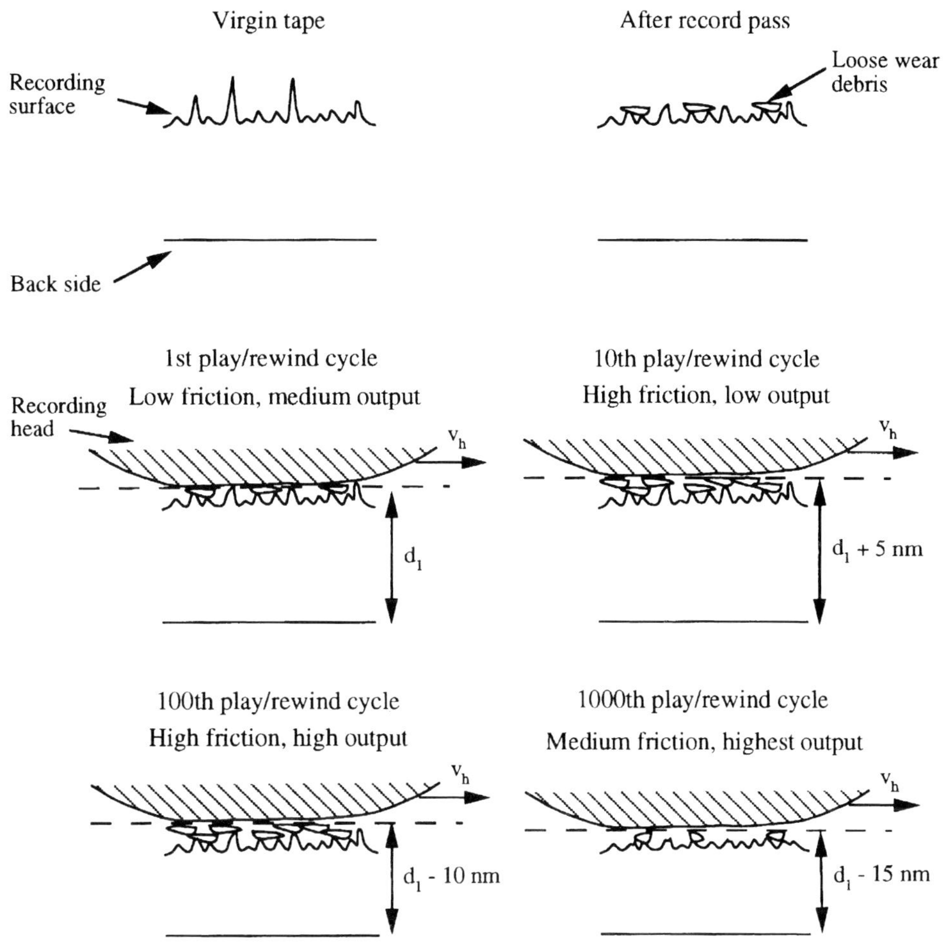

Fig. 12 Schematic illustration of the evolution of tape wear and the head-to-tape contact condition during play/rewind cycling with MP tape

interface can change the wear mechanism (**40**). Tension and normal load in the pause mode are about one-half of their values in the streaming mode, which may decrease wear or change the wear mechanism. In the streaming mode, the heads contact a larger surface area of tape than in the pause mode, and consequently wear of a given area of tape is influenced by wear debris from other areas of the tape surface. In the pause mode, if an abrasive particle was expelled from the rubbing track, then it could no longer abrade the tape surface, which is not true in the streaming mode where the tape moves.

A relationship between wear coefficients in the streaming and pause modes can be developed using the Archard wear equation. The Archard wear equation is

$$V = \frac{kWx}{H} \tag{2}$$

where V is the wear volume, k is the wear coefficient, W is the normal load, x is the sliding distance and H is the hardness of the softer material (in this case the tape). Equation (2) was developed for adhesive wear, but abrasive wear also follows it reasonably well (**39**, **41**). Using a simple analysis of a tape sliding over a curved surface, the expression for the normal load is

$$W = \frac{TL}{R} \tag{3}$$

and

$$\mu = \frac{F}{W} = \frac{FR}{TL} \tag{4}$$

where μ is the coefficient of friction, F is the friction force, R is the radius of curvature of the curved surface, T is the tape tension and L is the length of the chord which has end-points at either end of the tape contact region of the head surface. Optical micrographs in Fig. 9 show that the contact length did not change in going from the streaming mode to the pause mode based on the location of debris on the head surfaces (the contact area should be clean). Thus, the factor R/L will be the same in the streaming and pause modes. If it is assumed that there is little air bearing effect at the head-to-tape interface (the head and tape are in contact), then

$$\mu_s = \mu_p \tag{5}$$

where subscripts s and p refer to the streaming and pause modes respectively. Now, using equations (4) and (5),

$$\frac{F_s}{T_s} = \frac{F_p}{T_p} \tag{6}$$

Since F_s and F_p were measured directly, a relationship between T_s and T_p can be determined. For a given data track of width w and length l on the tape, the wear volume is

$$V = -wl\alpha\,\Delta d \tag{7}$$

where Δd is the change in physical spacing between the head and tape surfaces obtained from equation (1) and α is the fractional contact area. Note that α is an increasing function of normal load and consequently $\alpha_s > \alpha_p$ (**2**). The sliding distance x for the tape in equation (2) in the streaming and pause modes can be determined in the following way. The rotary drum of the VCR spins at 30 Hz. In the pause mode, there are three head passes per drum rotation; for instance, in a 7 min pause mode experiment there are 37 800 head passes. In the streaming mode, there are two head passes per play/rewind cycle; for instance, in 1000 play/rewind cycles there are 2000 head passes.

A relationship between the wear coefficients in the streaming and pause modes can now be determined. Equation (2) can be rewritten using equation (3) as

$$V = \frac{kTLx}{HR} \tag{8}$$

where H, R and L are the same in the streaming and pause modes. Therefore, from equation (8),

$$\frac{V_s}{V_p} = \frac{k_s T_s x_s}{k_p T_p x_p} \tag{9}$$

Using equations (6) and (7) in equation (9) and solving for k_s/k_p gives

$$\frac{k_s}{k_p} = \frac{\Delta d_s F_p x_p \alpha_s}{\Delta d_p F_s x_s \alpha_p} \tag{10}$$

Equation (10) will be evaluated for MP tape using pause mode data from Fig. 4 and streaming mode data from Fig. 7 and equation (1) to determine Δd_s, Δd_p, F_p and F_s. The entire 420 s of pause mode data will be used in equation (10), but only the first 300 play/rewind cycles of streaming mode data over which tape wear occurred due to an abrasive wear mechanism will be used in equation (10). This gives

$$\frac{(k_s)_{MP}}{(k_p)_{MP}} = \frac{(15\ \text{nm})(5.5\ \text{mN})(37\,800)\alpha_s}{(1.5\ \text{nm})(9\ \text{mN})(600)\alpha_p} \tag{11}$$

Using the fact that $\alpha_s > \alpha_p$ and solving for $(k_s)_{MP}$ gives

$$(k_s)_{MP} > 385(k_p)_{MP} \tag{12}$$

This result shows that the wear coefficient during the time that abrasive wear occurred in the streaming mode is over two orders of magnitude larger than the wear coefficient for adhesive wear in the pause mode. In the streaming mode, wear of a given area of the tape is influenced by loose wear debris from other areas of the tape that allowed abrasive wear to occur over the first 300 play/rewind cycles. However, abrasive particles expelled from the wear track in the pause mode could no longer contribute to tape wear and the wear mechanism was adhesive, which explains large differences in wear rates observed in equation (12).

For ME tape, application of the foregoing analysis is complicated due to removal of lubricant from the tape and unstable head output in the pause mode as shown in Fig. 4. The wear coefficient in the pause mode for ME tape depends on the time chosen to evaluate head output (physical spacing between the head and tape surfaces), whereas for MP tape there was no such problem because head output increased gradually throughout the experiment. The wear coefficient will be evaluated for ME tape in the pause mode up to the time corresponding to lubricant depletion from areas of the tape that were contacting the head surfaces which occurred at about 80 s. The wear coefficient in the streaming mode for ME tape will be evaluated over the first 200 play/rewind cycles during which an abrasive wear mechanism was operative. The result for ME tape is

$$(k_s)_{ME} > 4(k_p)_{ME} \tag{13}$$

The abrasive effect observed in the streaming mode with MP tape is also present in the streaming mode for ME tape. Lubricant removal from the tape and transfer to the head surfaces for ME tape in the pause mode as shown in Figs 9 and 10 allowed a decrease in the physical spacing between the head and tape surfaces and lessened the disparity between wear coefficients in the streaming and pause modes for ME tape as compared to MP tape.

3.2 Head wear and staining

Figure 13 shows three-dimensional AFM profiles and optical micrographs of MIG head surfaces after running against the lapping tape for 20 s and ME, MP and BaFeO tapes for 3 days. Metal core recession is defined to be the height difference between the ferrite and metal core surfaces at their interface. A well-defined uniform metal core recession can be seen for heads run against lapping and ME tapes and is shown schematically for the head run

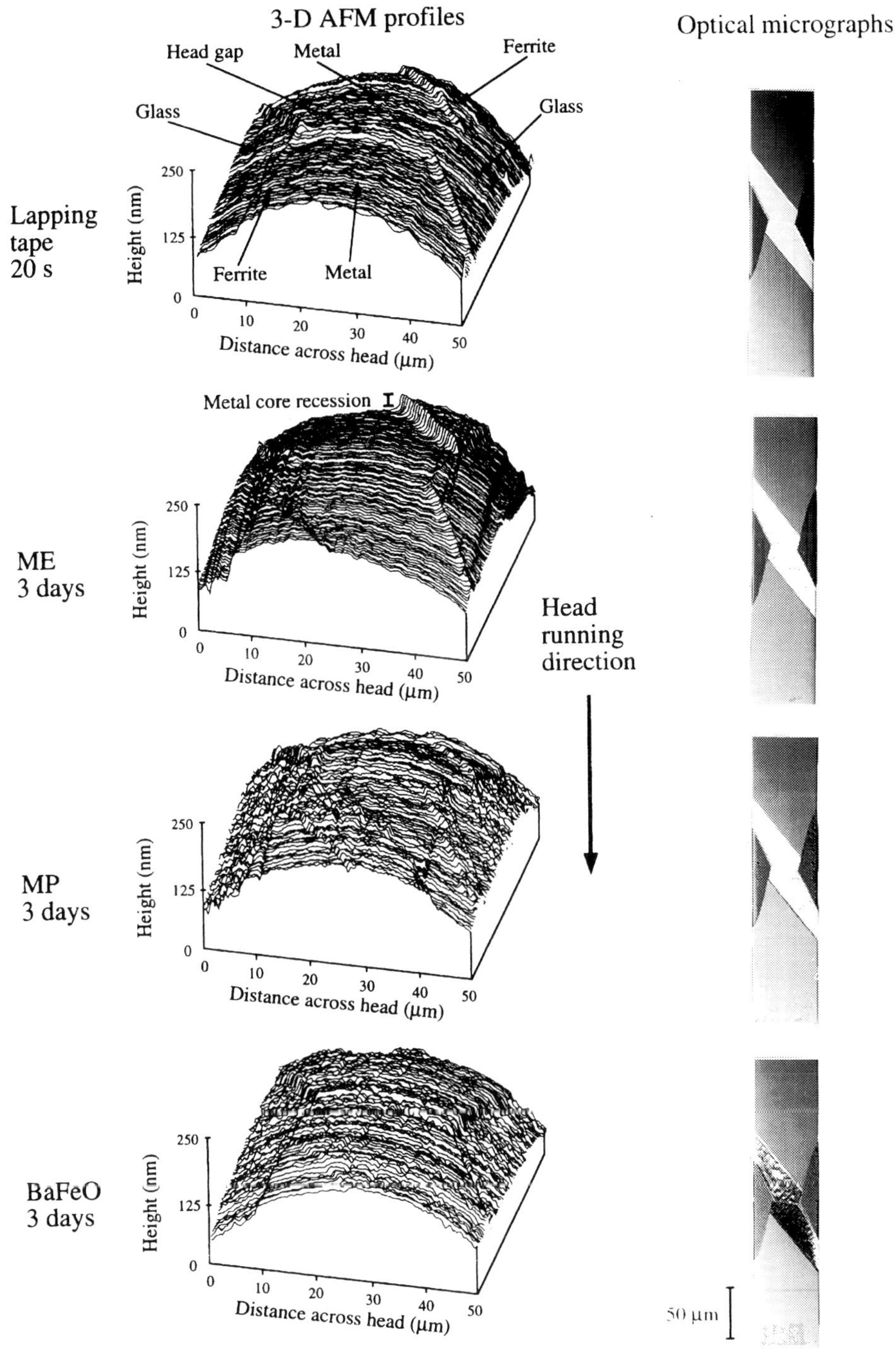

Fig. 13 Three-dimensional AFM profiles and optical micrographs of an MIG head surface after running against the lapping tape for 20 s and ME, MP and BaFeO tapes for 3 days

against ME tape. The presence of stain on the metal core for heads run against MP and BaFeO tapes made metal core recession somewhat less obvious in the three-dimensional profiles. Metal core recession was obtained by taking several two-dimensional cross-sections across the ferrite-to-metal core interface and then averaging values obtained from the cross-sections. The presence of stain on the metal core for heads run against MP and BaFeO tapes complicated the analyses in the sense that the experimenter had to be careful that the metal core itself, and not the stain on the metal core, was used as the metal core level. Metal core recession obtained with lapping, ME, MP and BaFeO tapes are 15, 25, 12 and 15 nm respectively. Thus, ME tape produced metal core recession about twice that produced by the other tapes, which gave similar values. These values are less than those reported in a previous study which used the same tapes and an NOP to measure metal core recession of CoNbZr MIG heads (**22**). This is somewhat surprising since Vickers hardness of CoNbZr was reported to be 700 kg/mm^2, which is comparable or even higher than that of Sendust, and both heads use a core of single-crystal Mn–Zn ferrite (**21**). The AFM correctly measured metal core recession, whereas the NOP apparently did not do so. Thus, a reliable method of measuring metal core recession of MIG heads has been developed which is similar to that used to measure pole tip recession in thin-film heads (**23**, **24**).

Optical micrographs in Fig. 13 show that stains formed on the metal core and glass surfaces of the heads run against MP and BaFeO tapes and that no staining formed on heads run against lapping and ME tapes. Staining was more severe with BaFeO tape than with MP tape. Stains are generally brown in colour, except for the case of the metal core run against BaFeO tape, where the stain appears to be black. Stains on the metal cores run against MP and BaFeO tapes are more prevalent towards the leading edges of the heads and their growth may nucleate at interfaces between dissimilar materials (see the MP stain in Fig. 13). Stains form more easily at lower tensions and lower contact pressures, and stain formations in Fig. 13 suggest that contact pressure in the vicinity of the head gap was lower on the leading side of the head, which is expected in a convergent channel type of bearing (**40**). This phenomenon has recently been observed by other researchers (**42**).

AFM imaging proved to be a valuable technique to determine stain morphology and thickness and yielded information that could not be obtained with an optical microscope. Stains are readily observed in the three-dimensional AFM profiles of Fig. 13 for heads run against MP and BaFeO tapes. Surface morphologies of stains formed on the metal core and glass surfaces are different, which may indicate that the mechanism of stain formation for the surfaces is different. Stains on the metal cores are smooth films with voids or pores, whereas stains on the glass are rough films with many high asperities. Thickness of the stains was determined by taking several two-dimensional cross-sections across the metal core, and then the height difference between the stain on the metal core and any exposed areas of the metal core was measured. Thickness of the stain was measured at several locations and averaged to obtain an average stain thickness. The average thicknesses of the stains on the metal cores run against MP and BaFeO tapes are about 15 and 25 nm respectively.

Figure 14 shows head contours formed after running against lapping tape for 20 s and ME, MP and BaFeO tapes for 3 days. Generally, lateral head contours formed by ME and BaFeO tapes had larger radii of curvature than those formed by MP tape, which is most evident along the line segments EF and IJ in Fig. 14. For line segment AB, stains on the heads run against MP and BaFeO tapes make interpretation of the data difficult. Based on tape stiffness data in Table 2, high TD stiffness of ME tape (compared to MP tape) caused lateral head contours formed by ME tape to have larger radii of curvature than those formed by MP tape. Lateral radii of curvature formed by compliant BaFeO tape are larger than expected.

Lateral head contours are symmetric with respect to the centre of the head along line segment AB for heads run against all of the tapes and along all line segments for heads run against compliant BaFeO tape. Lapping, ME and MP tapes appear to have been twisted from the leading to trailing sides of the head in going over the head since lateral contours for the heads run against these tapes along line segments EF (leading side) and IJ (trailing side) are the mirror image of each other. High contact pressure is expected near the head gap on the trailing side of the head. Transverse pressure profiles are influenced by hydrodynamic air-film formation above the rotating upper drum and the lack of such a film above the stationary lower drum. Pressure gradients should be larger on the stationary lower drum side, which may make lateral head contours asymmetric with respect to the centre of the head because local head wear should increase at higher pressures. Lateral head contours along line segment IJ for heads run against lapping, ME and MP tapes have higher gradients on the stationary lower drum side than on the rotating upper drum side due to larger pressure gradients on the stationary lower drum side. The surprising results of symmetric gradients for line segment AB and higher gradients on the rotating upper drum side for line segment EF are probably due to tape rigidity and their inability to deform and uniformly contact the rotary heads. The amount of asymmetry in lateral head contours is more for ME tape than for MP tape and is not present for compliant BaFeO tape. Table 2 shows that the stiffest tape is ME, followed by MP and BaFeO tapes.

Longitudinal head contours formed by MP and BaFeO tapes had smaller radii of curvature than those formed by ME tape, as shown in line segments CD, GH and KL in Fig. 14. The lower stiffness of MP and BaFeO tapes compared to ME tape may be responsible for this behaviour.

4 CONCLUSIONS

MP tape performance was superior to that of ME and BaFeO tapes when operated in the streaming mode of a rotary head recorder. Tribological and magnetic performance of ME tape was affected by waviness of the ME tape surface, and eliminating waviness should improve ME tape performance. Magnetic performance of MP tape improved to a saturated performance level as the tape was worn smooth. Loose wear debris on the tape surface caused high friction and accelerated tape wear. Different tapes were dissimilar in their tendencies to cause head stains and they formed different preferred head shapes due to varying mechanical properties.

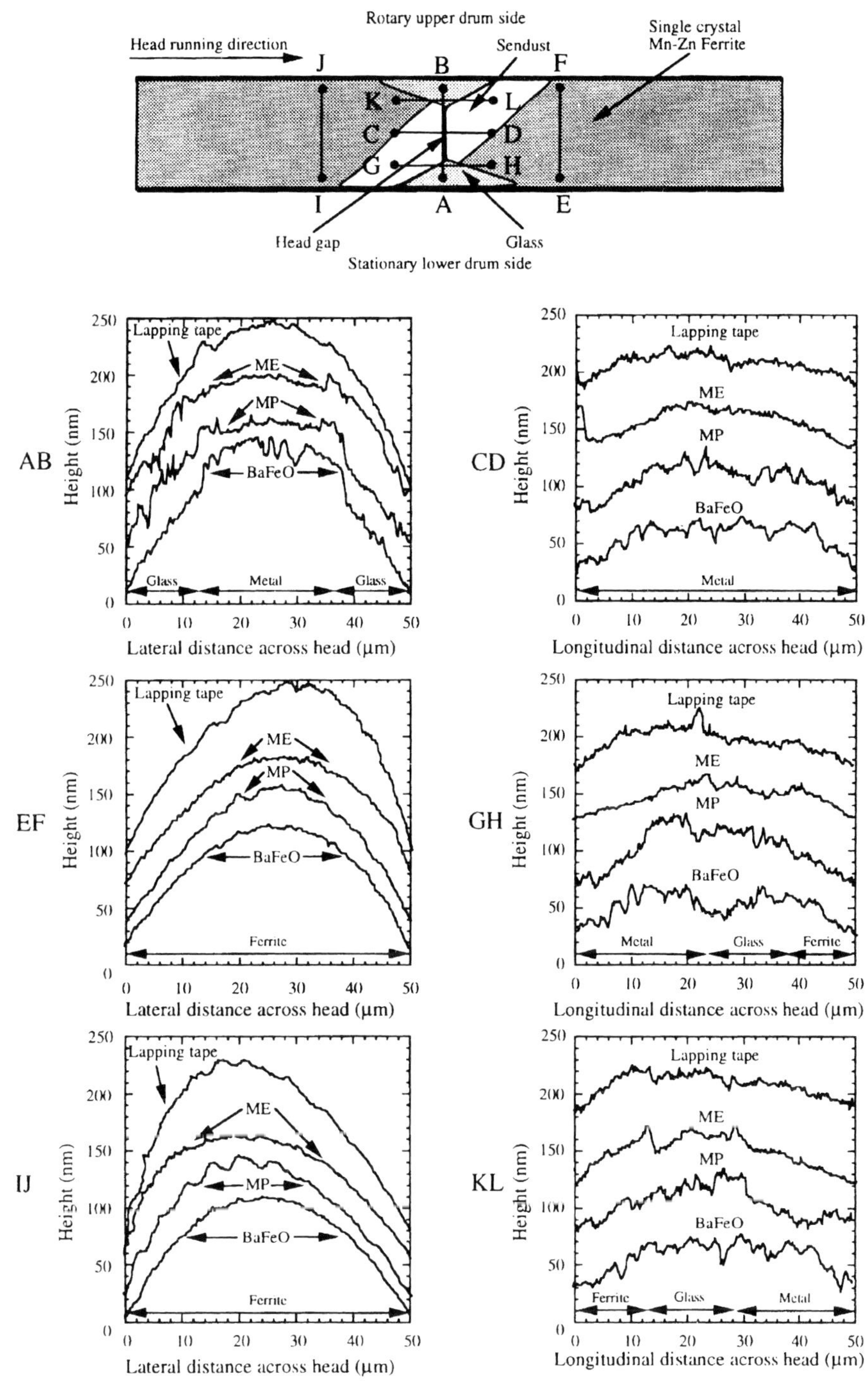

Fig. 14 Schematic of an MIG video head with line segments illustrating locations on the head surface where the contour was measured with an AFM, and head contours formed after running against the lapping tape for 20 s and ME, MP and BaFeO tapes for 3 days

ACKNOWLEDGEMENTS

The research was supported by the 3M/Imation Corp./Advanced Technology Program, National Institute of Standards and Technology, as part of co-operative agreement 70NANB5H1116. The authors express their gratitude to T. Schwarz of 3M Company for assistance in the procurement of the drop-out tester. The authors thank Dr Hirofumi Kondo of Sony Corp. Research Center, Yokohama, Japan, for supplying Hi-8 (120 min) ME and Hi-8 (120 min) MP professional video tapes, Dr T. Suzuki of Toshiba Corp. for supplying Hi-8 (120 min) barium ferrite professional video tape and Kazuya Irobe, Toshifumi Osawa, Hirofumi Murooka and Jerry Ishida of Teijin Limited for supplying ME tape substrate.

REFERENCES

1 **Bhushan, B.** *Mechanics and Reliability of Flexible Magnetic Media*, 1992 (Springer-Verlag, New York).

2 **Bhushan, B.** *Tribology and Mechanics of Magnetic Storage Devices*, 2nd edition, 1996 (Springer-Verlag, New York).

3 **Mee, C. D.** and **Daniel, E. D.** *Magnetic Recording Technology*, 2nd edition, 1996 (McGraw-Hill, New York).

4 **Kawana, T., Onodera, S.** and **Samoto, T.** Advanced metal evaporated tape. *IEEE Trans. Magn.*, 1995, **31**, 2865–2870.

5 **Shibata, S., Chino, N., Sato, N.** and **Chikamesa, H.** Simultaneous video tape dual-layer coating technology. *Fuji Film Res. Dev.*, 1992, **37**, 107–111.

6 **Richter, H. J.** and **Veitch, R. J.** Advances in magnetic tapes for high density information storage. *IEEE Trans. Magn.*, 1995, **31**, 2883–2888.

7 **Saitoh, S., Inaba, H.** and **Kashiwagi, A.** Developments and advances in thin layer particulate recording media. *IEEE Trans. Magn.*, 1995, **31**, 2859–2864.

8 **Sharrock, M. P.** and **Carlson, L. W.** The application of barium ferrite particles in advanced recording media. *IEEE Trans. Magn.*, 1995, **31**, 2871–2876.

9 **Speliotis, D. E.** Performance of MP++ and BaFe++ tapes in high density recording applications. *IEEE Trans. Magn.*, 1995, **31**, 2877–2882.

10 **Sugita, N., Maekawa, M., Ohta, Y., Okinaka, K.** and **Nagai, N.** Advances in fine magnetic particles for high density recording. *IEEE Trans. Magn.*, 1995, **31**, 2854–2858.

11 **Vom Wege, F.** and **Hornbogen, E.** Wear properties of metal evaporated tape. *Wear*, 1988, **128**, 291–305.

12 **Osaki, H., Fukushi, K.** and **Ozawa, K.** Wear mechanisms of metal-evaporated magnetic tapes in helical scan videotape recorders. *IEEE Trans. Magn.*, 1990, **26**, 3180–3185.

13 **Osaki, H., Oyanagi, E., Aonuma, H., Kanou, T.** and **Kurihara, J.** Wear mechanism of particulate magnetic tapes in helical scan video tape recorders. *IEEE Trans. Magn.*, 1992, **28**, 76–83.

14 **Patton, S. T.** and **Bhushan, B.** Effect of interchanging tapes and head contour on the durability of metal evaporated, metal particle and barium ferrite magnetic tapes. *Tribology Trans.*, 1995, **38**, 801–810.

15 **Patton, S. T.** and **Bhushan, B.** Friction and wear of metal particle, barium ferrite and metal evaporated tapes in rotary head recorders. *Trans. ASME, J. Tribology*, 1996, **118**, 21–32.

16 **Patton, S. T.** and **Bhushan, B.** Environmental effects on the pause mode performance of metal evaporated and metal particle tapes. *J. Appl. Phys.*, 1996, **79**, 5802–5804.

17 **Osaki, H., Kurihara, J.** and **Kanou, T.** Mechanisms of head-clogging by particulate magnetic tapes in helical scan video tape recorders. *IEEE Trans. Magn.*, 1994, **30**, 1491–1498.

18 **Hempstock, M. S.** and **Sullivan, J. L.** A study of the durability and lubricant performance of metal evaporated and metal particle tape. *IEEE Trans. Magn.*, 1996, **32**, 3723–3725.

19 **Patton, S. T.** and **Bhushan, B.** Tribological evaluation of the streaming mode performance of metal evaporated and metal particle tapes. *IEEE Trans. Magn.*, 1996, **32**, 3684–3686.

20 **Tsuchiya, T., Abe, M., Noro, Y.** and **Inada, K.** The gap recession reduced MIG head. *IEEE Trans. Magn.*, 1993, **29**, 3945–3947.

21 **Tsuchiya, T.** and **Bhushan, B.** Metal core recession and head stain studies of MIG heads sliding against cobalt-doped gamma iron oxide and metal particle tapes. *Tribology Trans.*, 1995, **38**, 941–949.

22 **Tsuchiya, T.** and **Bhushan, B.** Running characteristics of MIG heads against MP, barium ferrite and ME tapes. *IEEE Trans. Magn.*, 1994, **30**, 4182–4184.

23 **Bhushan, B., Patton, S. T., Sundaram, R.** and **Dey, S.** Pole tip recession studies of hard carbon-coated thin-film tape heads. *J. Appl. Phys.*, 1996, **79**, 5916–5918.

24 **Patton, S. T.** and **Bhushan, B.** Micromechanical and tribological characterization of alternate pole tip materials for magnetic recording heads. *Wear*, 1996, **202**, 99–109.

25 **Ota, H., Namura, K.** and **Ohmae, N.** Brown stain on VCR head surface through contact with magnetic tape. *Adv. Inf. Storage Systems*, 1991, **2**, 85–96.

26 **Stahle, C. M.** and **Lee, T. D.** Characterization of the deposits on helical scan heads. *Adv. Inf. Storage Systems*, 1992, **4**, 79–86.

27 **Bhushan, B.** and **Hahn, F. W.** Stains on magnetic tape heads. *Wear*, 1995, **184**, 193–202.

28 **Gupta, B. K., Bhushan, B., Zhou, Y., Winograd, N.** and **Krishnan, K.** Chemical analyses of stains formed on Co–Nb–Zr metal-in-gap heads sliding against oxide and metal particle magnetic tapes. *J. Mater. Res.*, 1995, **10**, 1795–1810.

29 **Mizoh, Y., Yohda, H., Kita, H.** and **Kotera, H.** Simulation of head wear and reproduction envelope by the finite element method. *Electronics and Communs in Japan, Part 2*, 1992, **75**, 91–101.

30 **Rogers, A. E.** and **Hinteregger, H. F.** Measurement of magnetic tape abrasivity by interchanging tape thickness. *Tribology Trans.*, 1993, **36**, 139–143.

31 **Mizoh, Y., Yohda, H., Inoue, K., Kotera, H.** and **Kita, H.** Influence of mechanical anisotropy of magnetic tape on head-to-tape interface in VCR. *J. Magn. Soc. Japan*, 1991, **15**, 709–714.

32 **Scott, W. W.** and **Bhushan, B.** Bending stiffness measurements of magnetic tapes and substrates. *Thin Solid Films*, 1997, **33**, 2513–2530

33 **Patton, S. T.** and **Bhushan, B.** Environmental effects on the streaming mode performance of metal evaporated and metal particle tapes. *IEEE Trans. Magn.*, 1997, **33**, 2513–2530.

34 **Oktay, S. T.** and **Suh, N. P.** Wear debris formation and agglomeration. *Trans. ASME, J. Tribology*, 1992, **114**, 379–393.

35 **Bhushan, B.** and **Patton, S. T.** Friction and wear of ultra high density magnetic tapes. *J. Appl. Phys.*, 1994, **75**, 5771–5773.

36 **Rabinowicz, E., Dunn, L. A.** and **Russell, P. G.** A study of abrasive wear under three-body conditions. *Wear*, 1961, **4**, 345–355.

37 **Timoshenko, S.** and **Young, P. H.** *Elements of Strength of Materials*, 1968 (Van Nostrand, New York).

38 **Hertzberg, R. W.** *Deformation and Fracture Mechanics of Engineering Materials*, 1989 (John Wiley, New York).

39 **Suh, N. P.** *Tribophysics*, 1986 (Prentice-Hall, Englewood Cliffs, New Jersey).

40 **Bowden, F. P.** and **Tabor, D.** *The Friction and Lubrication of Solids*, 1950 (Clarendon, Oxford).

41 **Archard, J. F.** Contact and rubbing of flat surfaces. *J. Appl. Phys.*, 1953, **24**, 981–988.

42 **Kim, S., Prabhakaran, V.** and **Talke, F. E.** Investigation of head wear and contamination in helical scan tape systems. *IEEE Trans. Magn.*, 1996, **32**, 3741–3743.

Effect of environment on drop-outs in rotary head–tape drives

R M Anderson and **B Bhushan**
Computer Microtribology and Contamination Laboratory, Department of Mechanical Engineering, The Ohio State University, Columbus, Ohio, USA

Abstract: The effect of environment on the drop-out performance of magnetic tapes in a rotary head–tape drive was studied. A capacitance probe, mounted on the rotating upper drum and positioned to measure tangential deflection of the head, was used to measure friction force. This technique, in conjunction with the measurement of head signal amplitude, was used to investigate the effect of environment on drop-outs. Streaming mode tests were conducted with two brands of S-VHS tapes at various combinations of temperature (6–38 °C) and relative humidity (10–80% RH). From the experiments, it was found that the propensity for intermittent drop-outs increases in a low relative humidity environment. A low relative humidity environment increases the propensity for debris to persist over the head gap and create a loss in the head signal. High temperature further increases the propensity for drop-outs in a low relative humidity environment. The propensity for intermittent drop-outs is reduced by increasing the relative humidity, changing magnetic tapes and to some extent by lowering the temperature of the environment. It was also determined that variability in the quality of magnetic tapes has a significant impact on the number of intermittent and permanent drop-outs encountered during playback.

Keywords: drop-outs, magnetic tapes, rotary head–tape drive

1 INTRODUCTION

A rotary head–tape drive and magnetic tape must be designed for optimum performance in the operating environmental envelope including extremely hot and dry or cold and wet without a loss in reliability. A decrease in reliability results in the loss of user data. A drop-out, a momentary loss of signal during playback of the recorded signal on the magnetic tape, is an indication of the reliability of the head–tape interface. A high number of drop-outs implies numerous data errors and poor reliability. Therefore, it is important to define an envelope for the operation of a tape drive in various environments in terms of drop-out performance and to study the drop-out mechanism in different environments, in order to determine which environments offer the best or worst data reliability.

Typical manufacturer specifications for reliable operation of tape drives and magnetic media are a temperature range of 5–40 °C and a relative humidity range of 10–80 per cent, non-condensing. The objective of this paper is to experimentally develop a safe operating envelope for rotary head–tape drives by studying changes at the head–tape interface during drop-outs in various environments at the limits of temperature (6–38 °C) and relative humidity (10–80% RH). It is of interest to determine the immediate effect of environment on the interface (within a day of testing), not over a long period. The effect of environment on the drop-out mechanism is investigated with the use of a new technique to measure friction on the head. A capacitance probe, mounted on the upper drum, is positioned to measure deflection of the head. This deflection is proportional to the friction force. The measurement of friction can signal changes occurring in the head–tape interface. An unusual variation in friction may signal a loss of optimum drive performance or poor quality in magnetic tapes.

In a previous paper (**1**), drop-outs were determined to be of three types: intermittent with correlation to friction force, permanent with correlation to friction force and permanent without correlation to friction force. Permanent drop-outs with correlation to friction force, repeating with each tape pass, typically result from tape defects and damage. Also, debris covering the head gap may prevent the ability to correctly write the signal during record, possibly resulting in a permanent drop-out without correlation to friction force during playback. Among other mechanisms, an intermittent drop-out may occur from

The MS was received on 11 September 1996 and was accepted for publication on 25 June 1997.

loose debris on the tape covering the head gap or debris, persisting over the head gap that prevents the read-out of the signal. To determine the effect of environment on drop-outs, the change in variable (intermittent) drop-outs in each test needs to be measured for different environments.

The macrodrop-outs, generally referring to a drop-out with duration on the order of milliseconds, investigated in this study are severe as the minimum loss in head signal is 6 dB and the duration is greater than 250 μs at the 6 dB loss level. A previous paper (**1**) investigated microdrop-outs, generally referring to a drop-out with duration on the order of microseconds, with a minimum loss of 2.5 dB and duration greater than 5 μs at the 2.5 dB loss level. The majority of microdrop-outs are permanent in a typical tape pass. Macrodrop-outs were chosen as the subject of this investigation as their occurrence results in a definite data error in the tape drive, a data storage device, used in the experiments.

Environment can play a significant role in the tribology of the head–tape interface and magnetic tapes. High humidity has been shown to lead to high friction and head wear (**2–6**). The long-term exposure of magnetic tapes to a hot and humid environment has been shown to result in tape degradation (**2**, **7–12**). The operation of a tape drive in low humidity generates stains and deposits on the heads (**2**, **13–18**). The effect of environment on the friction and wear of barium ferrite, metal particle and metal evaporated tapes in the pause mode has been studied (**19**, **20**), but little is known about the effect on drop-outs in the streaming (playback or reproduce) mode with particulate tapes. The results are the subject of this paper.

The sliding contact between the magnetic head and tape is known to generate rubbing noise (**21–25**). Other sources of noise in a tape drive include equipment, media and head noises. The frequency spectrum of the reproduced head signal generally consists of a distinct peak from the carrier frequency (for frequency modulation in magnetic recording) surrounded by frequencies from other noise sources. A very small amplitude white noise covers the spectrum which contains signals of uniform amplitude at all frequencies. Rubbing noise generally appears with a large amplitude and narrow band. The bandwidth for rubbing noise in the drive used in this study (a carrier frequency of about 12 MHz) is typically 10–20 MHz. Rubbing noise is typically severe at high head–tape speeds, low relative humidity and high head penetration. The level of rubbing noise can vary with different tapes and heads. It has been observed that at low humidity, rubbing noise may increase the level of noise in the signal by 2 or 3 times the typical noise level at ambient. This decreases the signal-to-noise ratio, thus increasing the probability of data error. To clarify the mechanism for rubbing noise, it is of interest to determine whether friction at the head–tape interface increases as the level of rubbing noise increases. Attempts were made to generate high levels of rubbing noise, but the results were inconclusive. However, rubbing noise is discussed in this paper.

2 DESCRIPTION OF EXPERIMENTAL APPARATUS

A tape drive, a data storage device [very large data store (VLDS) by Metrum Information Storage] using helical scan technology with S-VHS (Super VHS format) cassettes, was used in this study. In the drive, the read/write heads are mounted on a rotating upper drum that is positioned over a lower fixed drum. The drive parameters are summarized in Table 1. In forward mode, the tape is wrapped 186° around the fixed lower drum and the upper drum. The upper drum rotates at 3600 r/min. The supply arm maintains a supply tension of about 300 mN (30 g). The heads, shown in Fig. 1a, are single-crystal Mn–Zn ferrite and amorphous glass with a head gap creating a 70 μm wide track on the tape. The heads swipe over the tape at 6° relative to the direction of tape movement (helical angle) with a head–tape speed of 11.5 m/s. The head arrangement consists of two separate heads per window, one trailing and one leading, with two windows on the drum. Thus, there are four heads on the drum: two leading and two trailing. The head protrusion is 40 μm. In this experiment, only the two leading heads, diametrically opposite on the drum, are active during read/write operation.

In this study, a new technique is used to measure friction force at the head–tape interface (**1**). A capacitance probe is used to measure deflection of one of the leading heads. Thus, friction force is measured on only one of the four heads on the drum. The probe produces a 0–10 V signal for a range of 1.2–25 μm with a 10 nm resolution. A schematic with magnification of the probe mounted on the upper drum is shown in Fig. 1b. The probe is mounted to measure the deflection of the machined shoe. The deflection of the shoe is proportional to the friction force on the head. A groove is machined in the base of the leading head shoe. The shoe is reduced to a thickness of 0.43 mm at a distance of 6.09 mm from the head. This reduces the stiffness of the shoe which increases the measurability of shoe deflections. An aluminium mount is designed to securely attach the probe to the drum with an

Table 1 Rotary head–tape drive parameters

Drum diameter	62 mm
Upper drum rotating speed	3600 r/min or 11.7 m/s
Tape speed	0.1 m/s
Helical angle	6°
Relative head–tape speed	11.5 m/s
Wrap angle of tape over drum	186°
Supply tension	300 mN
Head length	1880 μm
Head width	140 μm
Head radius	11 mm
Head protrusion	40 μm
Effective track length	95.3 mm
Track width	70 μm
Track spacing (⊥ to tracks)	160 μm
Grooves on drum	5

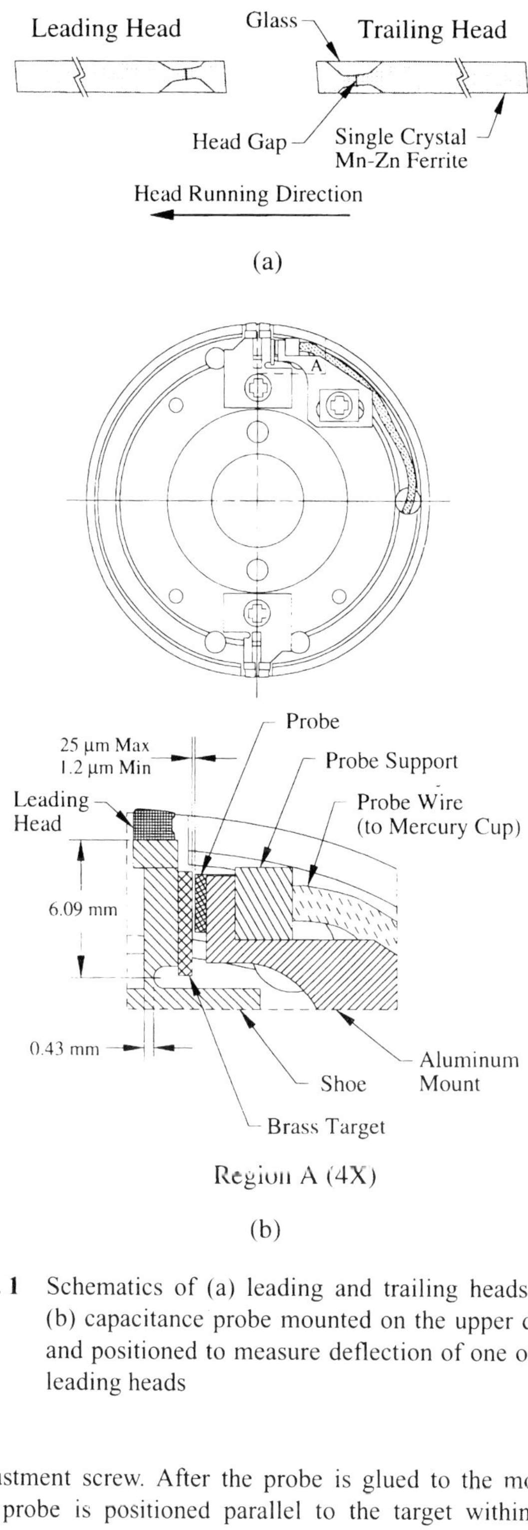

Fig. 1 Schematics of (a) leading and trailing heads and (b) capacitance probe mounted on the upper drum and positioned to measure deflection of one of the leading heads

adjustment screw. After the probe is glued to the mount, the probe is positioned parallel to the target within the range of 1.2–25 μm. The probe wire is passed through a hole in the drum and connected to a mercury cup that is fixed to the drive. This set-up is designed to channel the probe signal from the rotating drum while minimizing noise and vibration.

After the probe is mounted, the probe is calibrated, the drum is microbalanced and the response of the probe to an impact is determined. The probe is calibrated by standing the drum on its edge and adding weights to a basket hanging from the leading head. The displacement measured by the probe is linear over a range of weights from 0 to 40 mN, resulting in a calibration of 7.6 mV/mN. Since a 1 mV reading from the probe represents a 2.5 nm change in displacement (0–10 V for a range of 1.2–25 μm), the stiffness of the shoe measured at the probe is about 20 nm/mN. The distance from the probe to the head tip was measured as about 4.2 mm. Thus, the stiffness of the shoe at the head is about 45 nm/mN. The final voltage to friction force calibration after amplification is 0.41 V/mN. This static calibration is used to calculate friction force on the head, assuming that friction force is the stiffness of the shoe times the displacement of the head neglecting inertial forces of the small (about 0.1 g) mass of the movable shoe. The minimum resolvable friction force with this instrument is about 0.2 mN based on the low noise level of the probe signal and the authors' experience with the instrument. The response of the probe (after filtering and amplification) to a sudden impact on the head was determined with the use of a contact probe. The mechanical resonant frequency of the shoe, measured by the probe, is about 1400 Hz with an electronic delay in the probe response of 0.2 ms.

In order to capture the friction force and head signal amplitude during a drop-out in the streaming mode, the data acquisition system is designed to simultaneously measure friction force and head signal amplitude at a sampling rate of 10 μs per sample. To measure friction force, the signal from the probe amplifier is passed through a high-pass filter with a cut-off frequency of 10 Hz to remove the d.c. offset, thereby allowing analysis of the dynamic responses of the probe. The probe signal is amplified by 53, resulting in a final signal voltage to friction force calibration of 0.41 V/mN. To measure head signal amplitude, the equalized (reproduced) head signal from the drive is passed into a high-precision wide-band r.m.s.-to-d.c. converter with an averaging time constant of 67 μs, which converts the a.c. head signal to an r.m.s. signal.

To capture drop-outs at a 100 000 Hz sampling rate without gathering large volumes of data, an electronic trigger was developed. If the r.m.s. head signal falls below a specified dB loss level for a minimum specified duration and the instrumented head is swiping over the tape, a data recorder is triggered to simultaneously record friction force and head signal amplitude. A liquid crystal display counter is used to record the total number of triggers during a test.

During the reproduce mode in the drive, the two leading heads alternate in the task of reading data from the tape. The drive uses a digital head switch clock (clock in this paper) alternating from Hi to Lo to indicate when a head is swiping over the tape. This clock is a synchronous signal used by numerous systems in the drive. Each Hi or Lo of the clock represents 180° of drum rotation with a duration of 8.2 ms. The instrumented head is passing over the tape during the Hi of the clock.

Two commercially available S-VHS ST-120 tapes from two manufacturers were studied: A and B. Tapes A and B have a width of 12.7 mm and thickness of about 19 μm. The r.m.s. of the surface heights measured with an atomic force microscope for tapes A and B is about 10.3 and 7.9 nm respectively. The tapes are particulate magnetic tapes with a magnetic coating deposited on a flexible poly(ethylene terephthalate) substrate (**2**, **26**). Magnetic particles, polymeric binders, lubricants, a cross-linker or curing agent, a dispersant or wetting agent, and solvents form the magnetic coating. For the S-VHS tape, the magnetic particle is typically Co-modified γ-Fe_2O_3 with a length of 0.15–0.20 μm and aspect ratio of 5–10 (**2**, **26**).

Tapes A and B were tested for drop-out performance at the limits of temperature and relative humidity in the operating envelope. The testing was conducted in an environmental chamber. Tape A was tested in five environments: ambient (22 °C/45% RH), hot/wet (38 °C/80% RH), wet (22 °C/80% RH), hot/dry (38 °C/10% RH) and dry (22 °C/10% RH). Tape B was tested in these five environments plus cold/wet (6 °C/80% RH) and cold/dry (10 °C/10% RH). All environments were ±1 °C and ±5% RH. Tape A was not tested in all seven environments because the tape, as discussed in the results, performed well in all five environments tested. After the heads were run-in, an arbitrary signal set by the drive consisting of multiple frequencies from 2 to 11 MHz was recorded on a virgin tape at ambient. The tape was then played for one pass at ambient to determine the base level of permanent drop-outs.

It was also of interest to determine whether preconditioning the tape had an effect on drop-out performance. In two of the environmental tests for tape B, wet (22 °C/80% RH) and dry (22 °C/10% RH) tapes were removed from storage in an ambient environment and then preconditioned in the test environment for 4 and 6 days respectively. A tape was soaked at high humidity (> 80% RH) by using line air, a beaker and a hot plate (**6**). The line air was passed through a beaker of heated water and into a plastic enclosure containing the cassette. The air flowrate and water temperature were adjusted to obtain the correct relative humidity level without condensation. A tape was dried at low humidity (< 10% RH) by passing line air directly into a plastic enclosure. In both tests, the complete cassette was placed on edge in the enclosure to allow sufficient air flow on all sides. Prior to the start of the test, the reels were weighed. After the cassette was placed in the enclosure, the reels were weighed every 24 or 48 hour intervals.

Prior to the start of the environmental test, the drive and tape were conditioned in the test environment for about 1 hour. The heads and drum were cleaned with isopropyl alcohol prior to the start of the test to remove any possible sources of invalid drop-outs. The average head signal amplitude was estimated and the trigger level was set for a 6 dB loss with a minimum duration below the 6 dB loss level of 250 μs. The tests were conducted for 38 minutes in the streaming (reproduce) mode. The same segment of tape (about 220 m) was played in the drive for 11–20 passes. This resulted in a total test time of about 7–13 hours. For each pass, the number of triggers as counted and captured. After the test, the drum was removed and the heads were examined for debris and then photographed. The drum was removed for examination of debris to determine possible sources of the drop-outs.

3 RESULTS AND DISCUSSION

Table 2 gives the drop-out results from each environmental experiment on tapes A and B. A drop-out was recorded if it resulted in a loss of head signal during a single swipe of the instrumented head. A head signal loss of consecutive swipes would result in multiple drop-outs. The total number of intermittent drop-outs was divided by the number of tape passes to obtain the average intermittent drop-outs per pass (DP). The specific humidity (mass of water vapour/mass of dry air) from a psychometric chart indicates the amount of water vapour present for a given temperature and relative humidity and is given in Table 2. The DP results are rated

Table 2 Drop-out results for tapes A and B in each environment

Tape	Test condition	Specific humidity (approx.)	Tape condition prior to test	Total tape passes	Permanent drop-outs* per pass	Total intermittent drop-outs*	Average intermittent drop-outs* per pass
A	22 °C/45% RH	0.007	Ambient	16	1	2	0.1
	38 °C/80% RH	0.034	Ambient	11	0	0	0
	22 °C/80% RH	0.013	Ambient	20	0	2	0.1
	38 °C/10% RH	0.004	Ambient	11	1	1	0.1
	22 °C/10% RH	0.002	Ambient	20	0	0	0
B	22 °C/45% RH	0.007	Ambient	12	0	1	0.1
	38 °C/80% RH	0.034	Ambient	12	1	1	0.1
	22 °C/80% RH	0.013	4 day soak	16	4	1	0.1
	6 °C/80% RH	0.005	Ambient	12	0	0	0
	38 °C/10% RH	0.004	Ambient	12	4	7	0.6
	22 °C/10% RH	0.002	6 day dry	12	5	3	0.3
	10 °C/10% RH	0.001	Ambient	12	0	3	0.3

*Counted drop-outs exceed approximately 6 dB loss and duration greater than 250 μs at the 6 dB loss level.

graphically on a temperature–relative humidity envelope for tapes A and B in Fig. 2 in terms of good, fair and poor performances. Figure 3 shows the number of intermittent drop-outs per pass from each test. This section is divided into intermittent drop-outs at ambient, high humidity and low humidity, permanent drop-outs and rubbing noise for tapes A and B.

3.1 Intermittent drop-outs at ambient

Figure 4 defines an intermittent drop-out. An intermittent drop-out occurs during a single tape pass and correlates to a change in the friction force. In the seventh pass of tape A, the figure shows the r.m.s. head signal as the instrumented head swipes over the tape. The friction force rises sharply as the head contacts the tape in the inlet region. The friction force gradually rises across the head swipe to a maximum of about 3 mN. In the outlet region, the friction force drops as the head leaves the tape. The head passes around the back side of the drum (while the opposite leading head swipes over the tape) and the friction profile repeats after 8.2 ms. The opposite leading head refers to the head diametrically opposite to the instrumented leading head. The clock at the bottom of the figure alternates from Hi to Lo as the heads alternate in swiping over the tape. The

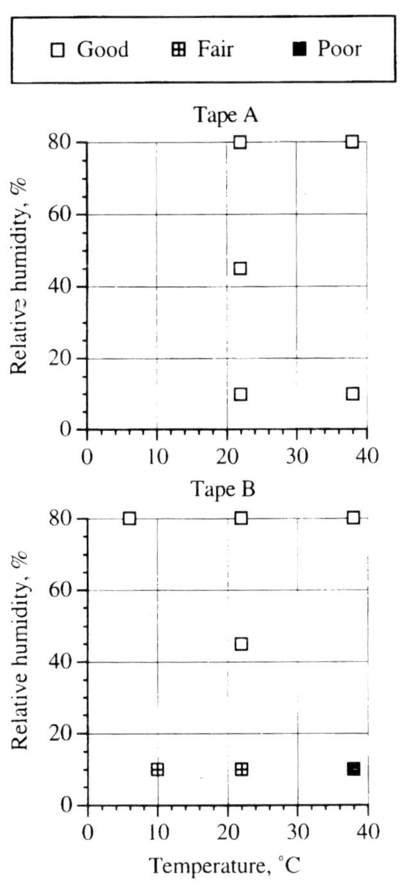

Fig. 2 Intermittent drop-out performance of tapes A and B in each environment

deep, repetitive drops (typically 6–12 dB) in the head signal at the beginning of the head swipe are a result of a high-frequency region recorded by the drive that is above the maximum frequency range of the r.m.s.-to-d.c. converter.

The steady state friction profile, the repeating profile measured as the instrumented head swipes over the tape, is associated with the tape flying height and deformation (**1**). In a rotary head–tape drive, a hydrodynamic air film is generated between the rotating upper drum and tape. The flying height between the drum and tape decreases from the inlet to the outlet regions. The flexible tape deforms and forms a tent around the fixed protruding head. As the head enters the inlet region, the tape flying height is maximum; thus the tape is easily deformed over the head. The resulting friction force is low. As the head continues to swipe over the tape, the flying height decreases. The amount of tape deformation increases and, correspondingly, the friction force increases.

In the eighth pass, a drop-out occurs and reaches a maximum loss of about 10 dB. The friction force spikes to a maximum of about 4.5 mN. The duration of the friction spike matches the duration of the head signal loss. This is an intermittent drop-out with correlation to friction force as it does not occur in the seventh or ninth tape pass.

As shown in Fig. 2, both tapes performed well at ambient with a DP of 0.1. For tape A, drop-outs occurred in the eighth and eleventh tape passes shown in Figs 4 and 5. For tape B, a drop-out occurred in the seventh pass as shown in Fig. 7. There is not a significant difference between these ambient drop-outs. The tape B drop-out persisted for a longer duration than the tape A drop-out. The friction force increased by 0.5 to 1 mN for the majority of the head swipe. As shown in the photographs of Figs 9 and 10, the contact region of the instrumented leading head and the trailing head for tapes A and B are clear of debris after testing. Even though the trailing head is not instrumented, debris that passes through the interface of the instrumented leading head may collect on the trailing head. Since more debris collects on the trailing head than the leading head, the photographs of Figs 9 and 10 also show differences in the debris generated by tapes A and B. A small amount of non-adherent debris, generally chunks and strands of binder material a few μm thick, collects on the non-contact region of the heads. This non-adherent debris is easily removed with a head-cleaning solvent.

In a previous paper (**1**), the authors found that loose debris on the tape surface which increases the head–tape spacing or debris generated at the head–tape interface that sticks to the head gap can result in intermittent drop-outs. The loose debris on the tape may originate from tape debris collected on the tape path guides, upper and lower drums or heads. Contaminants such as dust can also generate intermittent drop-outs. If the mechanism for intermittent drop-outs, as shown in Figs 4, 5 and 7, is the head passing over loose debris on the tape that results in an increase of the head–tape spacing and corresponding head signal loss

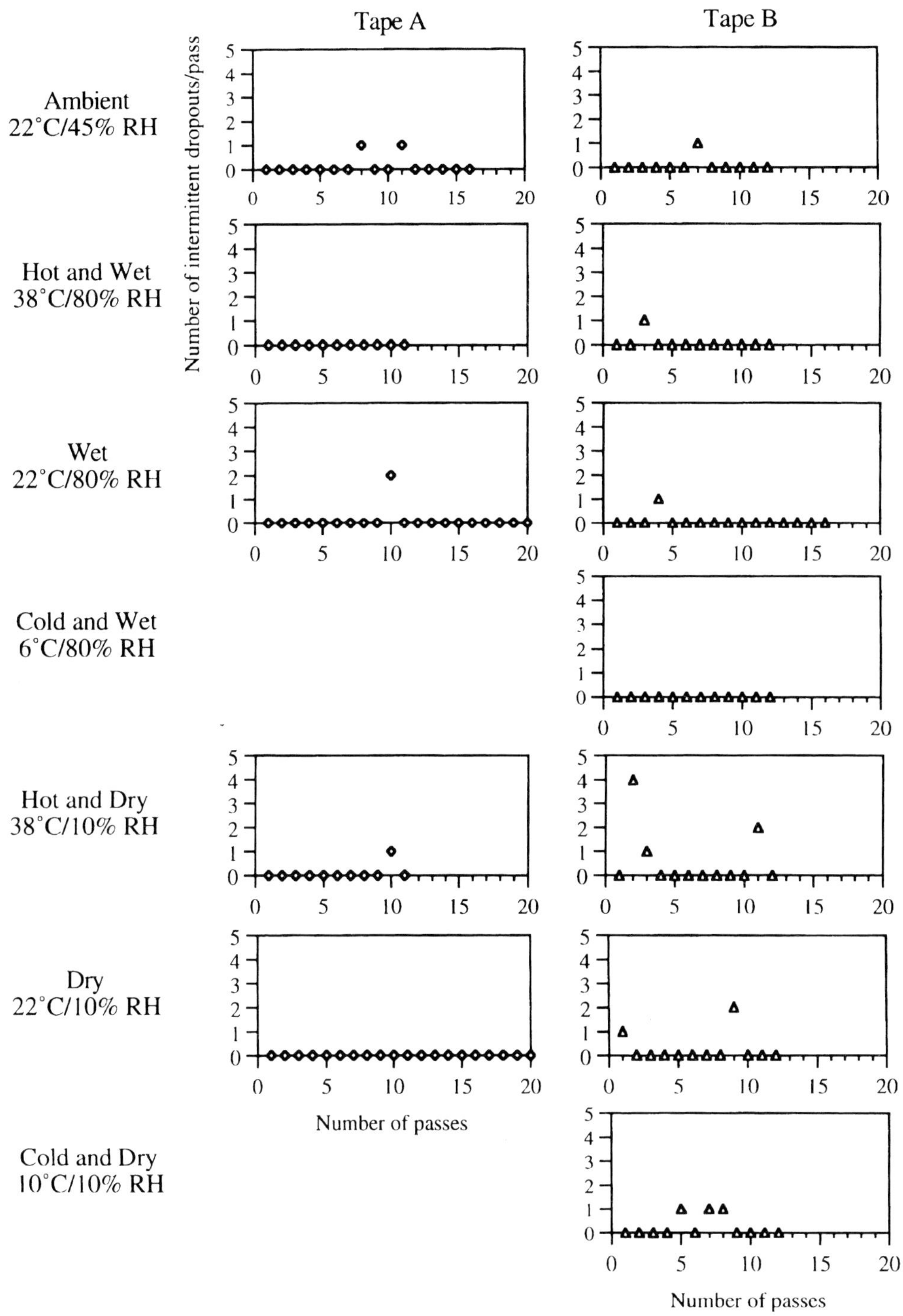

Fig. 3 Plots of number of intermittent drop-outs per pass for tapes A and B in each environment. Drop-outs have a minimum loss of 6 dB and duration greater than 250 μs at the 6 dB loss level

and friction spike, then the minimum size of the debris would be about 3 mm (head–tape speed times drop-out duration). From examinations of the tape and head surfaces, debris with a size of 3 mm is not typical. Therefore, it is more likely that these drop-outs are a result of loose debris on the tape surface or debris generated at the head–tape interface, sticking to the head gap and preventing the correct read-out of the signal. The friction force increases as the debris is slid against the tape surface. The debris persists until the interfacial friction is able to remove the debris from the head gap; then the head signal recovers.

Figure 5 shows typical and atypical intermittent drop-outs from ambient testing of tape A. The typical drop-out reached a maximum depth of about 9 dB and the friction force spiked for about 4 ms. In the atypical case, the head signal from the instrumented head drops below 6 dB for the

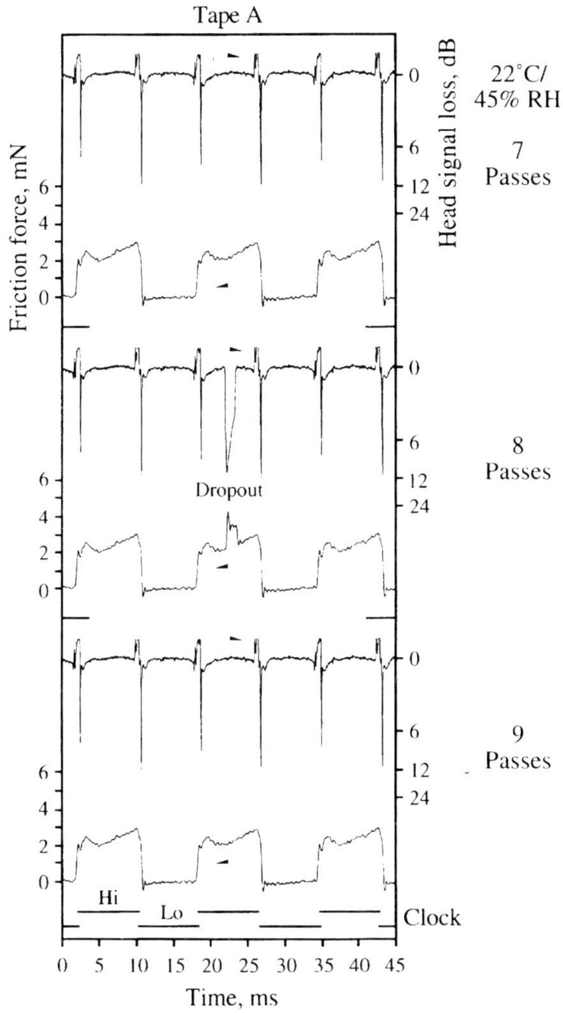

Fig. 4 An intermittent drop-out from the eighth pass of tape A at ambient

complete head swipe and the friction force increases by about 0.5 mN. This pattern repeats in the next swipe of the instrumented head. The swipe of the opposite head remains intact, therefore ruling out the possibility of substantial debris on the tape. The pattern actually repeated for 30 head swipes (0.5 s), resulting in 30 drop-outs or 30 tracks of lost user data. Debris stuck to the head gap at the beginning of a head swipe and persisted for 30 swipes, resulting in a severe head clog. The gap was sufficiently covered to reduce the head signal by 6 dB. This intermittent drop-out was extremely rare as ambient drop-outs are typically low frequency and usually not more than a head swipe in duration.

A problem encountered during ambient testing was variability in the samples of tape B. Figure 11 shows the results of 12 passes during an ambient test with a tape B sample. The number of intermittent drop-outs per pass increases significantly to 6 after 12 passes. These drop-outs were generally severe as the drop-outs typically persisted for consecutive head swipes and resulted in high friction

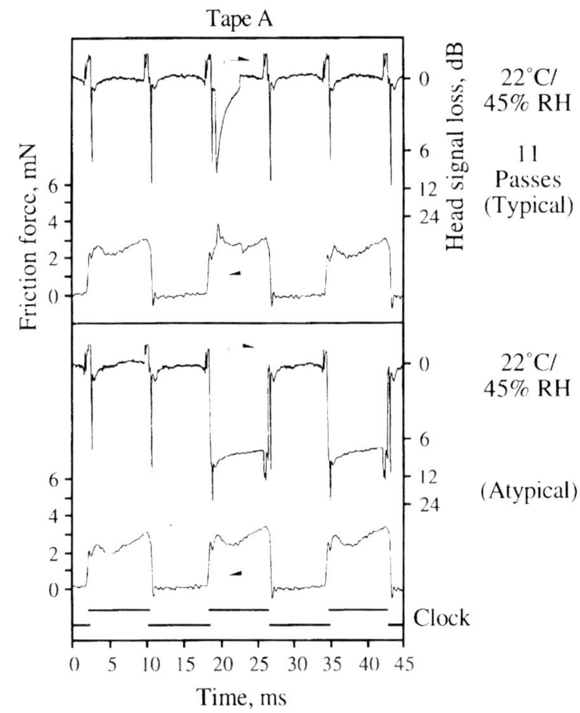

Fig. 5 Typical and atypical intermittent drop-outs from tape A at ambient

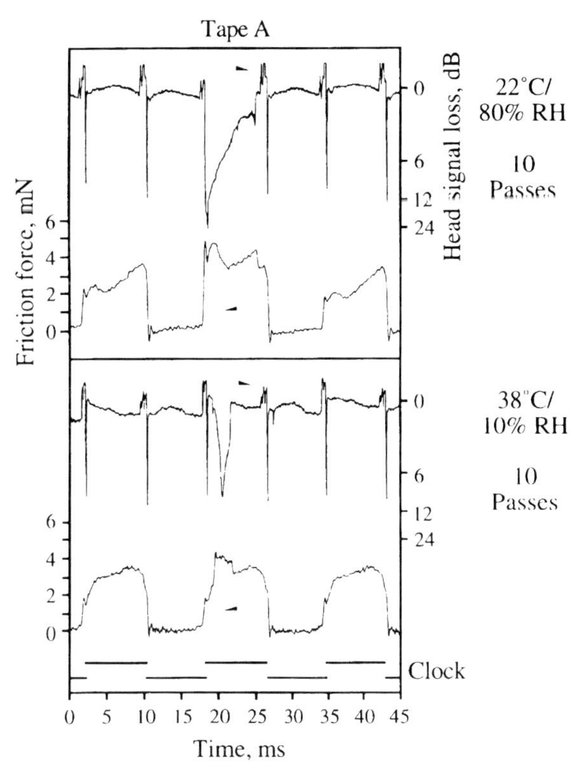

Fig. 6 Intermittent drop-outs from tape A in wet (22 °C/80% RH) and hot/dry (38 °C/10% RH) environments

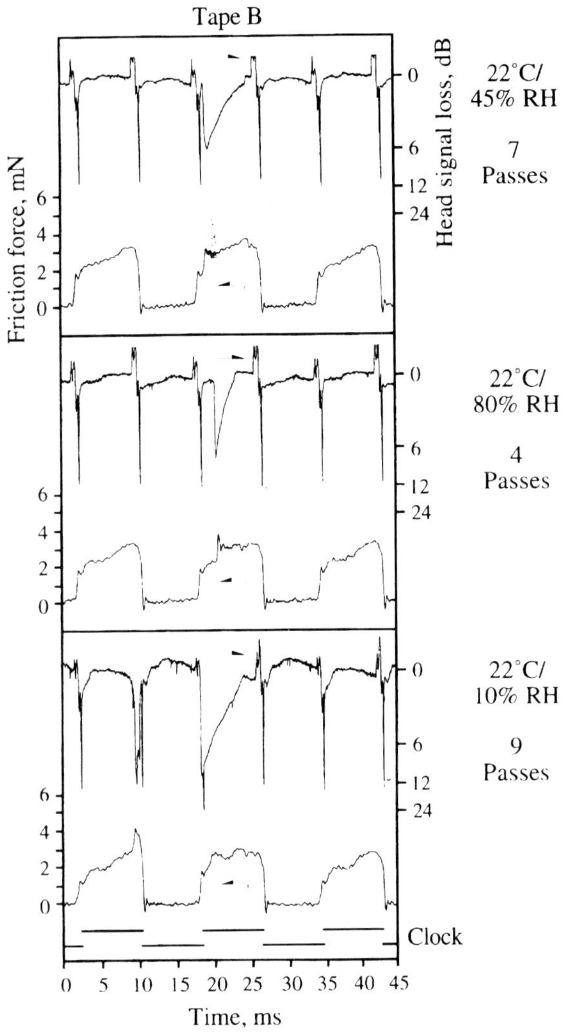

Fig. 7 Intermittent drop-outs from tape B in ambient (22 °C/45% RH), wet (22 °C/80% RH) and dry (22 °C/10% RH) environments

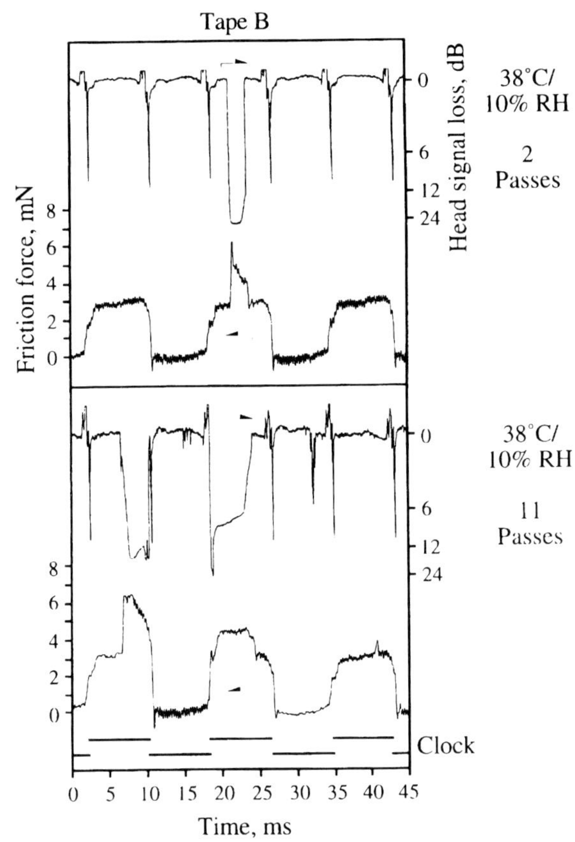

Fig. 8 Intermittent drop-outs from tape B in a hot and dry (38 °C/10% RH) environment

force spikes. The photographs reveal that heavy abrasion of the instrumented leading head surface has occurred, possibly due to sharp isolated asperities with a high contact pressure on the tape surface ploughing into the head material. The asperities may pull out from the tape, persist and create a drop-out or generate a rough head surface that increases the propensity for debris to stick to the gap. This problem was remedied by running a good-quality tape in the drive for about 5 hours at ambient.

3.2 Intermittent drop-outs at high humidity

The number of intermittent drop-outs at high humidity did not significantly change from ambient, as shown in Figs 2 and 3. The testing of tapes A and B at 38 °C/80% RH and 22 °C/80% RH did not have an effect on drop-outs and nor did testing tape B additionally at 6 °C/80% RH. The DP for the high-humidity experiments was either 0 or 0.1. The drop-out at 22 °C/80% RH from the tenth pass of tape A (Fig. 6) was more severe than the tape B drop-out at 22 °C/80% RH (Fig. 7). The friction force from the tape A drop-out reached a maximum of 5 mN and continued for the majority of the head swipe. The data are limited, but there is not a definite difference in the severity of the drop-outs at ambient or high humidity.

Although high humidity in combination with several temperatures did not effect drop-outs, high humidity is characterized by increased head wear and friction (**2**–**6**). An increase in wear was observed as the heads generally contoured to the tape faster at high humidity than ambient. Bhushan and Khatavkar (**6**) attribute the increase in friction to the formation of menisci bridges at asperity contacts. This increases the adhesive friction. An increase in the friction that would correlate to high humidity even at 38 °C/80% RH was not observed where the most water vapour was present. This may be due to a high head–tape speed of 11.5 m/s, which does not allow the formation of menisci. However, there was a 10–15 per cent increase in friction at cold and wet, and also at cold and dry. This may be due to increased stiffness of the magnetic tape at low temperatures.

Photographs of the instrumented leading head and the trailing head after each test are shown in Figs 9 and 10. It generally appears that, at high humidity, water tends to agglomerate the debris on the heads in the non-contact

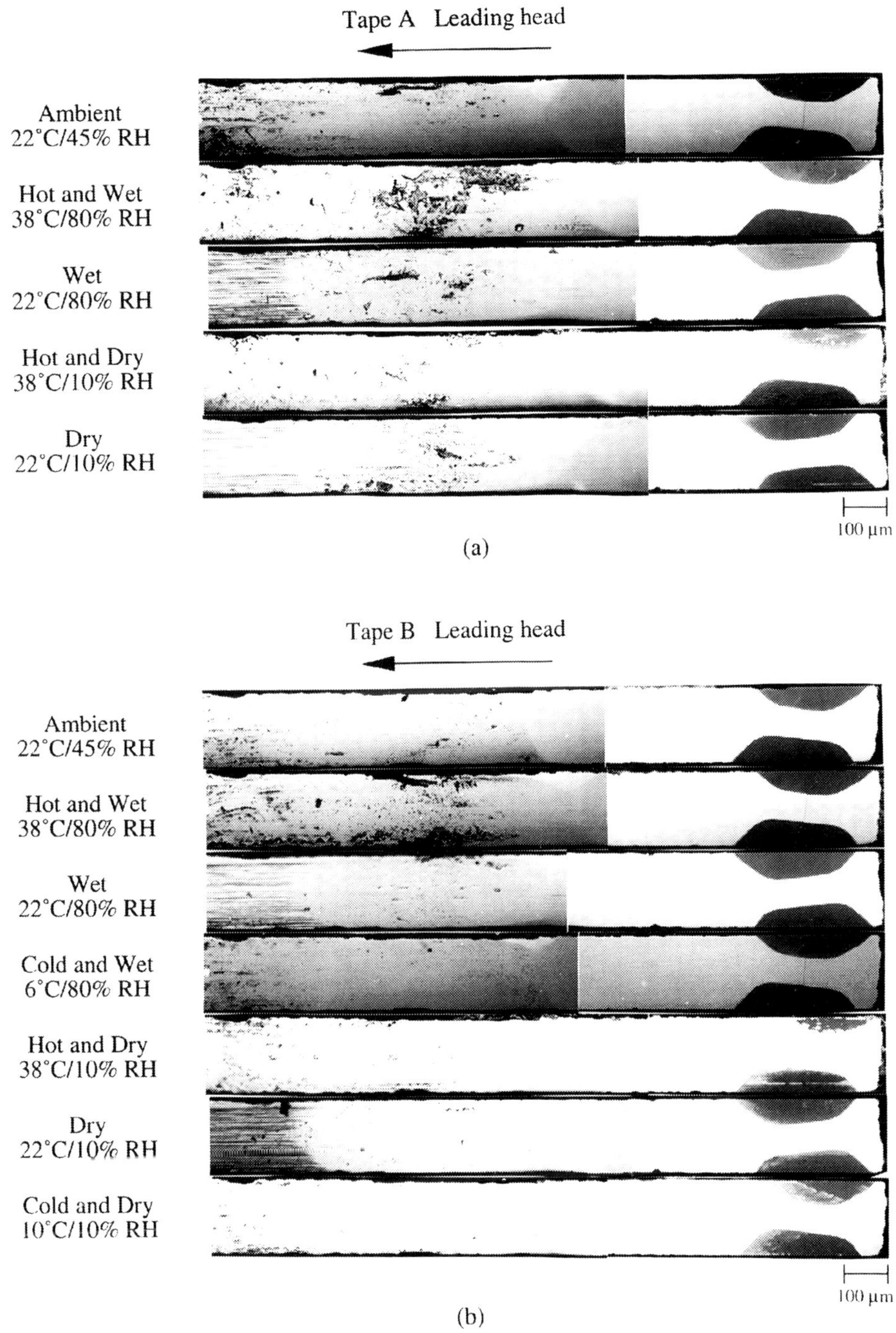

Fig. 9 Photographs of the instrumented leading head after testing of (a) tape A and (b) tape B in each environment. The arrows indicate the head running direction

region, but this does not significantly affect drop-outs. The photographs of the trailing head after testing of tapes A and B at 38 °C/80% RH show more agglomeration than the other environments, along with heavy flaking at 22 °C/80% RH for tape A and 38 °C/80% RH for tape B. The flake of thick debris on the trailing head after testing with tape A in the wet environment may have been transferred to the tape in subsequent passes, possibly resulting in a drop-out. Debris was not found near the head gap of the instrumented leading head after the tests, but the photograph (Fig. 12a) of the opposite leading head shows non-adherent debris near the head gap after the hot and wet test with tape A, which may have caused a drop-out in later tape passes.

It is of interest to determine whether preconditioning the tape prior to the start of the test has an effect on drop-outs. Tests were performed with tape B at 22 °C/80% RH after 4 days of soaking an ambient tape at >80% RH and at

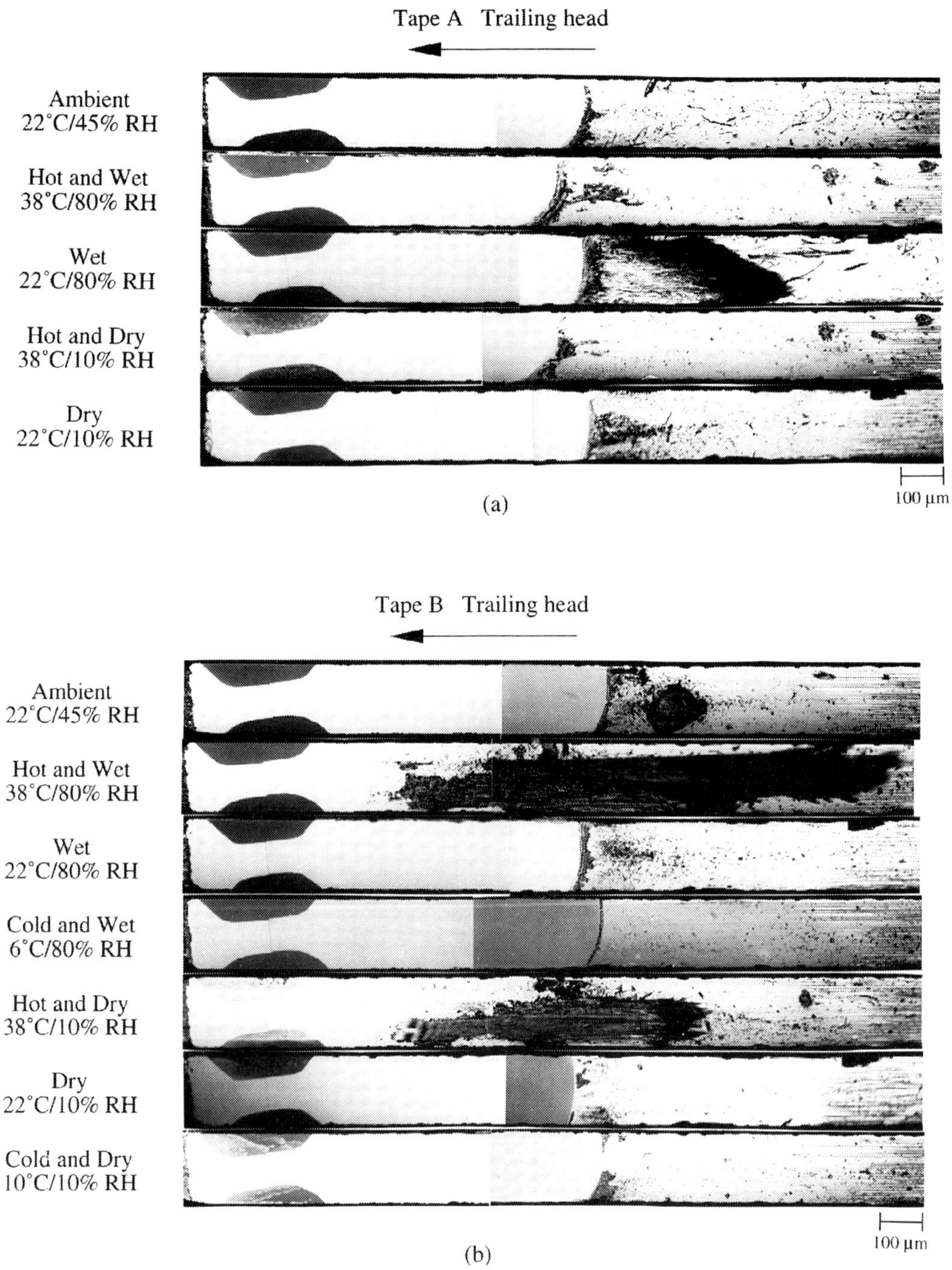

Fig. 10 Photographs of the trailing head (not instrumented) after testing of (a) tape A and (b) tape B in each environment. The arrows indicate the head running direction

22 °C/10% RH after 6 days of drying an ambient tape at < 10% RH. Figure 13 shows the percentage change in weight of the reels during soaking and drying of tapes A and B for 14 days. The amount of water absorbed or lost is reversible. The ambient weight of the reels prior to the start of the test, about 131 g for tape A and 125 g for tape B, is used as the reference weight. The data in Fig. 13 approximately fit with the equation:

$$\Delta W \approx A(1 - e^{-Bt})$$

where ΔW is the percentage change in weight of the reels and t is the soak/dry time in days. The saturation level, A, is 0.31 and −0.15 for soaking and drying respectively. The absorption rate, B, is 0.34 and 0.51 for soaking and drying respectively. The hygroscopic behaviour of magnetic tapes has been thoroughly studied by Vos *et al.* (**27**) and Cuddihy (**28**).

The drop-out test of soaked tape B at 22 °C/80% RH resulted in a DP of 0.1. This gives typical results at high humidity for both tapes; therefore preconditioning of the

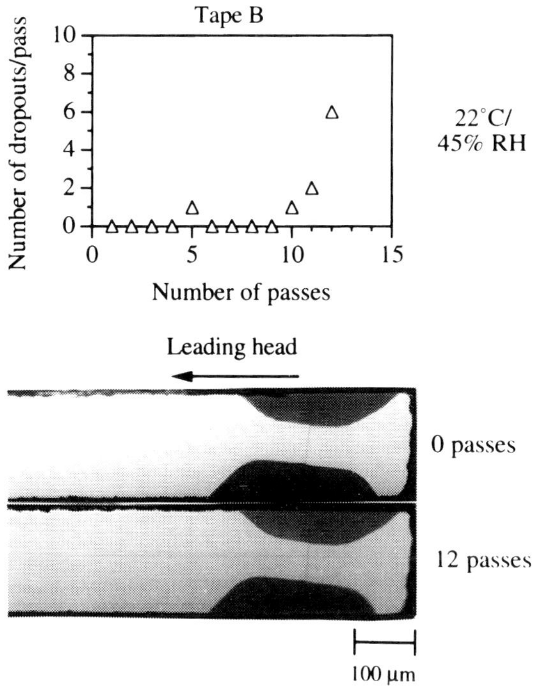

Fig. 11 A plot of the number of intermittent drop-outs per pass and photographs of the instrumented leading head at 0 passes and after 12 passes at ambient with tape B. The arrow indicates the head running direction

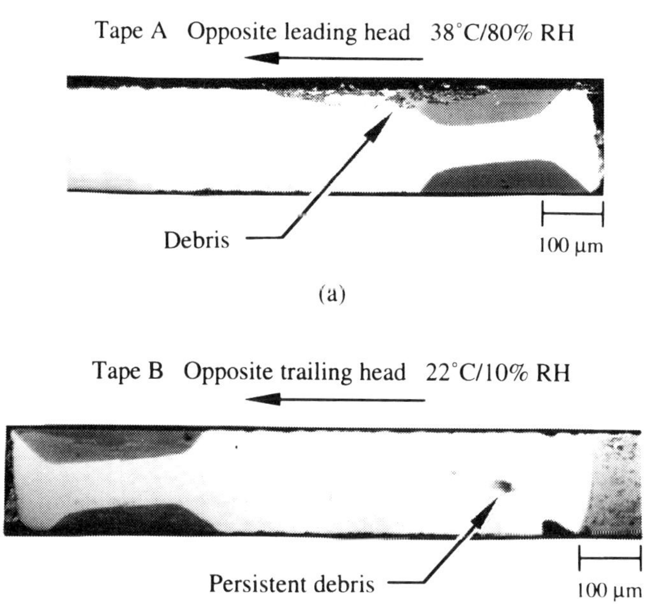

Fig. 12 Photographs of (a) non-adherent debris near the head gap of the opposite leading head after testing tape A in a hot and wet (38 °C/80% RH) environment and (b) persistent debris on the opposite trailing head after testing tape B in a dry (22 °C/10% RH) environment. The arrows indicate the head running direction

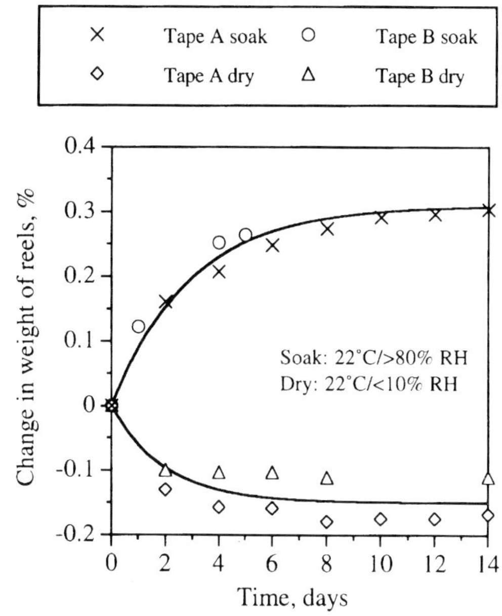

Fig. 13 Plot of percentage change in weight of the reels versus number of days for soaking and drying of tapes A and B. The reference weights of the reels for tapes A and B are about 131 and 125 g respectively

tape did not have an impact on drop-outs. The tape had been expected possibly to become gummy and sticky. This may have increased the persistence of debris on the head gap. The lack of drop-outs with the preconditioned tape may be due to the fact that once several tape passes have been completed in the test, the tape in the reels is exposed and acclimatizes quickly to the environment. The water content of an unwound tape, in general, requires about 0.6 min to equalize with each percentage change in relative humidity (**28**). Also, an increase in drop-outs may have been observed if the tape was soaked at an elevated temperature for about a week or more. Soaking along with baking a tape for about a week would most likely generate a substantial increase in drop-outs. An elevated temperature increases the ageing process of magnetic tapes. The process of tape degradation has been studied (**2**, **7**–**12**). Binder hydrolysis is known to cause a gummy and sticky binder layer that can create playback problems in magnetic tapes.

3.3 Intermittent drop-outs at low humidity

Tape A performed well in the low-humidity environments, as indicated in Fig. 2, with a maximum DP from the two tests of 0.1. Tape B, however, showed a significant increase in intermittent drop-outs at low humidity. The DP are 0.6, 0.3 and 0.3 for 38 °C/10% RH, 22 °C/10% RH and 10 °C/10% RH respectively. These values from the low-humidity tests are higher than values from the ambient or high-humidity tests. The test sample for tape B at 22 °C/10% RH was dried for 6 days prior to the start of the test, but this is most likely not attributable to the increase in drop-outs. As shown in Fig. 3, drop-outs resulted in the loss of user data five times in the first three tape passes at 38 °C/10% RH.

Figure 7 shows multiple drop-outs from the ninth pass of tape B at 22 °C/10% RH. A loss in the head signal occurred near the end of the first head swipe and then continued for the majority of the second head swipe. This resulted in two intermittent drop-outs which correlated to friction force. Figure 8 shows drop-outs from the hot/dry (38 °C/10% RH) test of tape B. For the drop-out from the second pass, a deep drop in the head signal to 24 dB correlated to a spike in the friction force that reached 6 mN. During the eleventh pass, a severe head clog occurs, which results in two deep drop-outs from 12 to 24 dB and friction spikes that reach 6.5 and 4.5 mN. As shown in the figures, the drop-outs from the low-humidity tests of tape B are generally more severe than drop-outs from ambient or high-humidity tests. The drop-outs from the hot and dry test are the most severe. Long duration and high friction spikes are observed in these drop-outs plus drop-outs occurring over consecutive head swipes. The authors believe that this is a result of the low-humidity environment increasing the propensity for debris to persist over the head gap due to the lack of water vapour acting as a lubricant. The persistence of debris is further increased by high temperatures in a low-humidity environment. However, high temperatures do not necessarily increase drop-outs as tape B did not show an increase in drop-outs at 38 °C/80% RH. Figure 12b is a photograph of the opposite trailing head after use of tape B at 22 °C/10% RH. A small piece of debris in the contact region is stuck to the head surface. This debris could not be removed with isopropyl alcohol.

Staining and deposits are known to form on heads in low-humidity environments (**2**, **13**–**18**). The photographs of the instrumented leading head and the trailing head (Figs 9b and 10b) after use in low humidity with tape B indicate that adherent deposits have formed in the amorphous glass region. The deposits are most severe at 38 °C/10% RH. There is no evidence of deposits from the photographs of ambient or high humidity. The deposits are most likely to be composed of tape materials, including magnetic particles and wear debris from the ferrite and glass regions of the head. The photographs of the heads (Figs 9a and 10a) after use at low humidity with tape A show significantly less formation of deposits than tape B. Also, the DP did not increase with tape A at low humidity. The deposits forming in the glass region are evidence of the tenacity of debris to persist on the head, although not necessarily being related to the mechanism responsible for the increase in drop-outs. The photographs also show that the formation of deposits can be reduced by using a different brand of tape, lowering the temperature of the environment and increasing the relative humidity. The most effective method to reduce the formation of deposits is to increase the relative humidity. The water vapour, as in the case of drop-outs at low humidity, acts as a lubricant to prevent the formation of deposits.

The drop-out performance of tape A is insensitive to the environment. Also, the propensity for the formation of deposits at low humidity was less in tape A than tape B. This may be due to a difference in binder chemistry.

It is interesting to note from the eleventh pass data in Fig. 8 that in the third friction profile a small friction spike, of about 1 mN, occurs without a loss in the head signal. This spike may have been a result of debris sticking to the contact region of the head but not covering the head gap; thus a drop-out does not occur.

In the severe environments of the drop-out tests, the reliability of the tape drive was often a greater concern than the reliability of the magnetic tape. The drive struggled in the cold environments (6 °C/80% RH and 10 °C/10% RH) where the tape path friction is high and greases/lubricants have a relatively low viscosity. Performance in the high-humidity environments (> 80% RH) was good as long as condensation was avoided.

3.4 Permanent drop-outs

Figures 14 and 15 are examples of typical permanent drop-outs encountered in environmental testing. The permanent drop-outs did not provide additional information for the environmental study, but can create a significant source of error in magnetic tapes. The number and severity of permanent drop-outs provide a good measure of tape quality. Permanent drop-outs fall into two categories: permanent drop-outs with correlation to friction force and permanent drop-outs without correlation to friction force (**1**).

A drop-out occurs in the first pass at 22 °C/45% RH with tape A (Fig. 14a). A friction force spike of about 2 mN correlates with a head signal loss. The drop-out repeats in the second pass. This is a permanent drop-out with correlation to friction force as it continually repeats with each tape pass and correlates to a change in friction force. Tape damage from a defect, mishandling or excessive use can generate this category of drop-out. Also shown in Fig. 14a, the drop-out in the swipe of the instrumented head is followed by a smaller drop-out in the swipe of the opposite leading head that starts at the same position in the head swipe. These drop-outs are possibly a result of damage along the machine direction of the tape. Figure 14b shows a permanent drop-out, which lasts for a substantial duration and correlates to friction force, probably due to a localized tape defect.

The second category of drop-outs (Fig. 15) is permanent drop-outs without correlation to friction force. During the first pass of tape B at 22 °C/80% RH, a drop-out occurs and the friction force remains unchanged. This is also the case in the fifth pass of tape B at 22 °C/10% RH, except that another drop-out occurs in the next swipe of the instrumented leading head. One possible mechanism for these drop-outs is debris covering the head gap during

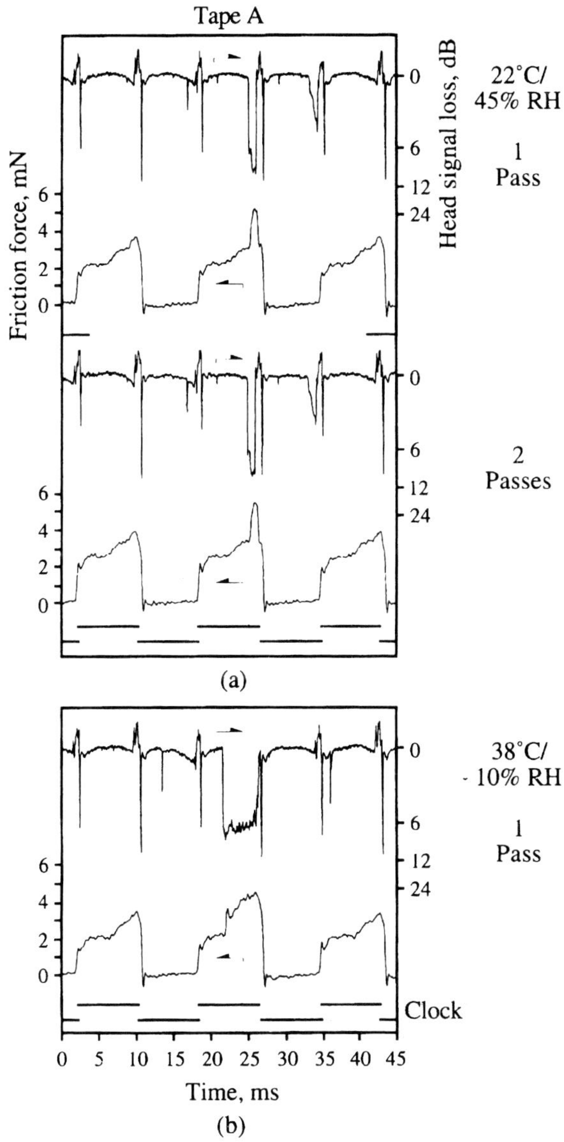

Fig. 14 Typical permanent drop-outs with correlation to friction force, (a) and (b), present in every pass and resulting in friction spikes encountered during testing of tape A

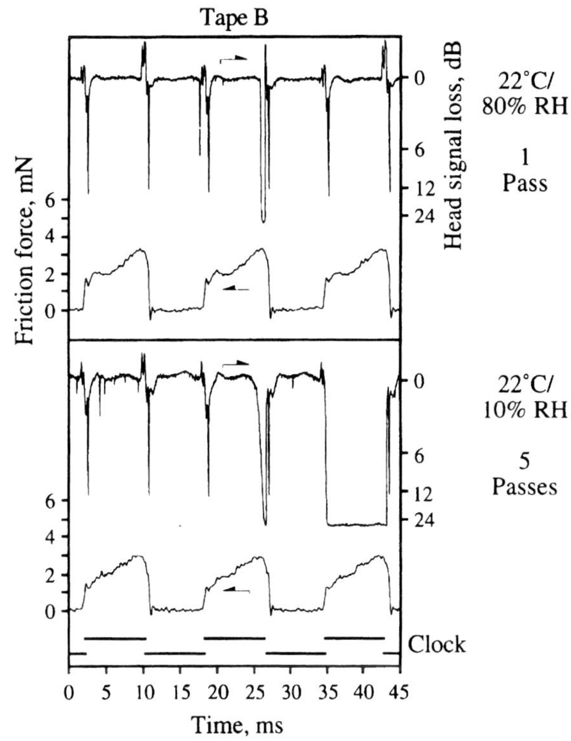

Fig. 15 Typical permanent drop-outs without correlation to friction force encountered during testing of tape B

recording, which prevents the ability of the head to correctly write the signal. On playback, a drop-out would appear without a change in the friction force as the tape remains undamaged. This mechanism can be checked by rewriting the track of data and then determining whether the drop-out has disappeared during playback. The permanent drop-outs at 22 °C/10% RH (Fig. 15) may have been the result of a head clog that persisted for several head swipes during recording, thus creating multiple permanent drop-outs in playback. Another possible mechanism is that magnetic particles are not completely dispersed in the magnetic layer, creating regions unable to sustain a recorded signal.

As shown in Table 2, there can be a considerable amount of variability in the number of permanent drop-outs in a tape. Two of the A tapes had one permanent drop-out while three of the B tapes had four or five permanent drop-outs. A good method to avoid a problem with permanent drop-outs in a tape is to screen the tape prior to use. Record the full tape and then check for drop-outs during playback. A permanent drop-out will repeat with each playback.

3.5 Rubbing noise

A problem that has arisen from the use of ferrite heads in magnetic recording is rubbing noise. Rubbing noise increases the noise level in the reproduced signal and is associated with sliding contact of the head and magnetic tape. The signal-to-noise ratio is reduced, thus decreasing clarity of the reproduced signal. Studies and theories of rubbing noise have been presented (**21**–**25**). Johnson and Huijer (**23**) describe rubbing noise as a result of the friction between the head and magnetic tape inducing mechanical vibrations in the head. Because of the dimensions and other properties of the head, the vibrations are generated in the MHz region at certain peak frequencies. Since Mn–Zn ferrite is magnetostrictive, the stress from the vibrations induces changes in the internal magnetization, thus varying the magnetic flux of head. Consequently, the variations in the flux generate noise in the head coils in the MHz region. A similar theory for rubbing noise is proposed by Mee and Daniel (**21**).

It has been observed that in a low-humidity environment, rubbing noise may generate a two- to threefold increase in the typical noise level of the reproduced signal. This severe rubbing noise has been possibly associated with a silver coloured deposit forming in the amorphous glass region of the head. It is of interest to determine whether a change in the interfacial friction occurs as the level of rubbing noise increases. In an attempt to initiate high levels of rubbing noise, tests were conducted in dry (22 °C/10% RH) and hot/dry (38 °C/10% RH) environments. Two brands of tape were tested and shuttled in the drive for 8 hours. The noise in the reproduced signal was monitored with an oscilloscope after recording a 'blank' signal on the tape. A wedge-shaped noise pattern was expected to develop where the noise level increased 2 to 3 times by the end of the head swipe. However, this was not observed in the test. As shown in Fig. 16, an atypical wedge shape pattern in the signal noise was encountered in one of the tapes similar to the expected rubbing noise pattern. This wedge-shaped noise pattern oscillated for short durations and repeated with each tape pass. The friction force at the head–tape interface did not change during the oscillations in noise. Therefore, it appears that a permanent region of the tape generated the excessive noise possibly caused by variations in the tape roughness, magnetic layer thickness, head–tape spacing, substrate thickness or head–tape velocity (**21**, **29**, **30**). It was expected that if the increase in noise was due to rubbing noise, the effect would be continuous at low humidity and could be reduced in an ambient environment. As shown in Figs 9 and 10, deposits were observed on the glass region. When rubbing noise is observed, the deposits may increase the roughness of the head surface, thus increasing head oscillations and generating rubbing noise. The fact that rubbing noise was not observed may have been due to the manufacturer of the head. Some ferrite heads generate more rubbing noise than others (**22**, **23**, **31**).

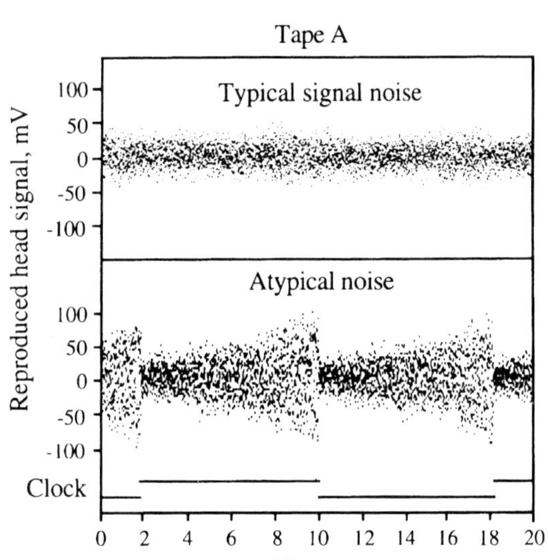

Fig. 16 Typical and atypical reproduced signal noises during playback of tape A

4 CONCLUSIONS

Environmental experiments were performed on S-VHS magnetic tapes in a rotary head–tape drive. The test results, summarized in Table 2 and Fig. 2, were used to determine the effect of environment on the drop-out performance of particulate magnetic tapes. The conclusions are as follows:

1. Tape A performed well in all test environments in terms of the number of intermittent drop-outs. The performance of tape B was fair in cold/dry (10 °C/10% RH) and dry (22 °C/10% RH) environments. The performance was poor in a hot/dry (38 °C/10% RH) environment.
2. The propensity for intermittent drop-outs increases in a low relative humidity environment. This increases the propensity for debris to persist over the head gap and create a loss in the head signal. High temperatures further increase the propensity for drop-outs in a low-humidity environment. Intermittent drop-outs were generally most severe in a hot and dry environment.
3. The propensity for intermittent drop-outs is reduced by increasing the relative humidity, changing magnetic tapes and to some extent by lowering the temperature.
4. The formation of deposits in the amorphous glass region is generally substantial in a hot and dry environment. The deposits generally do not form in an ambient or high-humidity environment.
5. Water vapour acts as a lubricant to reduce the propensity for drop-outs and deposit formation in a high relative humidity environment.
6. The duration of head signal loss matches the duration of friction force spike in intermittent drop-outs with correlation to friction force.
7. The variability in the quality of magnetic tapes, between brands and within a brand, can have a significant effect on intermittent and permanent drop-outs.

ACKNOWLEDGEMENTS

The authors would like to thank Dr Dan Soo for suggesting that a program be initiated to measure *in situ* friction force at the Ohio State University and for equipment and technical support during the course of this research. The authors would further like to thank Gary Simpson, Gerald Wade (formerly with Metrum) and John Schutte of Metrum Information Storage, Littleton, Colorado, for equipment and technical support. Tapes A and B were obtained from Maxell and 3M respectively. Financial support for this study was provided by the industrial membership of the Computer Microtribology and Contamination Laboratory.

REFERENCES

1 **Anderson, R. M.** and **Bhushan, B.** Concurrent measurement of *in situ* friction force and head signal amplitude during dropouts in rotary head tape drives. *Wear*, 1996, **202**, 35–49.

2 **Bhushan, B.** *Tribology and Mechanics of Magnetic Storage Devices*, 2nd edition, 1996 (Springer-Verlag, New York).

3 **Carroll, J. F.** and **Gotham, R. C.** The measurement of abrasiveness of magnetic tape. *IEEE Trans. Magn.*, 1966, **MAG-2**, 6–13.

4 **Mayer, D. H.** On the abrasivity of γ-Fe_2O_3 and CrO_2 magnetic tapes. *IEEE Trans. Magn.*, 1974, **10**, 657–659.

5 **Hegel, R. F.** Hygroscopic effects on magnetic tape friction. *Tribol. Trans.*, 1993, **36**, 67–72.

6 **Bhushan, B.** and **Khatavkar, D. V.** Role of water vapor on the wear of Mn–Zn ferrite heads sliding against magnetic tapes. *Wear*, 1996, **202**, 30–34.

7 **Cuddihy, E. F.** Aging of magnetic recording tape. *IEEE Trans. Magn.*, 1980, **MAG-16**, 558–568.

8 **Bertram, H. N.** and **Cuddihy, E. F.** Kinetics of the humid aging of magnetic recording tape. *IEEE Trans. Magn.*, 1982, **MAG-18**, 993–999.

9 **Wheeler, J.** Long-term storage of videotape. *SMPTE J.*, 1983, **92**, 650–654.

10 **Bradshaw, R. L.** and **Bhushan, B.** Friction in magnetic tapes III: role of chemical properties. *ASLE Trans.*, 1984, **27**, 207–219.

11 **Bradshaw, R., Bhushan, B., Kalthoff, C.** and **Warne, M.** Chemical and mechanical performance of flexible magnetic tape containing chromium dioxide. *IBM J. Res. Develop.*, 1986, **30**, 203–216.

12 **Edge, M., Allen, N. S., Hayes, M., Jewitt, T. S., Brems, K.** and **Horie, V.** Degradation of magnetic tape: support and binder stability. *Poly. Deg. Stab.*, 1993, **39**, 207–214.

13 **Bhushan, B.** and **Hahn, F. W.** Stains on magnetic tape heads. *Wear*, 1995, **184**, 193–202.

14 **Ota, H., Namura, K.** and **Ohmae, N.** Brown stain on VCR head surface through contact with magnetic tape. *Adv. Inf. Stor. Systems*, 1991, **2**, 85–96.

15 **Stahle, C. M.** and **Lee, T. D.** Characterization of the deposits on helical scan heads. *Adv. Inf. Stor. Systems*, 1992, **4**, 79–86.

16 **Gupta, B. K., Bhushan, B., Zhou, Y., Winograd, N.** and **Krishnan, K.** Chemical analyses of stains formed on Co–Nb–Zr metal-in-gap heads sliding against oxide and metal particle magnetic tapes. *J. Mater. Res.*, 1995, **10**, 1795–1810.

17 **Tsuchiya, T.** and **Bhushan, B.** Metal core recession and head stain studies of MIG heads sliding against cobalt-doped gamma iron oxide and metal particle tapes. *Tribology Trans.*, 1995, **38**, 941–949.

18 **Kim, S., Prabhakaran, V.** and **Talke, F. E.** Investigation of head wear and contamination in helical scan tape systems. *IEEE Trans. Magn.*, 1996, **32**, 3741–3743.

19 **Tsuchiya, T.** and **Bhushan, B.** Running characteristics of MIG heads against MP, barium ferrite and ME tapes. *IEEE Trans. Magn.*, 1994, **30**, 4182–4184.

20 **Patton, S. T.** and **Bhushan, B.** Environmental effects on the pause mode performance of metal-evaporated and metal-particle tapes. *J. Appl. Phys.*, 1996, **79**, 5802–5804.

21 **Mee, C. D.** and **Daniel, E. D.** *Magnetic Recording Technology*, 2nd edition, 1996 (McGraw-Hill, New York).

22 **Kawamata, T., Mizoh, Y., Ushigome, H.** and **Hagiwara, H.** Abrasivity effect of magnetic recording tape on rubbing noise (in Japanese). *Trans. Inst. Electronics, Inf. Commun. Engrs*, 1984, **J67-C II**, 62–69.

23 **Johnson, M. T.** and **Huijer, E.** Rubbing noise generated by acoustic vibrations in ferrite test heads. *IEEE Trans. Magn.*, 1988, **24**, 2398–2400.

24 **van Groenou, A. B.** On the interpretation of tape friction. *IEEE Trans. Magn.*, 1990, **26**, 144–146.

25 **Oehme, W. F.** and **Wei, N.** Investigation and modelling of helical scan tape recorder noise. Conference Publication 402, 1994, pp. 72–75 (Institution of Electrical Engineers, London).

26 **Bhushan, B.** *Mechanics and Reliability of Flexible Magnetic Media*, 1992 (Springer-Verlag, New York).

27 **Vos, M., Ashton, G., Van Bogart, J.** and **Ensminger, R.** Heat and moisture diffusion in magnetic tape packs. *IEEE Trans. Magn.*, 1994, **30**, 237–242.

28 **Cuddihy, E. F.** Hygroscopic properties of magnetic recording tape. *IEEE Trans. Magn.*, 1976, **MAG-12**, 126–135.

29 **Tarumi, K.** and **Noro, Y.** A theoretical analysis of modulation noise and d.c. erased noise in magnetic recording. *Appl. Phys. A*, 1982, **28**, 235–240.

30 **Daniel, E. D.** A preliminary analysis of surface-induced tape noise. *IEEE Trans. Commun. Electronics*, 1964, **CE-83**, 250–253.

31 **Watanabe, H.** and **Yamaga, I.** Low noise manganese–zinc single crystal ferrite heads. *IEEE Trans. Magn.*, 1972, **MAG-8**, 497–500.

Wear mechanisms of sandwich-type video heads

M J K Harrison[1], **J L Sullivan**[1] and **G S A M Theunissen**[2]
[1]Department of Electronic Engineering and Applied Physics, Aston University, Birmingham
[2]Philips Research Laboratories, Eindhoven, The Netherlands

Abstract: Differential wear between the various material components of composite magnetic recording heads is observed upon running against the magnetic medium. A developing level difference between the magnetic material and the medium bearing surface leads to signal degradation, exponential with the difference and the signal frequency. Experimental evidence is presented for a three-body erosion mechanism in the case of sandwich heads employed in commercial helical scan, Hi-8 video tape recorders. Two sintered ceramic substrates, $CaTiO_3$ and composite Zr-doped $CaTiO_3$, in combination with three soft magnetic materials, amorphous CoNbZr and the nanocrystalline iron alloys FeNbSiN and FeTaN, have been studied. They have been studied in conjunction with commercial metal evaporated (ME) and metal particulate (MP) tapes. Level difference has been measured by interferometry and AFM (atomic force microscopy), the latter also observing level differences between the various components of the sandwich: soft magnetic metal, ceramic laminations and the metallic bonding layer. AFM shows a pull-out of particles from the ceramic substrate with similar sized particles observed on the track. Scanning Auger micrographs have shown the transfer of ceramic particles on to the track and of metal on to the substrate. Wear rates and wear mechanisms have been identified for the materials in isolation by studying single-material dummy heads. This has been identified as plastic deformation controlled microabrasion, transforming to grain pull-out in the crystalline materials in some instances. It is speculated that these pull-outs are responsible for the wear of the soft metallic track when this is out of contact with the tape. The evidence for chemical wear is inconclusive. The involvement of transfer material from the tape to the head is speculated upon.

Keywords: magnetic recording head, wear, AFM, Auger, XPS, microabrasion, three-body abrasion

NOTATION

- C experimental tape factor
- d spacing between magnetic pole tips of the head and the tape (nm)
- h microhardness indent depth (μm)
- k_m wear coefficient, magnetic material
- k_s wear coefficient, substrate
- l length of long diagonal of Knoop microhardness indent (μm)
- p fractional porosity
- P_m head–tape contact pressure, magnetic material
- P_s head–tape contact pressure, substrate
- λ recording wavelength (μm)

1 INTRODUCTION

Ferrites, the traditional materials of choice for video recording heads do not meet the soft magnetic criteria for high-frequency magnetic recording such as Hi-8 and digital video recording for current and high-definition formats. The current solution is metal in gap (MIG) heads in which a soft magnetic material, most commonly Sendust, is deposited in the gap of a ferrite head using the tilted sputtered Sendust (TSS) method. With an alternative design—the 'sandwich head'—the manufacturing process is much simplified, both in terms of micromachining and sputter deposition. This design allows accurate determination of track width and the employment of laminate insulation layers to improve the magnetic performance of the head.

The main thrust of this work is to understand the principal mechanisms at work in the differential wear experienced in sandwich heads. This phenomenon, pole tip recession (PTR), also known as 'hollow-out', is common to all composite magnetic recording heads—MIG, thin film, hard disk sliders, etc. Recession of the magnetic flux guide at the pole tips relative to the tape bearing surface will cause a signal loss rising exponentially with the magnitude of PTR and the recording frequency:

$$\text{Signal} \propto \exp\left(\frac{-2\pi d}{\lambda}\right) \quad (1)$$

The MS was received on 17 January 1997 and was accepted for publication on 4 August 1997.

The loss for both recording and playback is about 1 dB for every 10 nm at a recording wavelength of 1 μm.

PTR is influenced by material combination, climatic conditions and tape abrasivity (**1**–**3**). PTR in sandwich heads is further complicated by differential wear of lamination layers within the soft magnetic sandwich, which may either be recessive or protrusive as compared to the soft magnetic material. Possible mechanisms for PTR have been modelled by Tsuchiya, Abe and Noro (**1**, **2**) and by Broese van Groenau *et al.* (**3**). The former find PTR to be broadly independent of the form of the composite head (MIG, sandwich, etc.) and of the contact pressure. They describe an experimental wear law for equilibrium PTR based on the individual wear coefficients of the magnetic material and the substrate, k_m and k_s respectively, and a 'tape factor', C:

$$\Delta d = C \ln\left(\frac{k_m}{k_s}\right) \tag{2}$$

The problem with both models is that they assume a mainly abrasive, linear (Archard) wear rate for the magnetic material and substrate, thus defining PTR with time:

$$\frac{d(\Delta d)}{dt} = k_m P_m - k_s P_s \tag{3}$$

with P_m and P_s being the contact pressures on the metal and substrate respectively. P_m and P_s need not be equal and an equilibrium PTR value will be approached if contact pressure on the metal falls with increasing Δd. However, no account is made in the models for other possible wear mechanisms responsible for PTR and transitions between mechanisms. Broese van Groenau and co-workers indeed identified corrosion and erosion in some cases. Tape abrasion is also highly unlikely to cause some of the higher values observed in PTR (values of 20–80 nm in 100 h). Broese van Groenau and co-workers calculated that standard tapes were unlikely to bend more than a few nm into a recess under a wide range of contact pressures (30–300 kPa). Impingement of the highest tape asperities into a recession-causing abrasion is unlikely as metal evaporated (ME) tape, which is smoother and more uniform than metal particulate (MP) tapes, is often observed to produce greater PTR (**4**). Thus the wear mechanism responsible for the propagation of PTR is likely to be different to that of the running tape against an individual head material component. To understand the pressures exerted upon the various components of the head, the influence of the air bearing and topical lubrication must be understood.

Of prime importance is the nature of the head–tape contact. Intimate contact is desired to minimize signal loss, as governed by equation (1). An air bearing is required to prevent catastrophic damage on repeated head–tape collision in the helical scan mode. Research by Wu and Talke (**5**), on linear streaming of cylindrical heads, has shown unlubricated head–tape separation to be limited by the tape roughness (usually greater than the head roughness). This places contact in the elastohydrodynamic regime on the Stribeck curve. Results presented in this paper suggest it is elastoplastohydrodynamic contact that occurs in the situations described. Mobile and boundary lubrication of the tapes, transferred to the head surface, serves to lower friction, i.e. reduce stresses parallel to head and tape surfaces as well as mediating perpendicular stresses, thus protecting higher asperities. Head staining is a well-documented but ill-understood phenomenon (**6**). Stains increase head–tape spacing, thus decreasing the signal, and disrupt the hydrodynamics of the contact. They may, due to chemical interaction or third-body erosion, increase wear or they may act as a barrier to wear.

In this paper wear data are presented for dummy heads of the two ceramic materials and for dummy heads with a surface coating of soft magnetic material. The purpose of this is to identify the wear mechanisms of the major components of sandwich heads in isolation and to provide for, in the case of the metal-coated heads, a sufficient surface area for spectroscopic analysis. Commercial recorders were used to provide realistic running conditions. Conditions of high relative humidity (90 per cent) at ambient laboratory temperature (20 °C) were used as these represent relatively severe tribological conditions for which chemical effects are most likely to be observed. The surfaces of unworn and worn heads were analysed topographically by atomic force microscopy (AFM) and chemically by X-ray photoelectron spectroscopy (XPS), Auger mapping and point Auger electron spectroscopy (PAES). Also presented are AFM and XPS/AES analyses of sandwich heads run under similar conditions.

2 EXPERIMENTAL PROCEDURE

2.1 The heads

The dummy and sandwich heads used are shown diagrammatically in Fig. 1. The ceramic heads are sawn from tiles and glued on to brass mounting plates. The long head curvature is laser cut. The short curvature is introduced by lapping. The head–tape conformation is fine tuned by running in against MP tape in Hi-8 recorders. A soft magnetic layer can then be sputter deposited to a thickness of 3 μm and annealed. Sandwich heads are prepared by depositing sequentially magnetic layers and insulating layers. The sandwich is formed by bonding two such tiles together using Mo–Au. Individual heads and their coil chambers and azimuthal gaps are sawn. These heads are then similarly mounted and conformed. Table 1 details the head materials and the head dimensions used. S2 is a zirconium-doped $CaTiO_3$ composite.

2.2 Wear tests

The heads, two to each drum, are mounted with a protrusion of 25 ± 5 μm, adjusted by measuring the

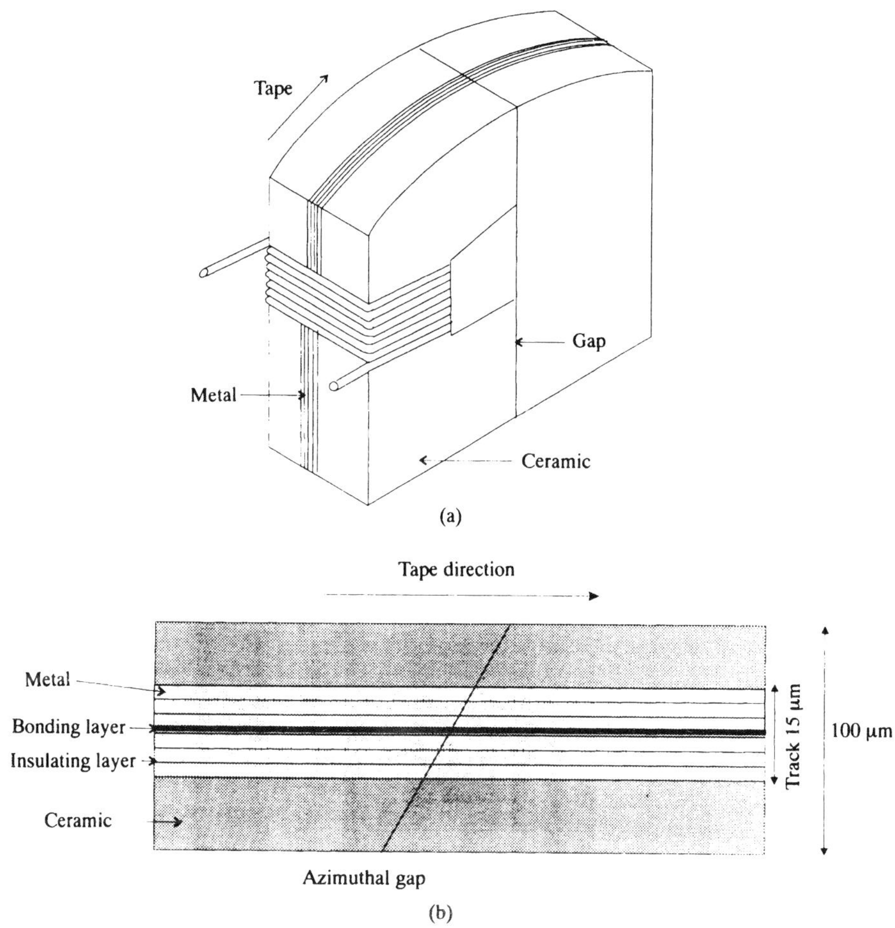

Fig. 1 (a) Dummy head used in cycling tests. (b) Surface view of sandwich head

height difference in the focal plane of the top of the head and the highest point on the drum. Rotation of the drum gives a head–tape velocity of 3.14 m/s in the play mode, corresponding to 5652 passes per hour for each head. The recorders are commercial Philips M870 camcorders, modified to allow computer control of the play, stop and rewind functions. The software employed allows the length and number of tape cycles to be preset. Commercial SONY Hi-8 60 min ME and MP tapes were used for cycling tests. A fresh tape was used for each set of heads and no more than 150 cycles were used as after this the tape signal was seen to degrade. High-humidity experiments were conducted in a perspex chamber with water-saturated air pumped in. The chamber was ventilated to maintain atmospheric pressure. A relative humidity (RH) of 90 ± 5 per cent was achieved.

Table 1 Dimensions of dummy heads used in cycling

	Designation	Material	Length (μm)	Width (μm)
Substrate Materials	S1	$CaTiO_3$	600	120 and 80
	S2	$Zr–CaTiO_3$	600	120 and 80
Magnetic Coatings	M1	CoNbZr	600 and 900	100
	M2	FeNbSiN	900	100
	M3	FeTaN	900	100

Temperature and humidity were monitored using an electronic thermometer/hygrometer.

Wear was measured by monitoring the change in depth of Knoop microhardness indentations with time. Knoop indentations, shown in Fig. 2a, are of a known geometry, with the following relationship between depth (h) and the length of the long diagonal (l):

$$h = \frac{l}{30.5} \tag{4}$$

The indentations were placed along the centre-line of the head along its entire length, shown in Fig. 2b, using a Buehler Micromet4 microhardness indenter. A separation of 50 μm between indentations ensured no mechanical interference between them. The long diagonal of the indentations was placed perpendicular to the direction of the tape to prevent scratching effects (**7**). An indenter force of 50 g and dwell time of 15 s gave indentations of $l \sim 30$ μm and $h \sim 1$ μm for each material. The materials were of similar hardness and the hardness of each material was recorded. The size of the indentations was constrained by the curvature of the heads. In the case of the ceramic heads, indentations were made after running in. The indentation length was measured using

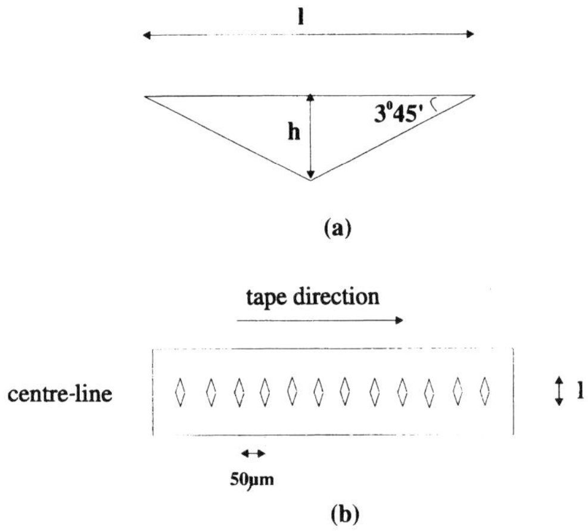

Fig. 2 (a) Geometry of the long diagonal of the Knoop indentation. (b) Distribution of the indents along the head surface

an optical microscope with graticule, giving a depth resolution of ± 0.05 μm.

The drum was removed after 5 hours of cycling, the head surfaces inspected optically and the indentations measured. The drum was then replaced if the wear was small and cycling continued. The indents were measured at intervals governed by experience of the expected wear rate until the indentations were nearly worn through.

The microhardness–indentation technique has been used by many workers (**7–10**). However, ultimate depth resolution has been identified as a major drawback by Bhushan and Lowry (**11**), who quoted a value of ± 30 nm from previous work (**9**), as compared to ± 50 nm in this paper. They employed nano-scratches instead, with depth changes measured by AFM. This is less obtrusive upon the head–tape interface and allows a much greater depth resolution (± 2 nm). The implication of low depth resolution for the microindentations is fairly long experiment times for the cycling tests. This may be troublesome when, in the case of a very low wear regime, no measurable wear is observed within the identified cycling lifetime of the tape.

2.3 Surface analytical techniques

Topographical information is gathered using a Topometrix Explorer SPM with SPMLab V.3.06.06 software operating in contact AFM and non-contact modes. Vacuum surface analytical techniques of small area X-ray photoelectron spectroscopy (XPS), point Auger electron spectroscopy (PAES) and scanning Auger microscopy (SAM) were performed using a VG ESCALAB 200D with the ECLIPSE software, operating at a base pressure of 1×10^{-9} mbar or lower. No special sample preparation was required. Only levelling algorithms were applied in obtaining quantitative data from the AFM images. XPS gives information on the chemical states of the elements in the top few atomic layers (< 100 Å) and thus can give information on the chemical bonding and show evidence of surface oxidation (corrosion) or reduction. AES gives details of elemental composition only, but as it employs an electron beam as an excitation source, elemental mapping may be obtained with micrometre resolution in rastering the beam across the surface in SAM.

3 RESULTS

3.1 Wear data

Microhardness data are presented in Table 2. For the ceramics this is after lapping, for the metals after annealing. Amorphous CoNbZr is the softest. The other materials are all of a similar hardness with the nanocrystalline Fe materials slightly harder than the ceramics.

A typical wear profile is shown in Fig. 3, showing the wear of an M2 head against both MP and ME tapes. The full length of the x axis represents the full length of the head. The MP tape causes the severest wear, but the profiles show a common shape observed optically on the metal heads as a highly polished elliptical wear scar: zero or almost zero wear within the first 200 μm from the trailing and leading edges, increasing over a distance of 100 μm to its maximum, sustained over the central 300 μm. The shorter ceramic heads at 600 μm wear more evenly over their entire length. The average maximum wear rate of each material is plotted in Fig. 4, this being the maximum observed on the wear curves of the type plotted in Fig. 3, averaged over all heads run under the same conditions. The metals wear more quickly than the ceramics, with the exception of S1 versus MP tape. M1, the softest material, is the greatest, with the iron materials showing wear rates similar to each other. Except for M3, wear is an order of magnitude less against ME tape than against MP tape, less than 10 nm/h for the metals against ME tape and of below-measurable limit for the ceramics. M3 versus ME displayed a much greater rate than the other materials against this tape type (> 25 nm/h). Between individual heads of the same material combination wear is greater for narrower heads and for those of greater protrusion, which is to be expected as both result in higher contact pressures.

Optical micrographs of the metal-coated heads show extensive tape debris to be collected about the periphery of wear scars and about the sides of the head. Debris around the edges is mainly in the form of 'polymer rolls'. These are less commonly observed on the head surface. The

Table 2 Knoop microhardness values of dummy heads prior to cycling

	Knoop microhardness	Standard deviation
S1	1082	31
S2	1111	57
M1	860	21
M2	1275	73
M3	1240	74

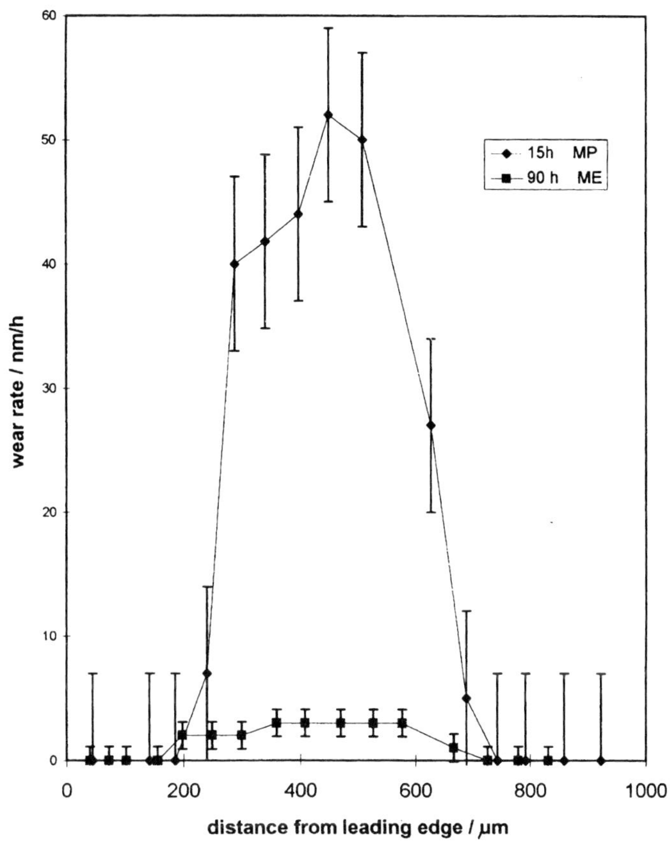

Fig. 3 Wear rate curves for S2 versus ME and MP tapes at 20 °C/90% RH

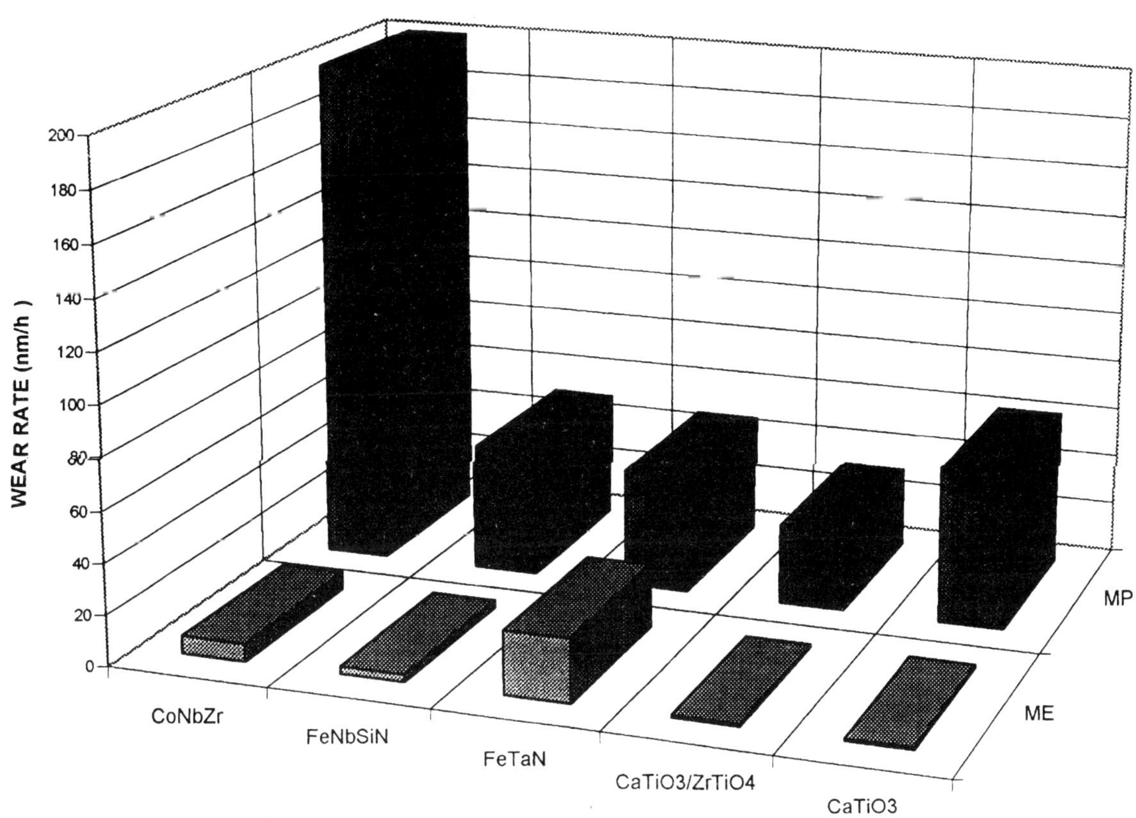

Fig. 4 Dummy head wear rates at 20–25 °C/90% RH versus ME and MP tapes

ceramic heads show virtually no surface debris. On the wear scars of some metals, e.g. for M3 versus MP, a wavy deposit is seen.

3.2 Surface analysis of ceramic heads

XPS spectra of ceramic heads run against the ME tape at high humidity are inconclusive as to the presence of a hydroxide component in the O 1s peak. Both S1 and S2 against the ME tape showed an F 1s peak, with two equal components at 687 and 688 eV that diminished in intensity upon exposure to the X-ray beam. This suggests a transfer film of the mobile component of the perfluoropolyether (PFPE) tape lubricant. The chemical shift is indicative of an organic fluoride, i.e. it was unreacted to the ceramic. Conversely, no chemical markers (N or Cl) typical of the binders found in the MP tape were found on head surfaces. Of interest is that of the Zr-doped $CaTiO_3$ composite ceramic heads run against ME at 80 per cent humidity; the O 1s peak shows a low oxidation state metal oxide component at 528 eV with a corresponding Zr $3d^5$ peak at 180.4 eV, indicative of the Zr^{II} oxidation state, in addition to the peaks at 530 and 182.2 eV of O and Zr^{IV} seen in pre-run and MP-run heads. Chemical states were obtained by reference to the NIST database (**12**). When comparing used and unused heads no significant differences were observed in the chemical state. This suggests that no tribochemical reaction has taken place.

Contact AFM on the wear surfaces shows two prime features superimposed upon the grain structure of the ceramic of both ceramic types: grooving and grain pull-out. The grooving is parallel to the direction of tape travel for

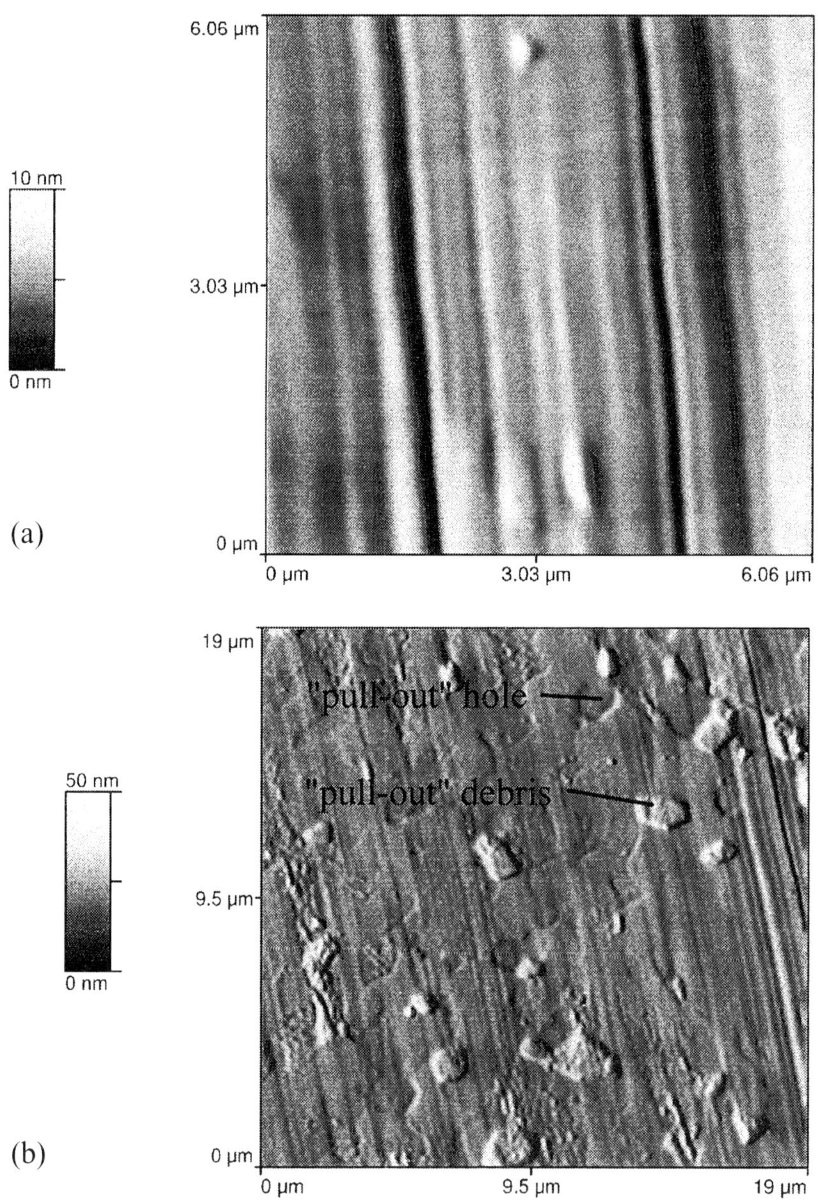

Fig. 5 AFM of substrate wear surfaces after running against ME tape at 20 °C/90% RH: (a) S1 showing plastic deformation and (b) S2 showing grain pull-out and deep and shallow grooving

ME and MP tapes. Two different types of grooving are seen for all substrate and tape combinations. The first type is that of deeper grooves, often visible optically, of 4–5 nm depth caused by ploughing by debris particles being dragged across the surface. The second type is that typical of plastic deformation, with r_a values typically < 1 nm, examples of which may be seen in Fig. 5a, S1 versus ME. Evidence of grain pull-outs and like-sized debris particles are shown in Fig. 5b, S2 versus ME, which also displays evidence of deep and shallow grooving. Depths of such features are typically 10–20 nm with diameters of 1–2 μm. The protruding debris may be unworn grains, rather than pull-out debris, and they are not removed by cleaning with compressed nitrogen, but they are present in the single-phase ceramic as well as the multiphase ceramic. The MP tape is a more effective polishing medium, removing more of the deeper scratching and particulate debris, resulting in lower r_a values for those heads cycled against MP tape as compared to those cycled against ME tape. Roughness values obtained from AFM of all head materials, together with those for tapes obtained by AFM, are tabulated in Table 3.

3.3 Surface analysis of metal-coated heads

Figure 6 shows the SAM of the wear scar observed on a CoNbZr head, mapping C, Co and O. These clearly show depletion of O and enrichment of Co in the wear scar, i.e. a thinner oxide layer. Point Auger analysis shows Cl and Fe to be present in the debris surrounding the wear scar, but not on the wear scar itself.

XPS depth profiling of the as-deposited metals (on S1) was carried out by Ar sputtering. The depth was reconstructed using sputtering rates calculated using the SUSPRE code (**13**). The chemical states and depths of the native oxide layers are summarized in Table 4. In addition to the NIST database (**12**), the chemical states of M3 were determined by reference to nitrogen ion implantation of metallic Ta. The native oxide of CoNbZr consists of oxides of all three metals, leading to a bulk composition of the elements in their metallic states. Figure 7 shows a synthesis of the N 1s/Ta 4p peak of uncycled FeTaN, after exposure to atmosphere. The presence of Ta oxide and nitride peaks at binding energies of 405.9 and 403.3 eV respectively is

Table 3 AFM r_a values (nm) of virgin tape, as deposited metal-coated heads and dummy heads after cycling

	Virgin/as deposited	ME	MP
ME tape	0.77		
MP tape	6.05		
S1		0.80	0.92
S2		2.20	1.86
M1	3.03	2.45	1.51
M2	2.87	5.58	1.46
M3	0.77	3.44*	1.66

* Surface deposit.

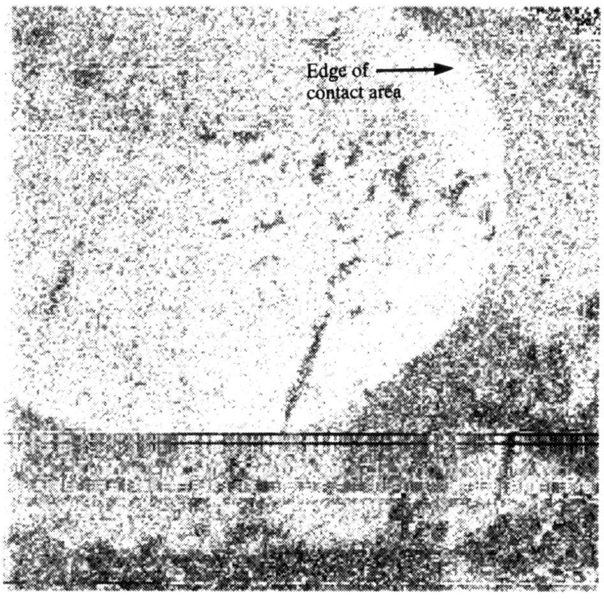

(a)

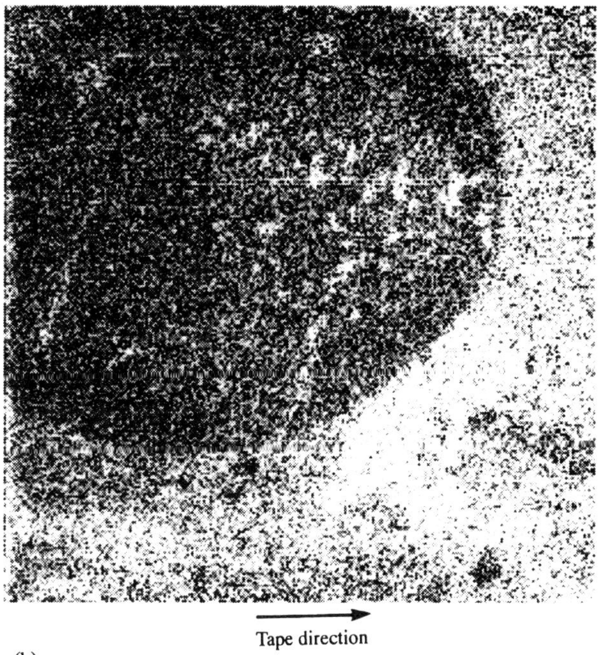

(b)

Fig. 6 SAM of the wear scar of M1 after running against MP tape at 20 °C/90% RH: (a) Co, (b) O

confirmed by the main Ta $4f^5$ peak, which also shows an absence of Ta in the metallic form. Therefore the peak at 400.8 eV is probably either an oxynitride N 1s peak, associated with Ta, or an N–O peak. The presence of Ta oxide with the absence of metallic Ta already suggests Ta–N bond breaking. The N 1s peak at 396.8 eV is typical of metallic nitrides. The Ta–O, Ta–N and N–O peaks persist

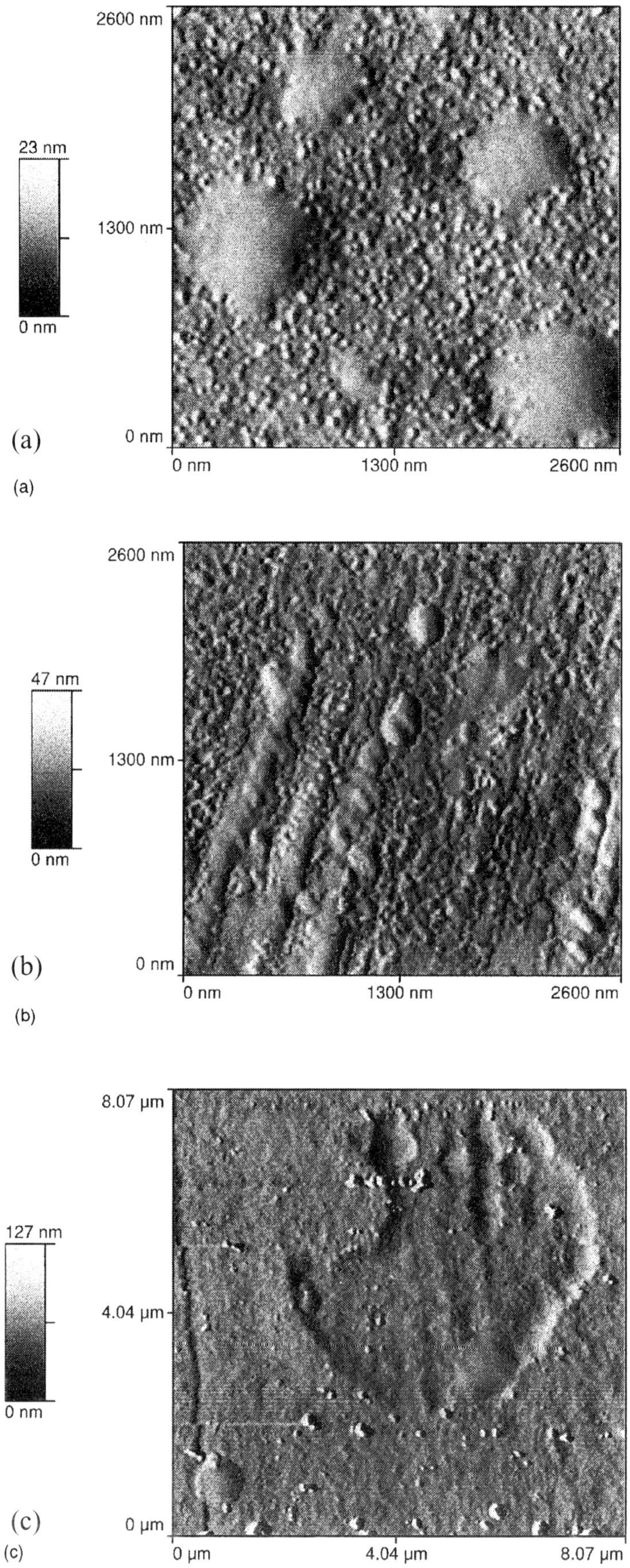

Fig. 9 AFM of M2 (a) as deposited, (b) after running against ME tape at 20 °C/90% RH and (c) after running against MP tape at 20 °C/90% RH. Note that (c) has a different scale

ME and MP tapes. Two different types of grooving are seen for all substrate and tape combinations. The first type is that of deeper grooves, often visible optically, of 4–5 nm depth caused by ploughing by debris particles being dragged across the surface. The second type is that typical of plastic deformation, with r_a values typically < 1 nm, examples of which may be seen in Fig. 5a, S1 versus ME. Evidence of grain pull-outs and like-sized debris particles are shown in Fig. 5b, S2 versus ME, which also displays evidence of deep and shallow grooving. Depths of such features are typically 10–20 nm with diameters of 1–2 μm. The protruding debris may be unworn grains, rather than pull-out debris, and they are not removed by cleaning with compressed nitrogen, but they are present in the single-phase ceramic as well as the multiphase ceramic. The MP tape is a more effective polishing medium, removing more of the deeper scratching and particulate debris, resulting in lower r_a values for those heads cycled against MP tape as compared to those cycled against ME tape. Roughness values obtained from AFM of all head materials, together with those for tapes obtained by AFM, are tabulated in Table 3.

3.3 Surface analysis of metal-coated heads

Figure 6 shows the SAM of the wear scar observed on a CoNbZr head, mapping C, Co and O. These clearly show depletion of O and enrichment of Co in the wear scar, i.e. a thinner oxide layer. Point Auger analysis shows Cl and Fe to be present in the debris surrounding the wear scar, but not on the wear scar itself.

XPS depth profiling of the as-deposited metals (on S1) was carried out by Ar sputtering. The depth was reconstructed using sputtering rates calculated using the SUSPRE code (**13**). The chemical states and depths of the native oxide layers are summarized in Table 4. In addition to the NIST database (**12**), the chemical states of M3 were determined by reference to nitrogen ion implantation of metallic Ta. The native oxide of CoNbZr consists of oxides of all three metals, leading to a bulk composition of the elements in their metallic states. Figure 7 shows a synthesis of the N 1s/Ta 4p peak of uncycled FeTaN, after exposure to atmosphere. The presence of Ta oxide and nitride peaks at binding energies of 405.9 and 403.3 eV respectively is confirmed by the main Ta $4f^5$ peak, which also shows an absence of Ta in the metallic form. Therefore the peak at 400.8 eV is probably either an oxynitride N 1s peak, associated with Ta, or an N–O peak. The presence of Ta oxide with the absence of metallic Ta already suggests Ta–N bond breaking. The N 1s peak at 396.8 eV is typical of metallic nitrides. The Ta–O, Ta–N and N–O peaks persist

Table 3 AFM r_a values (nm) of virgin tape, as deposited metal-coated heads and dummy heads after cycling

	Virgin/as deposited	ME	MP
ME tape	0.77		
MP tape	6.05		
S1		0.80	0.92
S2		2.20	1.86
M1	3.03	2.45	1.51
M2	2.87	5.58	1.46
M3	0.77	3.44*	1.66

* Surface deposit.

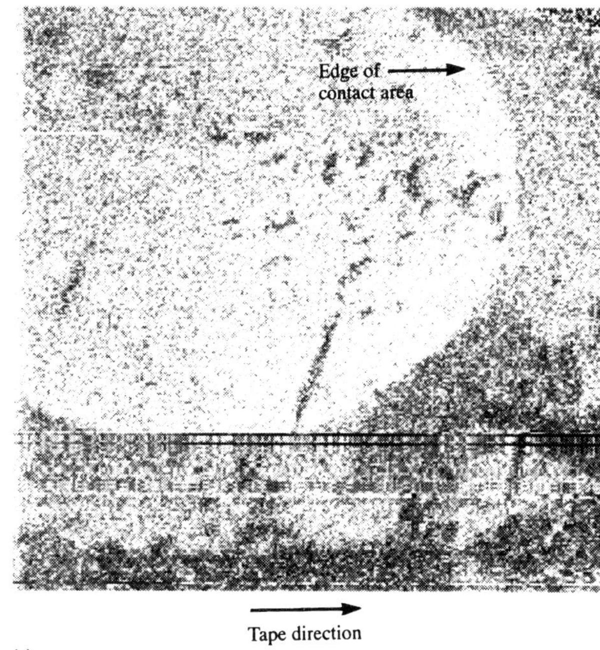

(a)

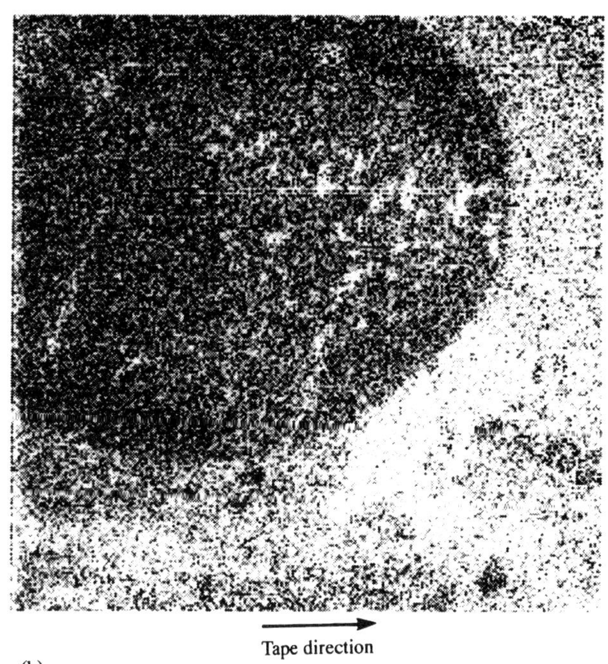

(b)

Fig. 6 SAM of the wear scar of M1 after running against MP tape at 20 °C/90% RH: (a) Co, (b) O

Table 4 Composition and thickness of native oxides by Ar ion XPS depth profiling

	Composition of native oxide	Depth of native oxide (nm)
CoNbZr	Co-mixed oxides Nb_2O_5 ZrO_2	20
FeNbSiN	Fe_2O_3, FeO_2 Fe–OOH Nb-mixed oxides N–O, N–metal	10
FeTaN	Fe_2O_3, Ta_2O_5, N–O	10

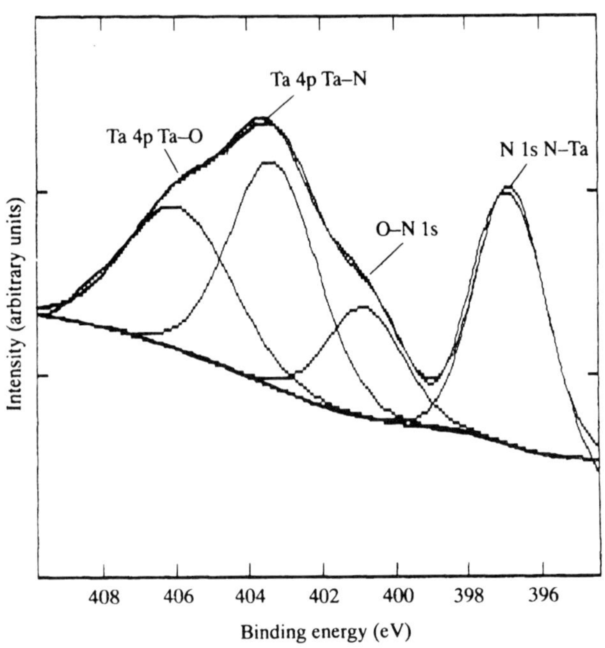

Fig. 7 Deconvolution of the Ta $4f^5$ peak of uncycled FeTaN, after exposure to air

into the bulk, which is predominantly metallic iron. The N 1s peak of M2 also shows evidence of a higher binding energy peak at 407 eV, as well as the nitride peak at 397.5 eV. Again, after cycling the tape, no significant differences were found in the chemical states of the metals.

Figures 8 to 10 show AFM micrographs of the unworn and cycled surfaces of M1, M2 and M3 respectively against ME and MP tapes. As deposited M3, Fig. 8a shows some pitting, perhaps due to precorrosion. The worn surfaces, like the ceramics, show microgrooving in the direction of tape travel. Unlike the ceramics, they show varying types of deposit on the worn surfaces which optically appear highly polished. The dimensions of the deposits are tabulated in Table 5. M1 versus ME and M2 versus MP, like the other heads, showed debris around the wear scar, but no characteristic deposits observed on the wear scar itself. The heights of these are typically of the order of 20 nm compared to the r_a values of < 2 nm for the tapes and the head surfaces. Microabrasion is evident under the ribbon-like deposits seen on M3 when run against MP tape (Fig. 10c). As deposited M2, Fig. 9a shows an interesting surface structure with larger spherical protrusions (0.3–1 μm in diameter), protruding (10–20 nm) above the smaller, underlying grain structure. Cycling against MP tape is seen to produce shallow (20 nm deep) delaminations, typically 5 μm in diameter. An example is shown in Fig. 9c which also shows the pull-out of the spherical protrusions, observed in as-deposited M2, within the delamination.

3.4 Sandwich heads

Figure 11a shows a Ti SAM of a simple sandwich head of S1 and M1 run against ME tape for 100 h, with a corresponding line scan seen in Fig. 11b. These clearly show ceramic particles in the soft metallic track.

Figure 12 shows AFM images of two examples of 100 h of cycling of sandwich heads at high humidity. Figure 12a, M1/S1, is typical of heads relatively free of debris and is from cycling against ME tape. The PTR of the soft magnetic track is 80 nm below the substrate. However, the bonding layer is not worn to such an extent, protruding 60 nm above the base of the track. Particles are observed on the track and substrate, and pull-outs from the ceramic are observed, which is consistent with the dummy results. Figure 12b, M3/S2, is with MP tape. No hollow-out is observed between the metal and the substrate (< 5 nm). The laminations are recessed about 5 nm below the depth of the metal, but the bonding layer protrudes ~45 nm above the surface of the head. Such a protrusion would be very bad for the head performance. The surface of the head is heavily coated in debris. Those particles on the metal track are about 1 μm in diameter and 90–150 nm deep. The relative wear seen between the different components of the sandwich is not always the same. Sometimes it is seen that the bonding layer (Mo–Au) is recessed relative to the metal (FeTaN), which in turn is worn more than the insulating layers (Si_3N_4). The reason for this variation in outcome is not yet fully understood.

4 DISCUSSION

4.1 Observed wear mechanisms

The wear of ceramics run against different tape types is not simple, but consists of two- and three-body microabrasion and several possible fatigue-based mechanisms. The manifestations of wear in the titanate ceramics used are plastically deformed grooves in the surfaces and grain pull-out. The grooves are evidence of two- and three-body microabrasion (plastic deformation) and the pull-outs of fatigue. Wang and Hsu (**14**) observed a transition from plastic deformation controlled wear in ceramics to micro-crack/grain pull-out controlled wear upon increasing tangential contact stresses for a variety of ceramics, with

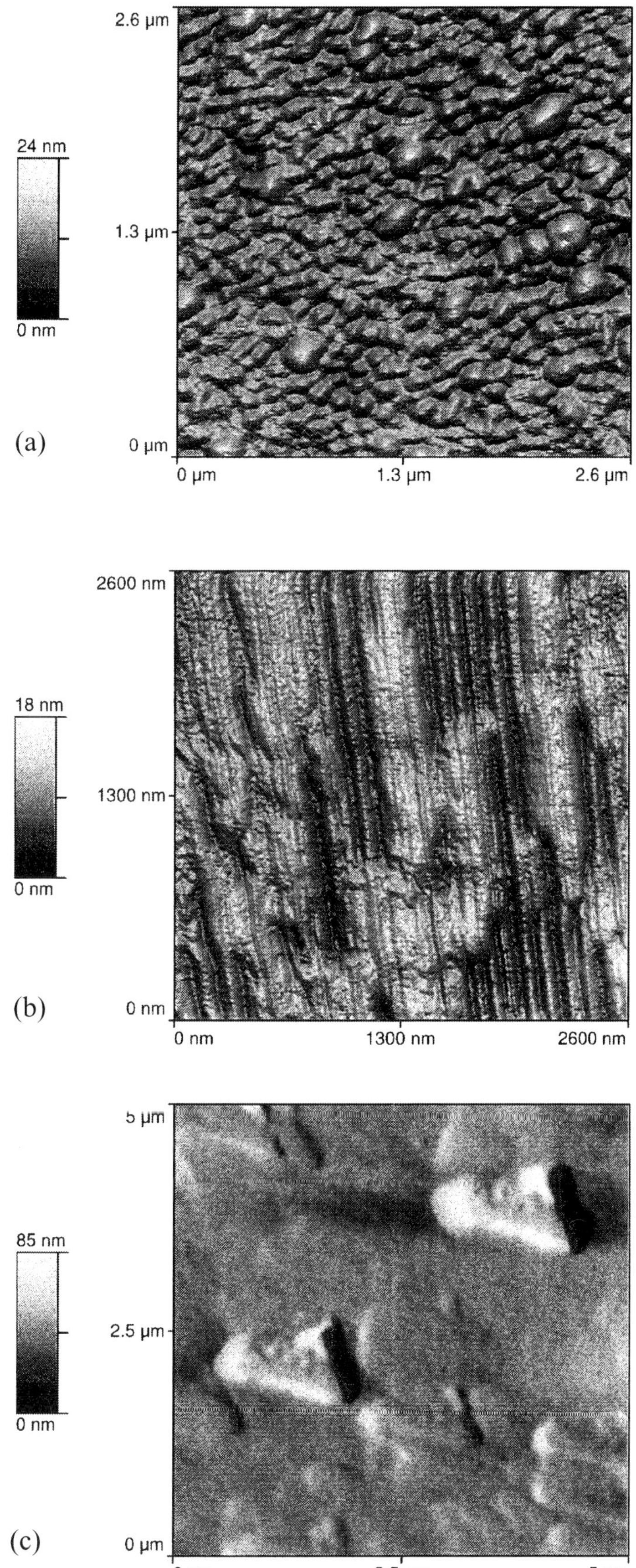

Fig. 8 AFM of M1 (a) as deposited, (b) after running against ME tape at 20 °C/90% RH and (c) after running against MP tape at 20 °C/90% RH. Note that (c) has a different scale

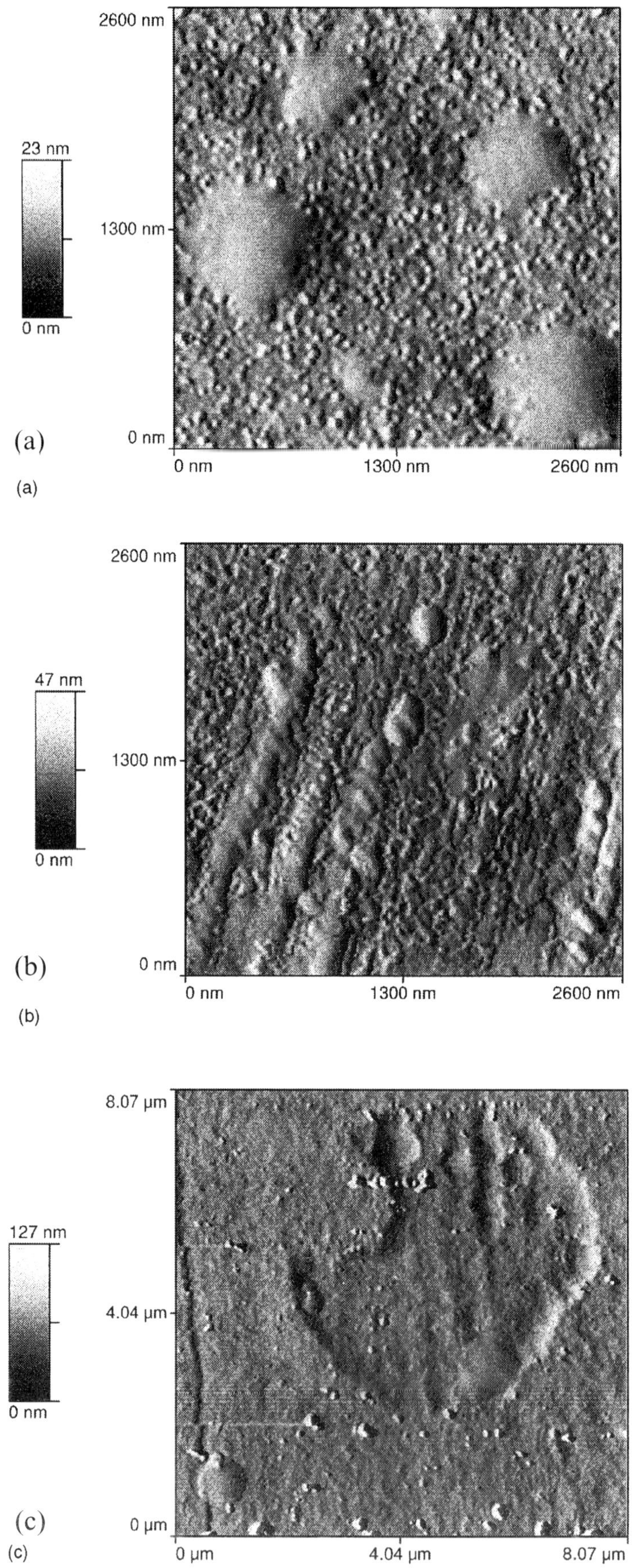

Fig. 9 AFM of M2 (a) as deposited, (b) after running against ME tape at 20 °C/90% RH and (c) after running against MP tape at 20 °C/90% RH. Note that (c) has a different scale

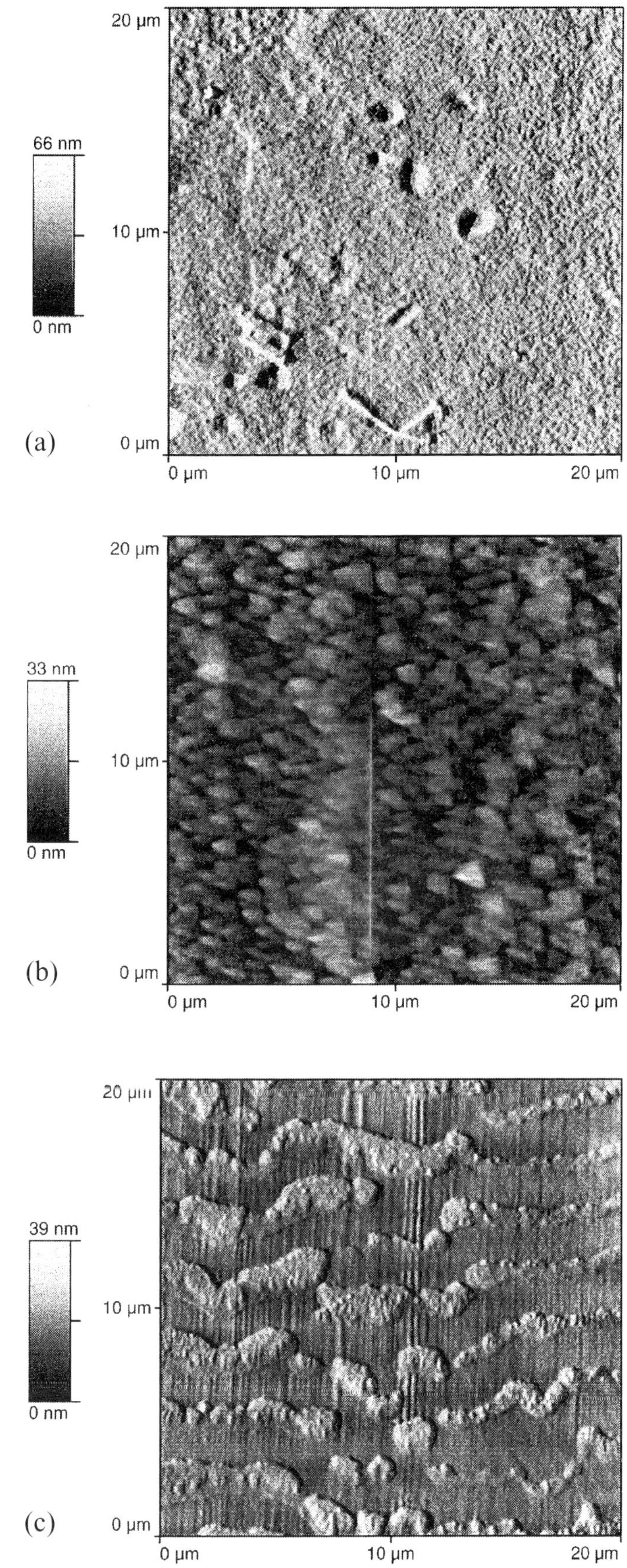

Fig. 10 AFM of M3 (a) as deposited, (b) after running against ME tape at 20 °C/90% RH and (c) after running against MP tape at 20 °C/90% RH

Table 5 Deposits observed on metal-coated dummy heads after cycling. M1 and M2 do not show debris deposits against ME and MP respectively

	Tape	Shape	Diameter (μm)	Height (nm)
M1	MP	Triangular	1–2	60–100
M2	ME	Smearing in direction of tape travel		100
M3	ME	Ribbon	Length 2–10 Width 2	20
M3	MP	Triangular	0.5–1.6	6–20

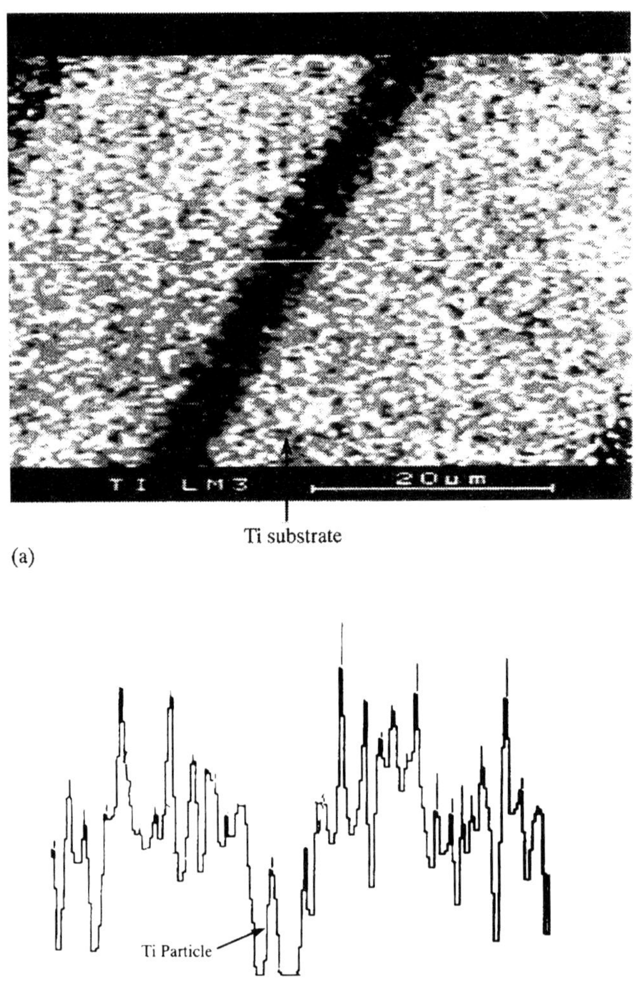

Fig 11 (a) SAM of sandwich head (S1/M1) after running against ME tape at 20 °C/90% RH for 100 h. (b) Line scan, of line marked in (a), noting the Ti particle in the soft magnetic track

the critical microcrack stress falling with the number of loading cycles, e.g. for Si_3N_4 it falls from 11 GPa for one loading to 2 GPa for 100 cycles. The contact pressures at the head–tape interface are typically very low, in the order of 100 kPa, but head–tape contact is continuous and hence the number of loading cycles may be regarded as very high, thus allowing for a microcrack controlled wear. Third-body ploughing by the pull-outs or of tape debris particles as they are dragged across the surface is a secondary source of wear and will increase both abrasion and subsurface crack initiation and propagation.

The dominant wear mechanism for the ceramics depends on the tape type against which it runs. In the case of ME tape the tape surfaces are very smooth and the head–tape interface is separated by a continuous boundary lubricant film, usually a perfluoropolyether; hence microabrasion is minimal and the dominant wear mechanism is due to fatigue-induced crack initiation and propagation, which eventually leads to grain pull-outs. This is illustrated by the almost zero wear rate observed with the ceramics run against ME, but with both materials displaying increasing areal density of pull-outs with running time. With this mechanism the wear rate might be expected to increase with contact time, due to the increased number of particles removed from the surface becoming available for three-body processes. Though some of the present experiments of ME run against ceramic heads seem to support this, the increasing wear rate can also be a result of running-in effects or profile errors.

When the ceramics are run against MP tape the surfaces are far more abrasive due to magnetic pigment and to head cleaning agents. Further to this, surface lubrication is not continuous (**15**); thus microabrasion is dominant although pull-outs still occur. The wear rate due to microabrasive processes is much greater than that due to fatigue-initiated pull-outs and again this is illustrated in the present comparisons of wear rates of materials worn against ME and MP tape types. Against MP, S1 wore the most and this ceramic is known to be the more porous and to exhibit a greater wear rate increase with greater humidity (**16**). The effect of humidity on ceramics is a very complex problem (**17**). In many ceramics there is a tribochemical reaction at the wearing surface which may produce hydroxides. This effect can either reduce wear, by providing an easily sheared low surface energy film at the surface, or, if the film is removed too rapidly due to the wearing conditions, can increase wear. In the case of heads examined no evidence was found that the enhanced wear rate was due to a chemical effect. However, because of the greater porosity, a mechanical process due to water molecule ingress seems to be more likely. This effect is a physical one due to the non-compressibility of water and is well known in the pitting of metals (**18**). Sintered ceramic materials are always porous to a certain extent. The greater fractional porosity, p, of S1 also leads to lower critical bending stresses. These are proportional to the elastic modulus, which in turn is proportional to $\{\exp(-p)\}$. Thus S1 will also be more susceptible to microabrasive wear.

For the soft magnetic materials, amorphous CoNbZr and crystalline FeTaN appear to wear principally by plastic deformation controlled microabrasion. In addition to this, nanocrystalline FeNbSiN also exhibits delamination and grain pull-out. As shown in the previous section, there is no

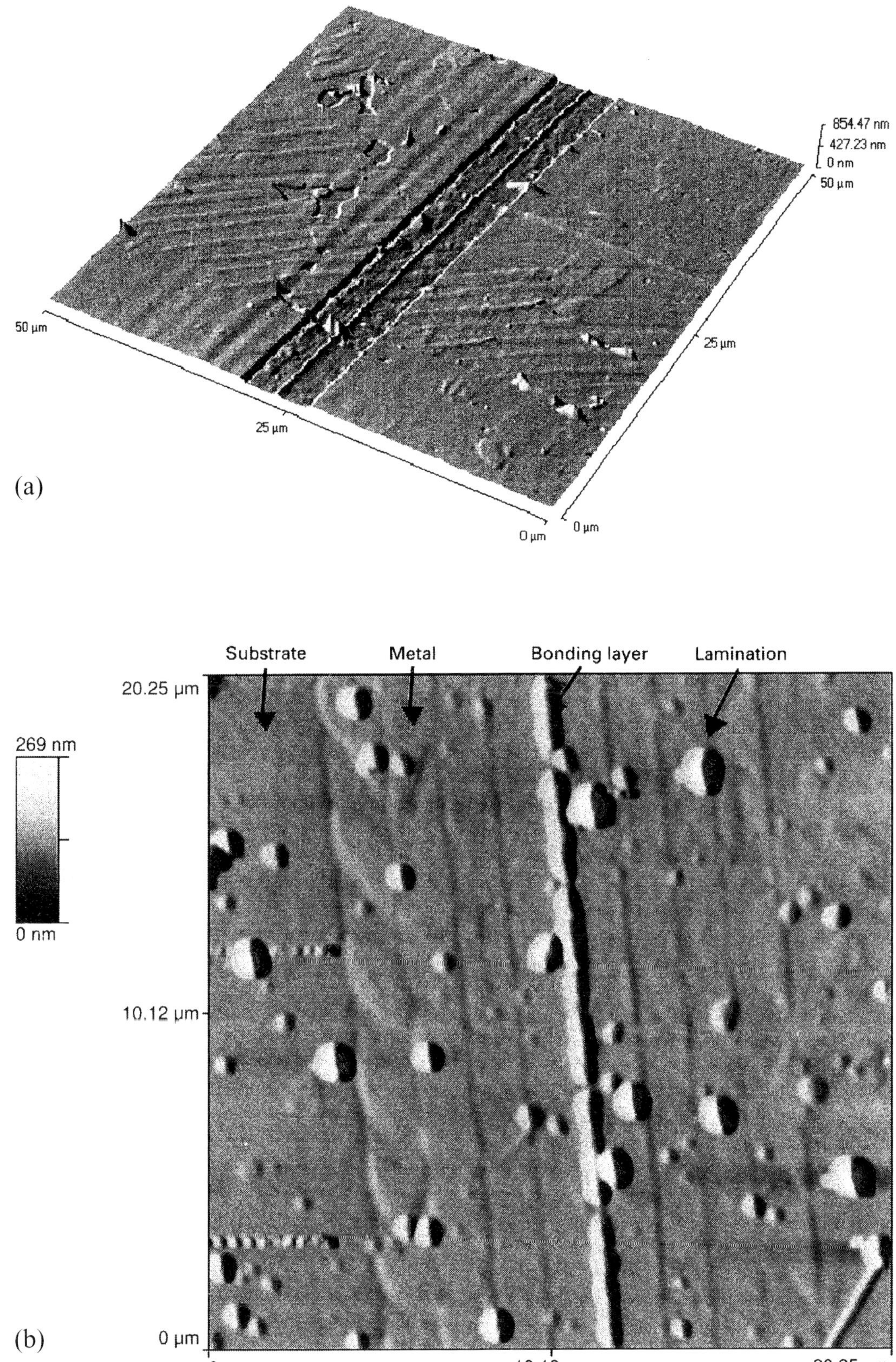

Fig 12 AFM of sandwich heads after cycling at high humidity: (a) M1/S1 versus ME tape, noting the pull-out from the ceramic substrate and the protruding bonding layer; (b) S2/M2 versus MP tape, noting little PTR, and substrate, magnetic material, lamination and bonding layers

hard evidence for tribochemical wear. The fact that CoNbZr exhibited a significantly lower Knoop microhardness than all the other materials and suffered a much higher wear rate tends to confirm microabrasion as the dominant mechanism in the case of these materials. Some caution must be exercised here, however, since observed wear rates of the metals were generally less than 0.1 μm per hour and the Knoop indents were about 1 μm deep. Hence, the usefulness of microhardness (as opposed to nanohardness) as an indicator of abrasion resistance may be questioned. Indeed, the Knoop microhardness values found for M3, quoted in Table 3, are somewhat lower than the nanoindentation data presented by Deng *et al.* (**19**). Native oxide layers were found to be 10 and 20 nm for as-deposited FeNbSiN and CoNbZr respectively, compared to *Ra* values of 2.9 and 3.0 nm. Thus, initially, at least, plastic deformation and abrasion is occurring within the oxide layer. Re-oxidation rates of the fresh metal surface during cycling and at rest could influence equilibrium wear rates and it is possible that a form of oxidative wear has been observed. The SAM of CoNbZr in Fig. 6 shows the oxide layer within the wear scar to be thinner as a result. In FeTaN, all three elements seem to be prone to oxidation, though the AFM results tend to suggest an abrasive mechanism. Though FeNbSiN is similar in structure to FeTaN (**20**), it shows a transition to microcrack-controlled wear in a similar way to the ceramics.

4.2 Implications for the hollow-out mechanism

AFM images of sandwich heads show that hollow-out is a complex problem of differential wear of all components in the sandwich. The depth of recession observed in some cases, especially those where the Mo–Au bonding layer down the centre of the track wears much less than the metal in the tracks (e.g. see Fig. 12a and b), shows that in these cases wear cannot be caused by tape bending into the recess and by tape contact abrasion. Abrasion by sporadic high asperities on the tapes may also be ruled out as a lower wear rate would be expected for the hard silicon nitride lamination layers and a higher wear rate for the softer Mo–Au layer, when in fact both exhibit higher, similar or lower recession than the soft magnetic material. The most likely mechanism for significantly large recession where tape contact is impossible is erosion caused by third bodies becoming entrapped in the soft magnetic region of the head. The present results show that such entrapment is much more likely in the metal regions. These third bodies may originate from grain pull-outs observed as a major wear mechanism in all of the sintered ceramic regions of the heads or wear particles originating from the tape surfaces. Both of these particle types have been observed in the present analyses. It has been shown that tape debris is collected on all heads after cycling. For the particulate tape, when analysed, this is shown to contain binder marker elements (Cl, N) and magnetic particles (Fe). The ceramic surfaces generally appear 'clean', but a lubricant transfer layer has been shown to exist. On the metal surfaces, however, microscopic debris, probably from the tape, was seen to be adhering to the surface, typically 10–20 nm high, increasing the roughness of the surface and leading to greater head–tape separation.

In this case particles entrapped in the area are given kinetic energy by the movement of the media across the head and produce erosive wear. The amount of wear will depend on the kinetic energy and density of the impinging particle stream, as well as on such factors as the impact angle and particle size (**21**). The same material properties are required for the resistance to erosive wear as those required for resistance to abrasive wear, but here fracture toughness is also an important factor. In cases where the microabrasive wear of the ceramic is relatively high, this type of recession is not likely to occur.

5 CONCLUSIONS

1. The wear rates of the major components of sandwich heads in isolation have been determined. The MP tape always resulted in greater individual component wear than the ME tape.
2. The principal wear mechanisms of these materials against ME and MP tapes under contact pressures experienced in Hi-8 video recording have been isolated. In sintered simple $CaTiO_3$ and Zr-doped $CaTi0_3$ composite ceramics, plastic deformation controlled microabrasion + subsurface crack propagation leading to grain pull-out occurs. In amorphous CoNbZr and nanocrystalline FeTaN, plastic deformation controlled microabrasion is dominant. For nanocrystalline FeNbSiN, plastic deformation controlled microabrasion is dominant with the ME tape, but subsurface crack propagation leading to delamination and grain pull-out is evident with the MP tape.
3. Deep recession of the soft magnetic region of the heads examined is probably due to an erosive mechanism caused by third-body entrapment in this region. This mechanism in turn must be related to the type and rate of formation of head and tape debris.

REFERENCES

1 **Tsuchiya, T., Abe, M.** and **Noro, Y.** The gap recession reduced MIG head. *IEEE Trans. Magn.*, 1993, **29**(6), 3945–3947.

2 **Abe, M., Tsuchiya, T.** and **Noro, Y.** A study of recess by tape running in metal and ceramics hybrid heads. *Dig. ICF-6*, 1992, 451.

3 **Broese van Groenau, A., Meulenbroeks, H. J. J. C.** and **de Jong, M.** Level differences in hybrid heads after contact with various tapes. *IEEE Trans. Magn.*, 1990, **26**(1), 153–155.

4 **Theunissen, G. S. A. M** and **de Jongh, M.** Unpublished work.

5 **Wu, Y.** and **Talke, F. E.** The effect of surface roughness on the head–tape interface. *Trans. ASME, J. Tribology*, 1996, **118**, 376–381.
6 **Bhushan, B.** and **Hahn Jr, F. W.** Stains on magnetic tape heads. *Wear*, 1995, **184**, 193–202.
7 **Reeves, L. A.** and **Sullivan, J. L.** The measurement of in-situ wear of video heads. *Tribology Int.*, 1990, **23**(4), 251–257.
8 **Hayashi, K., Hayakawa, M., Ishikawa, W., Ochiai, Y., Iwasaki, Y.** and **Aso, K.** Soft magnetic properties of FeRuGaSi alloy films: SOFTMAX®. *J. Appl. Phys.*, **64**(2), 772–779.
9 **Bhushan, B.** and **Lowry, J. A.** Friction and wear studies of various head materials and magnetic tapes in a linear mode accelerated test using a new nano-scratch wear measurement technique. *Wear*, 1995, **190**, 1–15.
10 **Bhushan, B.** and **Martin, R. J.** Accelerated wear test using magnetic particle slimes. *Tribology Trans.*, 1980, **31**, 228–238.
11 **Katori, K., Ohmori, H., Shoji, M.** and **Hayashi, K.** Wear and corrosion resistance of Fe–N based micrograin soft magnetic thin films. *IEEE Trans. Magn.*, 1994, **30** (2), 331–336.
12 **Wagner, C. D.** *NIST X-ray Photoelectron Spectroscopy Database, Version 1.0*, 1989.
13 SUSPRE V1. Surrey University Sputter Profile Resolution from Energy Deposition.
14 **Wang, Y.** and **Hsu, S. M.** Wear and wear transition mechanisms of ceramics. *Wear*, 1996, **195**, 112–122.
15 **Sullivan, J. L.** The tribology of flexible magnetic recording media. *J. Magnetism and Magn. Mater.*, 1996, **155**, 312–317.
16 **Theunissen, G. S. A. M** and **de Jongh, M.** Unpublished work.
17 **Fischer, T. E.** and **Mullins, W. M.** Chemical aspects of ceramic tribology. *J. Phys. Chem.*, 1992, **96**, 5690–5701.
18 **Sullivan, J. L.** and **Middleton, M. R.** The pitting and cracking of SAE 52100 sliding rolling contact in the presence of an aqueous lubricant. *ASLE Trans.*, 1985, **28**(4), 431–438.
19 **Deng, H., Inturi, V. R.** and **Barnard, J. A.** Nanoindentation and tribological studies of magnetically soft high moment FeTaN thin films. *IEEE Trans. Magn.*, 1995, **31**(6), 2697–2699.
20 **Hasegawa, N.** and **Saito, M.** Soft magnetic properties of micro crystalline Fe–(V, Nb, Ta) films with high thermal stability. *IEEE Trans. J. Magn. Japan*, 1991, **6**, 91–94.
21 **Finnie, I.** Erosion of surfaces by solid particles. *Wear*, 1960, **3**, 87–103.

Technical Note

Full-field interferometric measurement of head–tape contact spacing

R Swain*, **E Baugh** and **F E Talke**
Center for Magnetic Recording Research, University of California, San Diego, California, USA

Abstract: Interferometry is the preferred measurement method for the contact spacing of the head–tape interface, due to its ease of use and high accuracy. Full-field techniques provide spacing data at all points in the field of view as opposed to a single measurement point. Dynamic interferometric measurement necessitates the use of a strobed illumination source, and thus prohibits the alteration of the fringe field to resolve the ill-constrained nature of the analysis problem. In addition, the head–tape contact problem is further complicated by the low phase gradient in the region of head–tape contact, which is often large in comparison to the field of view.

In this paper two algorithms are compared which do not require alteration of the fringe field and therefore can easily be applied to the dynamic head–tape interface. The first technique is a two-dimensional implementation of a biquadratic fit to the fringe extremes, while the second is a pseudo-spatial heterodyne technique using the two-dimensional Fourier transform. Theoretical implementation and results are presented.

Keywords: interferometry, head–tape interface, Fourier techniques, Bernoulli interface

NOTATION

$a \ldots f$	coefficients of biquadratic fit
$a(\boldsymbol{r})$, $A(\boldsymbol{r})$	background fringe intensity
$b(\boldsymbol{r})$, $B(\boldsymbol{r})$	fringe intensity modulation
$c(\boldsymbol{r})$	complex form of cosine
$c^*(\boldsymbol{r})$	conjugate complex of cosine
f	carrier frequency
$h(\boldsymbol{r})$	head–tape spacing
i	$\sqrt{-1}$
$I(\boldsymbol{r})$	detected intensity
k	extinction coefficient
$\max(\boldsymbol{r})$	maximum fringe intensity
$\min(\boldsymbol{r})$	minimum fringe intensity
n	index of refraction
$\boldsymbol{r}$	(x, y) pixel coordinates on CCD image
z	fringe envelope value
λ	wavelength of illumination
ϕ	phase shift upon reflection
$\Psi(\boldsymbol{r})$	phase function

The MS was received on 2 October 1996 and was accepted for publication on 4 August 1997.

**Present address: Western Digital Corp., Irvine, California.*

1 INTRODUCTION

The read/write element of a tape head must be as close as possible to the magnetic medium to maximize the recording bit density. Any new head design requires verification of the spacing, commonly referred to as 'contact spacing', since it is limited by media surface roughness asperities. The spacing measurement should be done on many points, since the design engineer requires information on the coupling mechanism between the head and tape for the complete head–tape interface, i.e. a two-dimensional spacing measurement technique is highly desirable. The head, tape or both are in motion, producing the additional constraint of the simultaneous acquisition of the entire data field. Pulsed interferometry is generally used for this purpose, with either the head or tape being replaced by a transparent replica. The fringe pattern from the interferometer must be numerically processed to extract the spacing, which is the purpose of this paper.

2 HEAD–TAPE INTERFEROMETRY

Interferometry is a technique that can produce a visual representation of the contact area of the head–tape interface. Figure 1 shows one of the methods of measuring the

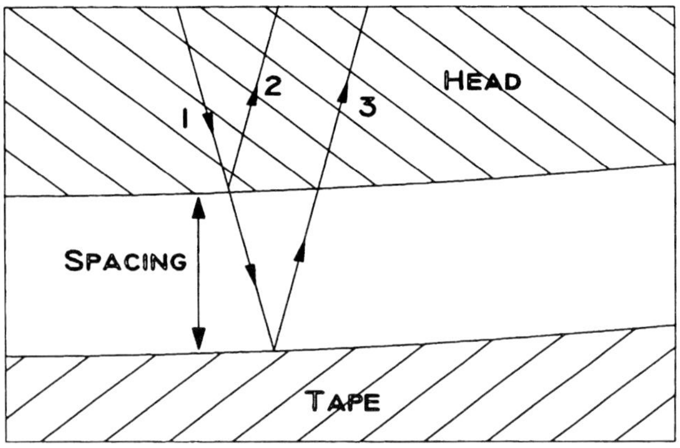

Fig. 1 Interferometry through a transparent head

head–tape spacing using interferometry, in this case through a transparent head (drawn at exaggerated incidence). The incident irradiation (ray 1) is delivered through a microscope objective lens at near normal incidence, causing interference to occur between rays 2 and 3. The phase difference between these rays is a measure of the head–tape spacing. The interference pattern is imaged through the microscope and digitized with a CCD (charge coupled device) camera and framestore. For near normal incidence, the fringe pattern intensity is given by (single reflection theory)

$$I(\boldsymbol{r}) = A(\boldsymbol{r}) + B(\boldsymbol{r})\cos\left[\frac{4\pi h(\boldsymbol{r})}{\lambda} - \phi\right], \qquad \boldsymbol{r} = \begin{pmatrix} x \\ y \end{pmatrix} \tag{1}$$

where vector $\boldsymbol{r}$ represents the coordinates of a pixel (x, y) on the CCD detector, $A(\boldsymbol{r})$ and $B(\boldsymbol{r})$ are the background and modulation terms respectively, λ is the wavelength of the illumination and $h(\boldsymbol{r})$ is the head–tape spacing. The constant phase term inside the cosinusoid, ϕ, is the phase shift upon reflection for ray 3, computed from the refractive index and extinction coefficient, $n + ik$, measured using ellipsometry. It is seen that equation (1) is ill-constrained since there is one equation with three unknowns, so determination of $h(\boldsymbol{r})$ from $I(\boldsymbol{r})$ requires supplementary information.

3 DATA EXTRAPOLATION

The basic interferometric equation [equation (1)] has three unknown terms: $A(\boldsymbol{r})$, $B(\boldsymbol{r})$ and $h(\boldsymbol{r})$. The $A(\boldsymbol{r})$ and $B(\boldsymbol{r})$ terms must be calculated separately before the spacing term $h(\boldsymbol{r})$ can be determined. Where the argument to the cosine term in equation (1) reaches an integer multiple of π, the cosine takes on a value of ± 1 and the intensity $I(\boldsymbol{r})$ is equal to either $A(\boldsymbol{r}) + B(\boldsymbol{r})$ or $A(\boldsymbol{r}) - B(\boldsymbol{r})$. At these points, the intensity reaches a maximum or minimum value (the fringe intensity extrema) and the spacing can be calculated immediately using simple algebra since the cosine argument is known within modulo (2π). There are two limitations to this technique. Firstly, the spacing is obtained only at fringe intensity extrema and, secondly, the head–tape contact spacing is generally below the spacing corresponding to the first extrema value.

Experimental techniques cannot be used to obtain $A(\boldsymbol{r})$ and $B(\boldsymbol{r})$ independently of $h(\boldsymbol{r})$ in many head–tape interface problems since the required mechanical perturbation of the interface is not always possible (**1**). Alternatively, numerical methods are used (**2**). In this procedure, the intensity of the fringe extrema $A(\boldsymbol{r}) + B(\boldsymbol{r})$ [the maximum value denoted by $\max(\boldsymbol{r})$] and $A(\boldsymbol{r}) - B(\boldsymbol{r})$ [the minimum value denoted by $\min(\boldsymbol{r})$] are numerically interpolated from the existing fringes into the region between the fringe intensity extrema. The interpolated max and min values are known as the 'calibration envelopes'. From these values, the $A(\boldsymbol{r})$ and $B(\boldsymbol{r})$ terms at any position in the interface are given by

$$A(\boldsymbol{r}) = \frac{\max(\boldsymbol{r}) + \min(\boldsymbol{r})}{2}, \qquad B(\boldsymbol{r}) = \frac{\max(\boldsymbol{r}) - \min(\boldsymbol{r})}{2} \tag{2}$$

For a fringe pattern with fringe curvature in two dimensions, a bivariate surface can be used as the interpolation function. A functional form with undetermined coefficients for the surface is chosen and the least-squares method used to determine the coefficients that best match the chosen surface to the experimental data. A functional form must be chosen which can capture the modulation behaviour of the existing fringes and interpolate that behaviour into the contact region. The detected intensity is affected by the head curvature, tape angle and imaging lens numerical aperture. Since the heads in helical scan and Bernoulli interfaces have two radii of curvature, a biquadratic surface was chosen as it can locally approximate two radii of curvature (**3**). The functional form is

$$z_i = a_i + b_i x + c_i y + d_i x^2 + e_i y^2 + f_i xy, \qquad i = \text{upper/lower} \tag{3}$$

and the coefficients $(a_i, b_i, \ldots, f_i)$ are found by selecting representative fringe intensity extrema from the entire image and performing the least-squares technique once for each envelope. After the coefficients are determined for the upper and lower envelopes, the values of $A(\boldsymbol{r})$ and $B(\boldsymbol{r})$ can be found anywhere in the image by using the interpolation function. The spacing calculation then proceeds in a straightforward fashion by inverting equation (1) and taking into account the fringe order, if necessary.

4 FOURIER TRANSFORM METHOD

The Fourier transform method of fringe analysis was first presented (in one dimension) by Takeda *et al.* (**4**) in 1982. Macy (**5**) extended the method into two dimensions, by considering a fringe image to be made of slices, each of which could be analysed using the one-dimensional algorithm. Genuine two-dimensional processing of fringe patterns using a two-dimensional Fourier transform was first presented by Bone *et al.* (**6**), and has been implemented by numerous researchers since (**6**–**12**).

Consider a fringe pattern imaged by a sensor. Assuming the capture device is linear, the two-dimensional spatial intensity distribution can be expressed in simplified form as (**12**)

$$i(\boldsymbol{r}) = a(\boldsymbol{r}) + b(\boldsymbol{r})\cos[\Psi(\boldsymbol{r})], \qquad \boldsymbol{r} = \begin{pmatrix} x \\ y \end{pmatrix} \tag{4}$$

Equation (4) [which is essentially equation (1) rewritten] assumes that only the phase of the measurement quantity modulates the intensity (if it is assumed that the background and depth of modulation are relatively constant). Ideally, the interferometer would be spatially heterodyned by tilting one of the interfering wavefronts, generating 'carrier' fringes (in much the same way as audio information modulates the frequency of a carrier in FM systems). In this case, the additional carrier fringes would increase the spatial frequency of the fringe pattern, which allows the modulating phase to be selectively placed in the spatial frequency domain (see Fig. 2). Equation (4) would be modified to

$$i'(\boldsymbol{r}) = a(\boldsymbol{r}) + b(\boldsymbol{r})\cos[2\pi \boldsymbol{f}\boldsymbol{r} + \Psi(\boldsymbol{r})] \tag{5}$$

where vector $\boldsymbol{f}$ is the carrier frequency. Equation (5) can be written as

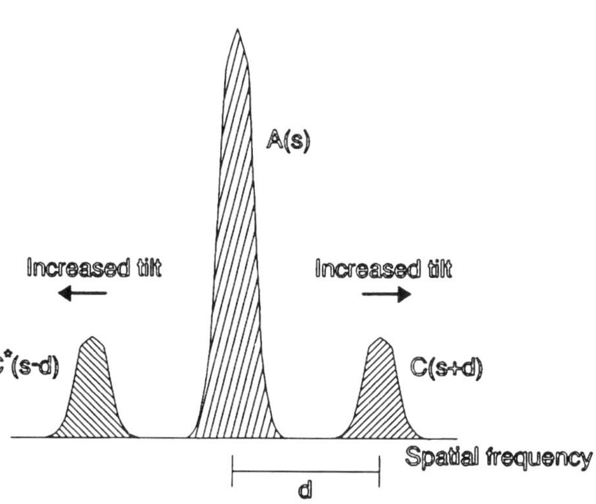

Fig. 2 Graphical representation of equation (8)

$$i'(\boldsymbol{r}) = a(\boldsymbol{r}) + c(\boldsymbol{r})\,e^{2\pi i \boldsymbol{f}\boldsymbol{r}} + c^*(\boldsymbol{r})\,e^{-2\pi i \boldsymbol{f}\boldsymbol{r}} \tag{6}$$

where

$$c(\boldsymbol{r}) = \tfrac{1}{2} b(\boldsymbol{r})\,e^{i\Psi(\boldsymbol{r})} \tag{7}$$

and $c^*(\boldsymbol{r})$ represents the complex conjugate of $c(\boldsymbol{r})$. These functions carry identical information about the phase function $\Psi(\boldsymbol{r})$. Computation of the two-dimensional Fourier transform (denoted by capital letters) of equation (6) gives the Hermitian (conjugate symmetric) function:

$$\begin{aligned} I'(\boldsymbol{s}) &= A(\boldsymbol{s}) + C(\boldsymbol{s}+\boldsymbol{d}) + C^*(\boldsymbol{s}-\boldsymbol{d}) \\ \boldsymbol{s} &= \begin{pmatrix} u \\ v \end{pmatrix}, \qquad \boldsymbol{d} = \begin{pmatrix} p \\ q \end{pmatrix} \end{aligned} \tag{8}$$

where $\boldsymbol{s}$ is the spatial frequency variable and $\boldsymbol{d}$ is the spatial frequency of the carrier. It is now seen that selective filtering in the Fourier domain enables destruction of the Hermitian data and extraction of the phase. Applying a pass-band filter to one side (either) of the Fourier spectrum gives

$$I''(\boldsymbol{s}) = C(\boldsymbol{s}+\boldsymbol{d}) \tag{9}$$

Spatially shifting the spectrum by $\boldsymbol{d}$ and computing the inverse Fourier transform gives

$$i''(\boldsymbol{r}) = c(\boldsymbol{r}) = \tfrac{1}{2} b(\boldsymbol{r})\,e^{i\Psi(\boldsymbol{r})} \tag{10}$$

The phase information is now available from

$$\Psi(\boldsymbol{r}) = \tan^{-1}\left\{\frac{\mathrm{Im}[c(\boldsymbol{r})]}{\mathrm{Re}[c(\boldsymbol{r})]}\right\} \tag{11}$$

where Re and Im are the real and imaginary operators. It should be noted that since the phase is computed from an arctangent function, it will be supplied modulo 2π (wrapped). The phase must be corrected (unwrapped) for the 2π phase jumps (**13**–**17**). With application to the head–tape interface, which uses a 'division of amplitude' interferometer, no tilt can be applied. This is not a problem if a sufficiently large number of fringes are available in the image, such that these fringes 'behave' like a carrier term. This scheme does not require any calibration of the head–tape interface during the data acquisition stage, nor does it require numerical interpolation to obtain the envelope. The algorithm is numerically more complex than the other methods, especially when computed in two dimensions.

5 RESULTS

Figure 3 shows a section of a fringe pattern taken from a Bernoulli removable disk drive head–disk interface; the dark vertical bar on the right of the image is a transverse air

Fig. 3 Bernoulli fringe pattern

Fig. 5 Wrapped phase map from the fringe pattern

slot, designed to bleed away the hydrodynamic air bearing. Figure 4 shows a one-dimensional trace of the upper and lower envelopes computed by extrapolation of the fringe extrema from higher-order fringes over the zero-order region. In this analysis, two-dimensional biquadratics were fitted to the fringe peaks:

$$\begin{aligned} z_{\max} &= 179.69 + 0.303x + 0.080y - 0.0006x^2 \\ &\quad - 0.0010y^2 + 2.2 \times 10^{-6}xy \\ z_{\min} &= 154.43 - 0.511x - 0.145y + 0.0011x^2 \\ &\quad + 0.0002y^2 - 5.9 \times 10^{-6}xy \end{aligned} \tag{12}$$

Displaying such surfaces is not easy, hence the one-dimensional representation in Fig. 4. Figure 5 shows the wrapped phase map from the fringe field computed using the two-dimensional Fourier transform method of fringe analysis. In this image, the grey scales map to the relative phase (e.g. a fringe extremum, as described in Section 3) not spacing, with black representing the zero phase and white representing a phase of 2π. By computing the absolute spacing at a single point, it is possible to map the relative phase to absolute phase. Figure 6 is the head–tape spacing for a 130 pixel × 100 pixel section of the central region of the image computed using the biquadratic fit algorithm, and Fig. 7 is the spacing computed using the two-dimensional Fourier transform method. It is seen that agreement between the two methods is excellent.

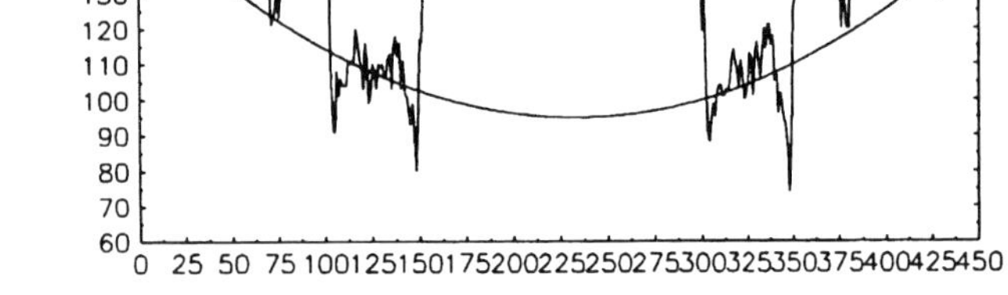

Fig. 4 Trace along a line of the fringe pattern

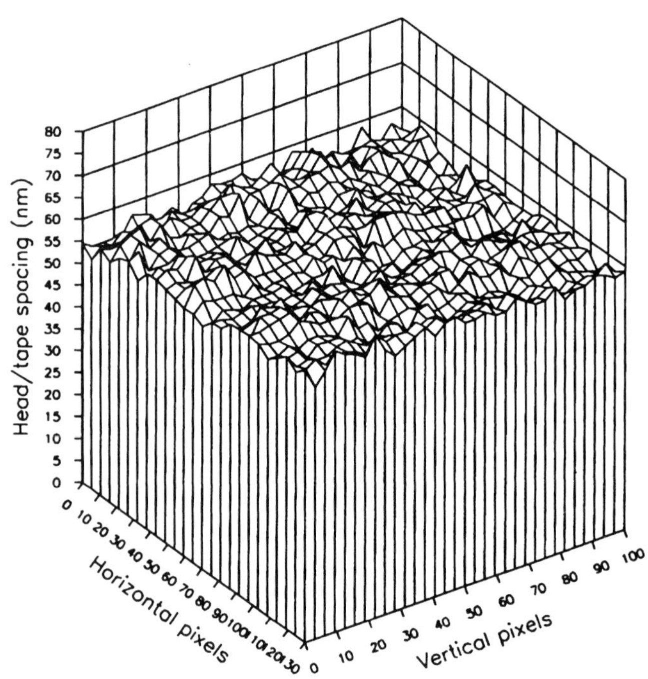

Fig. 6 Spacing obtained from the biquadratic fit (unsmoothed)

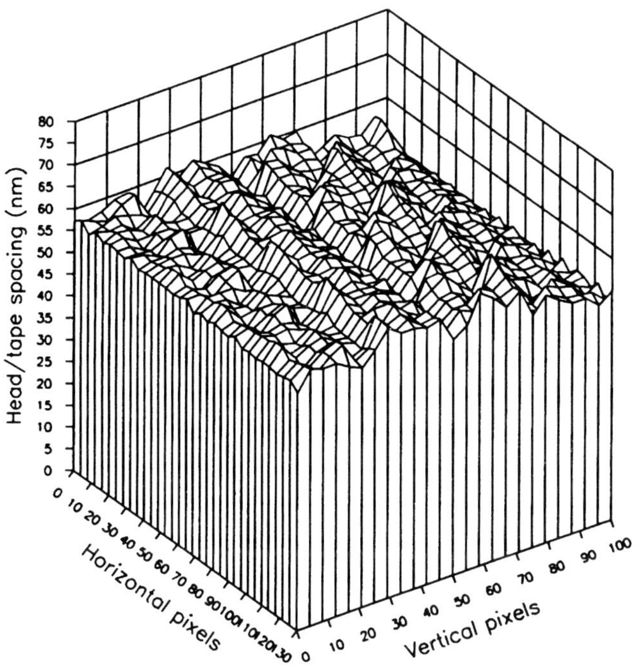

Fig. 7 Spacing obtained from the Fourier transform method (unsmoothed)

6 CONCLUSION

The two-dimensional envelope functions that are used to 'bridge' the contact region of the fringe pattern produce a head–tape spacing measurement that is within 10 nm of that obtained using the two-dimensional Fourier transform method. In the contact region, which occupies most of the field of view, the change in phase is small, and hence the two methods have a considerably lower signal-to-noise ratio. By adjusting the ratio of the contact area to the field of view, the signal-to-noise ratio can be easily improved. Simple smoothing can be used to remove the high spatial frequency content of the data, revealing the underlying low-frequency information about the contact mechanics. In addition, both methods have the added advantage of the prospect of complete automation and could therefore be used in production testing environments.

The choice of technique could be determined by several factors. For the Fourier method to work, a minimum number of regularly spaced fringes is required. In addition, sharp intensity changes such as those caused by bleed slots would cause the phase unwrapping to fail. While the data extrapolation method is less restrictive of the fringe pattern and is unaffected by the bleed slots, as these can be masked out, it is more sensitive to intensity measurement errors, such as discretization, and the calculation of the calibration envelopes is a dominant source of error. Finally, the data extrapolation method is simpler to implement and faster to execute.

REFERENCES

1 **Lacey, C. A.** and **Talke, F. E.** Measurement and simulation of partial contact at the head/tape interface. *Trans. ASME, J. Tribology*, 1992, **114**(4), 646–652.

2 **Mizukawa, M., Hosaka, H.** and **Hara, S.** Study on spherical foil bearings. *Bull. Jap. Soc. Mech. Engrs*, 1985, **28**(243), 2105–2111.

3 **Baugh, E.** and **Talke, F. E.** Biquadratic surface fits for two-dimensional interferometric head/tape spacing. *IEEE Trans. Magn.*, 1994, **30**(6), 4188–4190.

4 **Takeda, M., Ina, H.** and **Kobayashi, S.** Fourier-transform method of fringe-pattern analysis for computer based topography and interferometry. *J. Opt. Soc. Am.*, 1982, **A72**(1), 156–160.

5 **Macy Jr, W. W.** Two-dimensional fringe-pattern analysis. *Appl. Optics*, 1983, **22**(23), 3898–3901.

6 **Bone, D. J., Bachor, H. A.** and **Sandeman, R. J.** Fringe-pattern analysis using a 2-D Fourier transform. *Appl. Optics*, 1986, **25**(10), 1653–1660.

7 **Roddier, C.** and **Roddier, F.** Interferogram analysis using Fourier transform techniques. *Appl. Optics*, 1987, **26**(9), 1668–1673.

8 **Kreis, T. M.** Digital holographic interference-phase measurement using Fourier transform techniques. *J. Opt. Soc. Am.*, 1986, **A3**(6), 847–855.

9 **Kujawiska, M.** and **Wójciak, J.** High accuracy Fourier transform fringe pattern analysis. *Optics and Lasers in Engng*, 1991, **14**, 325–339.

10 **Reid, G. T.** Automatic fringe pattern analysis: a review. *Optics and Lasers in Engng*, 1986, **7**(1), 37–68.

11 **Schwider, J.** Advanced evaluation techniques in interferometry. In *Progress in Optics*, Vol. 28, 1990, pp. 273–359 (Elsevier Science Publishing Company, Oxford).

12 **Spik, A.** Two-dimensional phase decoding from bounded fringe patterns by using the Fourier-transform method. *Optica Applicata*, 1987, **17**(4), 349–354.

13 **Huntley, J. M.** Speckle photography fringes analysis: assessment of current algorithms. *Appl. Optics*, 1989, **28**(20), 4316–4322.

14 **Creath, K.** Phase-measurement interferometry: beware these errors. *Proc. Soc. of Photo-Opt. and Instrum. Engrs*, 1991, **1553**, 213–220.

15 **Maas, A. A. M.** Phase shifting speckle interferometry. PhD thesis, Delft University of Technology, Netherlands, 1991.

16 **Huntley, J. M.** Noise-immune phase unwrapping algorithm. *Appl. Optics*, 1989, **28**(15), 3268–3270.

17 **Ghiglia, G. A., Mastin, G. A.** and **Romero, L. A.** Cellular-automata method of phase unwrapping. *J. Opt. Soc. Am.*, 1987, **A4**, 267–280.

Contact modelling

Contact mechanics of rough surfaces in tribology: multiple asperity contact

B Bhushan
Computer Microtribology and Contamination Laboratory, Department of Mechanical Engineering, The Ohio State University, Columbus, Ohio, USA

Abstract: Contact modelling of two rough surfaces under normal approach and with relative motion is carried out to predict real area of contact and surface and subsurface stresses affecting friction and wear of an interface. When two macroscopically flat bodies with microroughness come in contact, the contact occurs at multiple asperities of arbitrary shapes, and varying sizes and heights. Deformation at the asperity contacts can be either elastic and/or elastic–plastic. If a thin liquid film is present at the interface, attractive meniscus forces may affect friction and wear. Historically, statistical models have been used to predict contact parameters, and these generally require many assumptions about asperity geometry and height distributions. With the advent of computer technology, numerical contact models of three-dimensional rough surfaces have been developed, particularly in the past decade, which can simulate digitized rough surfaces with no assumptions concerning the roughness distribution. In this article, a comprehensive review of modelling of multiple asperity contacts in dry and wet conditions is presented. Contact models for homogeneous and layered, elastic and elastic–plastic solids with and without tangential loading are presented. The models reviewed in this paper fall into two groups: (a) analytical solutions for surfaces with well-defined height distributions and asperity geometry and (b) numerical solutions for real surfaces with asperities of arbitrary shape and varying size and height distributions. Implications of these models in friction and wear studies are discussed.

Keywords: contact mechanics, tribology, asperity contact, surface deformation, contact area, rough surfaces

1 INTRODUCTION

Solid surfaces, irrespective of their method of formation, contain irregularities or deviations from a prescribed geometrical form. The high points on the surfaces are referred to as asperities, peaks, summits, or hills and the low points are referred to as valleys. When two nominally flat surfaces are placed in contact, surface roughness causes contact to occur at discrete contact spots (junctions). Deformation occurs in the region of the contact spots, establishing stresses that oppose the applied load. The sum of the areas of all the contact spots constitutes the real (true) area of contact, and for most materials with applied load, this will be only a small fraction of the apparent (nominal) area of contact (that which would occur if the surfaces were perfectly smooth). The real area of contact is a function of the surface topography, material properties and interfacial loading conditions. The proximity of the asperities results in adhesive contacts caused by interatomic interactions. When two surfaces move relative to each other, the adhesion of these asperities and other sources of surface interactions contribute to friction force. Repeated surface interactions and surface and subsurface stresses, developed at the interface, result in the formation of wear particles and eventual failure. Thus, the contact modelling of two rough surfaces is of considerable interest in the study of friction and wear (**1**–**3**).

Roughness maps of a ceramic substrate measured using three different measuring instruments with different sampling intervals (lateral resolutions) are shown in Fig. 1. The figure shows that roughness is found at scales ranging from millimetre to nanometre (atomic) scales. Although surface roughness is *intrinsic*, measured roughness is extrinsic. Instruments using different sampling intervals measure features with different length scales. It can be concluded that a surface is composed of a large number of length scales of roughness that are superimposed on each other (**4**–**6**). Therefore, 'It is not that different asperities come in different sizes but it is that one asperity comes in different sizes'. Distribution of size and shape of asperities is dependent on the measurement technique. When the sampling interval at which the

This paper was first published in Tribology Letters, 1998, **4**, *pp.* 1–35, *and is reproduced here by permission of J. C. Baltzer AG, Science Publishers.*

surface is examined is reduced, the number of asperities detected and their curvature appear to rise without a limit down to the atomic scale. This means that an asperity is not a 'definite object'. Attempts have been made to identify a correct sampling interval which yields the relevant number of summits and the relevant curvature for a particular application (functional filtering) (**7, 8**). An asperity relevant for contact mechanics is defined as that which makes a contact in a particular application (contacting asperity). As the two surfaces are brought in contact, nanoscale asperities are the first to come into contact, instantly deforming plastically, and merging to form micro- to macrocontacts, which upon further application of load may deform by elastic or elastic–plastic deformation (**9**). Thus, size and number of contacting asperities are dependent on the interface conditions. Both friction and wear are strongly influenced by the degree of plastic interaction of the contacting asperities.

Modelling of the contact of rough surfaces is difficult and has been treated in many papers using a number of approaches over the past three decades. The difficulty in the development of a theoretical model is that the surface is a random structure and may be anisotropic, so that stochastic models must be used. The classical statistical model for a combination of elastic and plastic contacts between rough surfaces is that of Greenwood and Williamson (**10**), which assumes that surfaces are composed of hemispherically tipped asperities of uniform radius of curvature, with their heights following a Gaussian distribution about a mean plane. Their model was

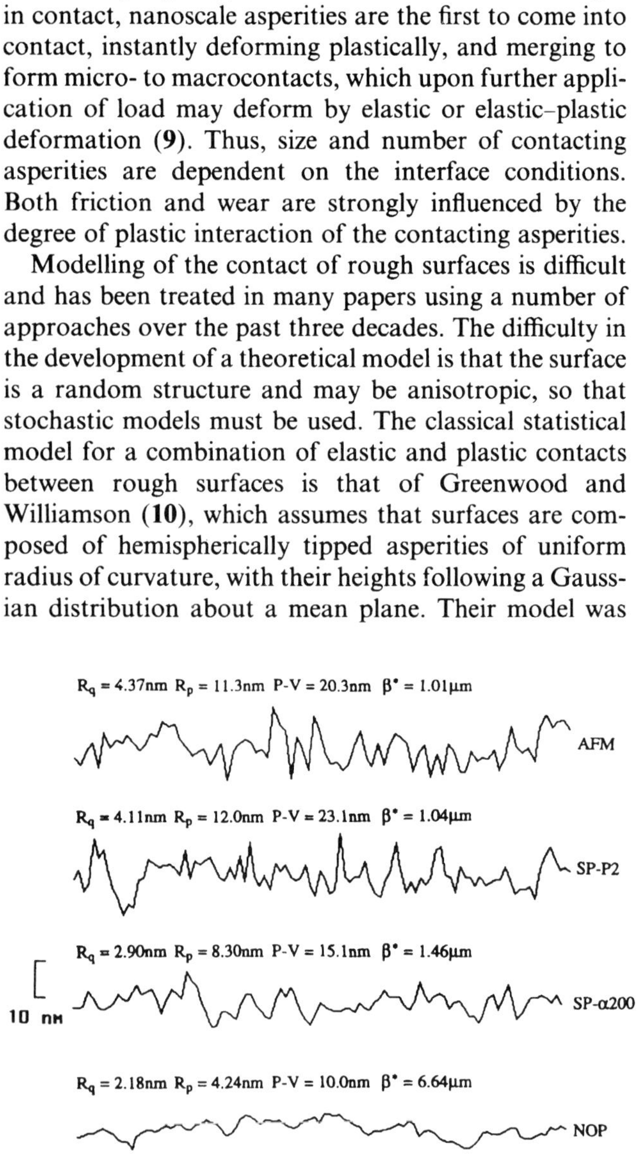

Fig. 1 Surface roughness profiles of a glass-ceramic substrate measured using an atomic force microscope (lateral resolution ~15 nm), stylus profiler (SP) P2 with a stylus tip of 0.2 µm radius (lateral resolution ~0.2 µm), and Alpha-step 200 (α-200) with a stylus tip of 5 µm radius (lateral resolution ~2 µm), and non-contact optical profiler (NOP) (lateral resolution ~1 µm) (**29**)

modified by Whitehouse and Archard (**11**) and Onions and Archard (**12**). Nayak (**13**), Bush *et al.* (**14, 15**), Gibson (**16**) and McCool (**17**) developed an elastic contact model that treated asperities as elliptical paraboloids with randomly oriented elliptical contact areas. Slight modifications to these models have been presented in several papers [e.g. see references (**7, 18–21**)]. Kotwal and Bhushan (**22**) developed a statistical model for non-Gaussian surfaces. A non-Gaussian surface may exhibit skewness and kurtosis. Skewness represents asymmetric spread and kurtosis represents peakedness of the distribution (**3**). Although these statistical models result in simple relationships, the assumptions are unrealistic.

A surface with Gaussian distribution of heights and exponential autocorrelation function can be characterized by two parameters: the standard deviation of surface heights (σ or R_q or r.m.s.) and correlation length (β^*) (**11**). Other parameters that are commonly measured are peak-to-mean distance (R_p) and peak-to-valley (P–V) distance. Due to the multiscale nature of surfaces, it is found that the surface roughness parameters depend strongly on the resolution of the measuring instrument or any other form of filter, and hence are not unique for a surface; for an example see Fig. 2. Therefore, predictions of the contact models based on conventional roughness parameters may not be unique to a pair of rough surfaces. Majumdar and Bhushan (**4**), Bhushan (**6**) and Ganti and Bhushan (**5**) have shown that the roughness of engineering surfaces can be characterized by fractal geometry. Majumdar and Bhushan (**23**) and Bhushan and Majumdar (**24**) developed a new fractal theory of elastic and plastic contact between two rough surfaces which uses fractal parameters for surface characterization. [Also see reference (**6**).]

Although the statistical models mentioned earlier so far can predict important trends in the effect of surface properties on the real area of contact, their usefulness is very limited because of over-simplified assumptions about asperity geometry and height distributions, the difficulty in determination of statistical roughness parameters, and the neglecting of interactions between adjacent asperities. With the advent of computer technology, a measured surface profile can be digitized and used for computer simulation (**25–28**). Digital maps of pairs of different surfaces can be brought together to simulate contact inside the computer and contours of contacts can be predicted. In computer simulations, the resulting contour maps can be analysed to give contact parameters for various interplanar separations of the rough surfaces. The common approach has been to convert the complex stress–deformation analytical expressions to a system of linear algebraic equations, which are generally solved, based on iterative numerical methods, with a computer. Most of these analyses do not require assumptions of surface isotropy, asperity shape, or distribution of asperity heights, slopes and curvatures. However, one still has to select scan size and

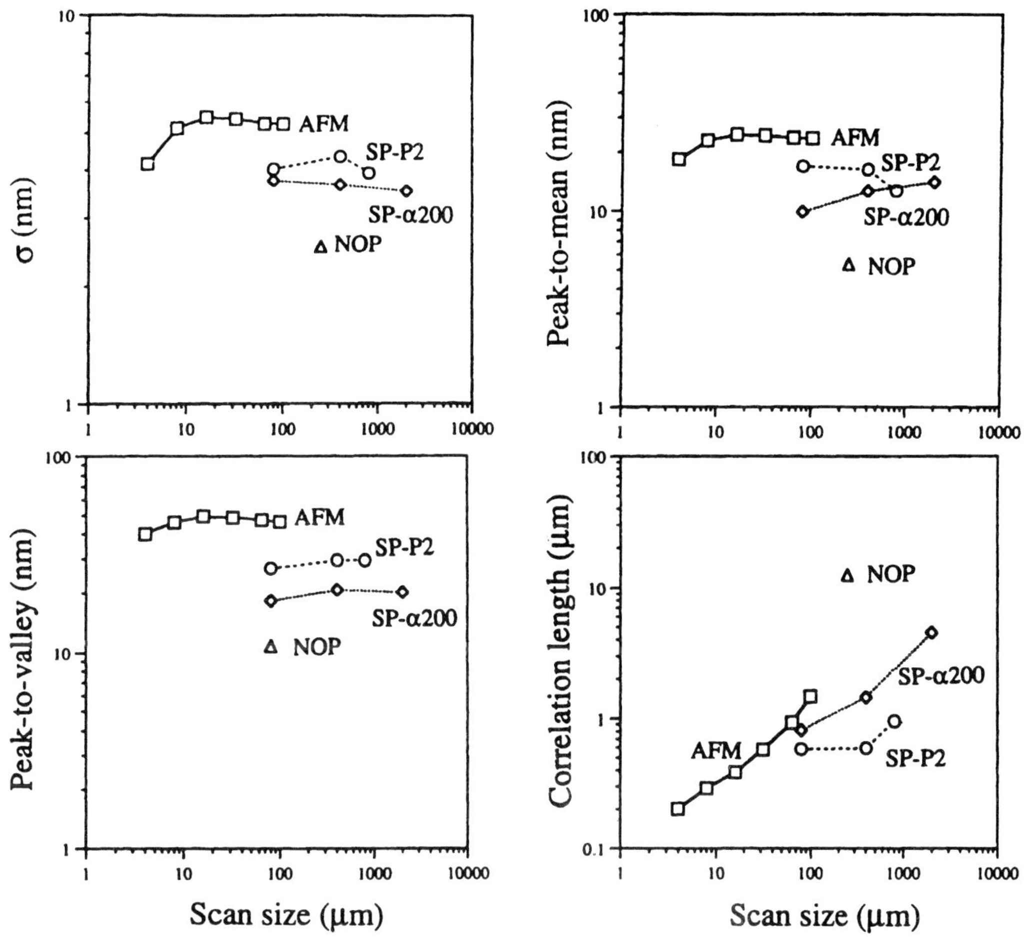

Fig. 2 Scale dependence of standard deviation of surface heights (σ), peak-to-mean (R_p), peak-to-valley (P–V) distance and correlation length (β^*) for a glass-ceramic substrate, measured using AFM, stylus profiler (SP-P2) and non-contact optical profiler (NOP) (**29**)

lateral resolution of the instrument relevant for the interface problem on hand (**29**).

Several numerical techniques have been developed and used to solve contact problems of rough surfaces numerically. Finite element methods (**30**) have been used to solve plane-strain elastic contact problems with only a few cylindrical asperities in contact (**31**). For three-dimensional rough surface contact problems with many asperities of arbitrary shape, requirements of a large number of mesh elements make the finite element approach unfeasible. One of the techniques provides a deterministic solution to stresses and areas for the approach of two real rough surfaces (**32–38**). This technique takes full account of the interaction of deformation from all contact points and predicts the contact geometry of real surfaces under loading. It provides useful information on the contact pressure, number of contacts, their sizes and distributions, and the spacing between contacts. For contacts with a moderate number of contact points, both elastic and elastic–plastic analyses of two rough surfaces can be carried out (**35, 38, 39**). Sayles and co-workers [e.g. reference (**35**)] have used an iterative process with a conventional matrix inversion technique to solve the contact problem for contact pressure and contact area. However, with an increase in contact points of rough surfaces, the so-called influence matrix will become very large, and possibly ill-posed. The large, ill-posed influence matrix will not only require a great deal of time for the conventional matrix inversion method to complete a large number of iteration cycles, but also, in some cases, cause serious non-converging problems.

Instead of using the conventional matrix inversion approach, Tian and Bhushan (**38**) utilized the approach on the basis of the variational principle. The use of this principle leads to a standard quadratic mathematical programming problem after an infinite-to-finite dimension transformation. In the variational method, the real area of contact and contact pressure distributions are those that minimize the total complementary potential energy. The use of the variational principle for elastic contact problems was first introduced in the early 1970s (**40, 41**), but the numerical difficulty of using a standard simplex-type algorithm to search for the minimum of the potential energy of contact with a large number of contact points has kept the variational approach from general applications in engineering practice. The variational approach developed by Tian and Bhushan (**38**) uses a direct quadratic mathematical programming method which guarantees the uniqueness of the solution of rough surface contact problems. It is also able significantly to reduce computing time since it does not need an additional iteration process which has been required by the conventional matrix inversion technique, and makes it feasible to solve the three-dimensional rough surface contact problem with a larger number of contact points.

Subsurface stress fields and the effect of friction on the contact stresses and contact area have been analysed by Bailey and Sayles (**42**) and Yu and Bhushan (**43**). Merriman and Kannel (**44**) and Cole and Sayles (**45**) developed a two-dimensional model to analyse the contact of layered rough bodies. A three-dimensional model of layered rough bodies has not been reported to date. Oden and Martins (**46**) and Martins *et al.* (**47**) have presented a continuum model of sliding of two rough surfaces using the finite element method.

In the presence of a thin liquid film with a small contact angle (wetting characteristics), such as a lubricant or an adsorbed water layer at the contact interface, curved menisci form around contacting and non-contacting asperities. The attractive meniscus force arises from the negative Laplace pressure inside the curved meniscus as a result of surface tension (**48–50**). This intrinsic attractive force may result in high static friction (stiction), kinetic friction and wear. This problem is particularly important in the computer data storage industry (**3, 51**). Several models have been developed to predict meniscus forces developed at a wet interface using a statistical approach (**52–55**). Kotwal and Bhushan (**55**) have extended statistical models to study non-Gaussian surfaces with and without a liquid film present at the interface. More recently Tian and Bhushan (**56**) developed a numerical model that uses digitized real rough surface maps without any arbitary assumption of the shape and size of asperities. The model accounts for the effect of surface roughness, thin liquid film, environment and operating conditions. Bhushan and Chilamakuri (**57**) and Chilamakuri and Bhushan (**58**) used this numerical model to study interaction of non-Gaussian surfaces with skewness and kurtosis.

It is evident that substantial on-going contributions, particularly in the last decade, are being made in the modelling of multiple asperity contacts under both dry and wet conditions. In this perspective article, the author presents various contact models for two rough surfaces in static and sliding contact. Homogeneous and layered elastic and elastic–plastic solids, with and without a liquid film present at the interface, will be considered. Emphasis in this paper has been on the results and implications of the contact models. The paper is primarily intended for a newcomer or individual not very familiar with this field. The paper starts out with a short review of modelling of single asperity contact followed by a comprehensive review of the modelling of multiple asperity contacts under both dry and wet conditions.

2 SINGLE ASPERITY CONTACT

If the asperity tip is assumed to be spherical then the contact problem reduces to a sphere in contact with a flat surface. A comprehensive review of modelling of single asperity contact has been presented by Bhushan (**59**). Here, the important results of contact stresses and contact areas for homogeneous and layered elastic and elastic–plastic solids in both static and sliding contacts are summarized. These results are of interest to understand the modelling of multiple asperity contacts more fully.

2.1 Elastic contact

The first case to consider is a generalized case of a spherical body in a sliding contact with a layered elastic half-space with a layer of thickness h, Fig. 3 (**60, 61**). In the following results, Poisson's ratio was taken as 0.3 for both layer and substrate. Poisson's ratio has little effect on stresses. Plastic deformation will be incorporated later. Various stress profiles on and beneath the contacting interface for homogeneous and layered elastic solids in static and sliding contacts are presented in Figs 4 to 6. Figure 4 presents the pressure profile beneath the indenter at various E_1/E_2. The pressure is normalized by p_0, which is the maximum pressure under the centre of indenter for a homogeneous medium (E_1/E_2) when the radius of contact $a_0=h$. Note that the so-called Hertzian pressure distribution is parabolic with a maximum at the centre of contact. The radius of the circular contact zone decreases and the maximum pressure

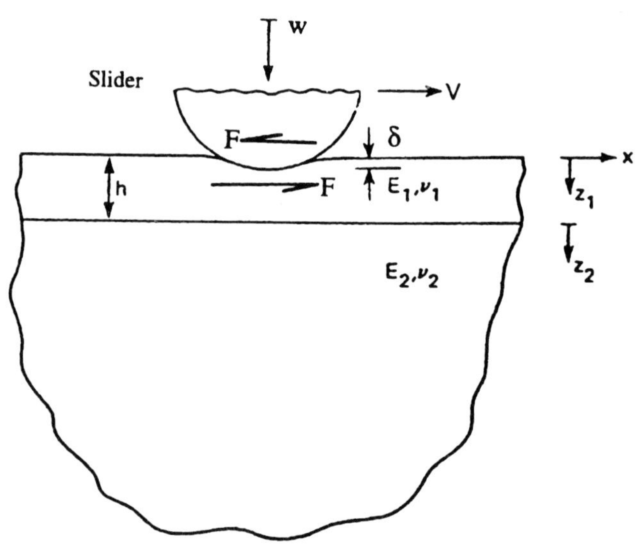

Fig. 3 Schematic diagram of a spherical rigid body in a sliding contact with a layered elastic half-space

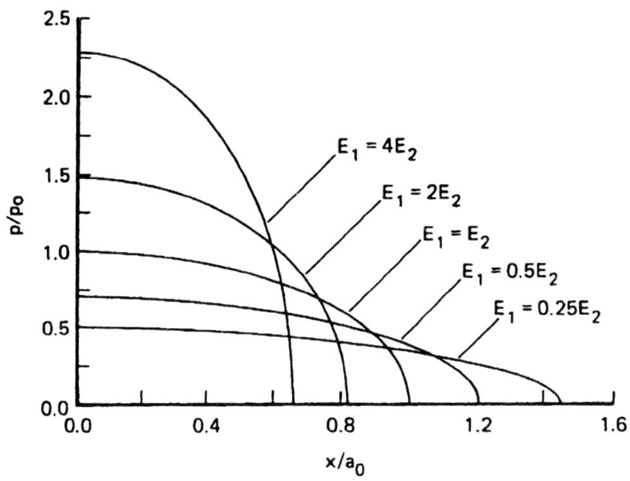

Fig. 4 Normal contact pressure profile beneath the rigid spherical indenter along the x axis for a layered elastic half-space with different values of E_1/E_2 and $\mu=0$ and when the radius of contact (at E_1/E_2) $a_0=h$ (**61**)

increases with an increase in the value of E_1/E_2. Figure 5 presents the three non-zero stress components σ_z, σ_x and τ_{xz}, as a function of depth in the layer and substrate under the centre of the indenter ($x=y=0$) for a coefficient of friction $\mu=0.25$ and for various values of E_1/E_2. Note that σ_x drops off rapidly through the body (along the z axis) (at $z/a_0\sim 1.3$ for homogeneous solids) whereas the σ_z drops off slowly (at $z/a_0>3$ for homogeneous and non-homogeneous solids). Maximum values of these stresses increase with an increase in friction and E_1/E_2. The stress component σ_x is tensile at the interface for a stiffer layer, which is significant for cracks at the base of the layer and orthogonal to the interface. The shear stress component is also enhanced by the stiffer layer, although it decays rapidly into the bulk. High interfacial shear stress adversely affects the adhesion of the layer to the substrate. For layers that are more compliant than the substrate, both the maximum value of normal stresses and the interfacial shear stress are reduced. Figure 6 shows the normal stress profile on the contacting surface along the x axis. Note that for a homogeneous solid, the normal stress (in all radial directions) is tensile outside the loaded circle. It reaches its maximum value at the edge of the circular contact. This is the maximum tensile stress occurring anywhere and it is held responsible for the ring cracks which are observed to form when brittle materials such as glass are pressed into contact (**62**). Normal stresses are compressive inside the contact. A stiffer layer and friction both increase the maximum tensile and compressive stresses on the surface. As the coefficient of friction increases, σ_x becomes unsymmetrical, compressive at the leading edge of the contact area ($x=a$) and maximally tensile at the trailing edge ($x=-a$). A stiffer layer and friction can thus degrade the brittle failure characteristics

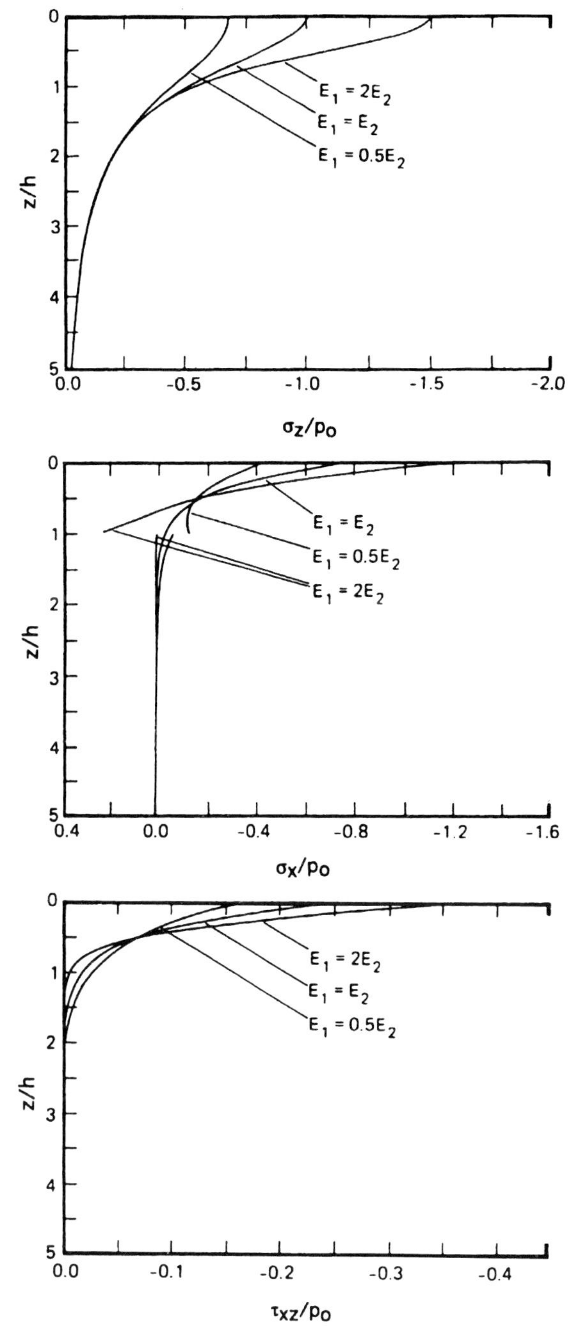

Fig. 5 Normal and shear stresses under the centre of indenter along the z axis for a rigid spherical indenter acting on a layered elastic half-space with different values of E_1/E_2 when $\mu=0.25$ and when the radius of contact (at E_1/E_2) $a_0=h$ (**61**)

of a layered medium, whereas a more compliant layer can be beneficial. Based on the normal and shear stresses at the interface and maximum tensile normal stresses at the surface, layers that are more compliant (e.g. solid lubricants such as Ag, MoS_2, graphite) than the substrate for multilayered body are preferred. However, high wear resistance may require harder (which are

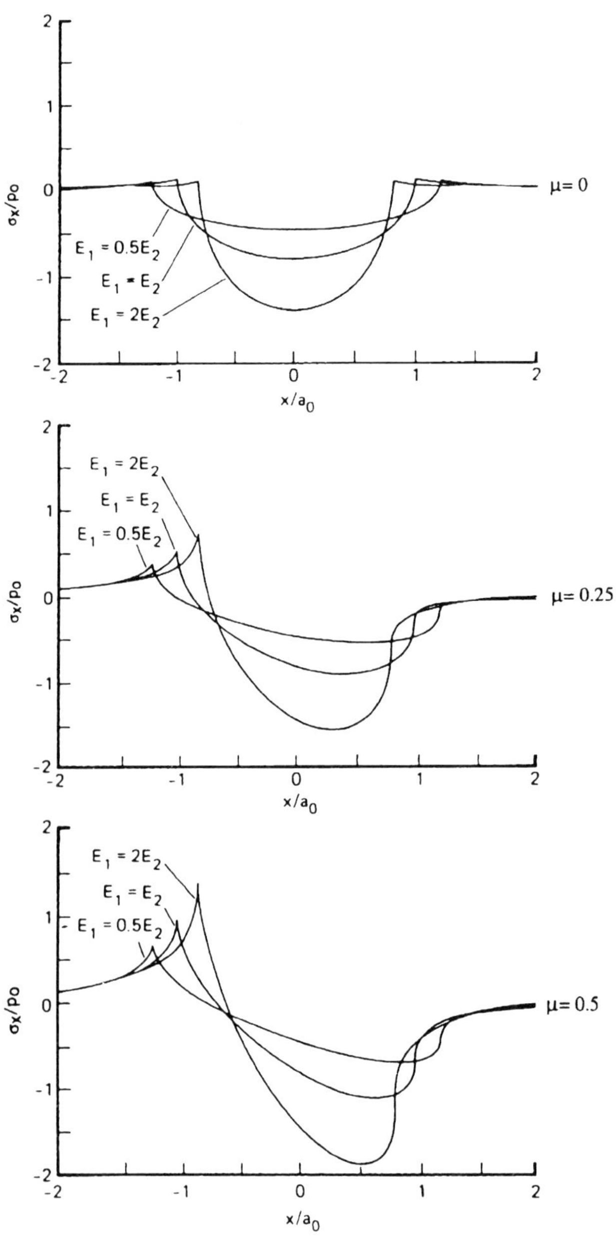

Fig. 6 Variation of normal stress on the surface beneath the rigid spherical indenter along the x axis for an indenter acting on a layered elastic half-space with different values of E_1/E_2 and μ and when the radius of contact (at $E_1 = E_2$) $a_0 = h$ (**61**)

generally stiffer) layers. Low values of coefficient of friction are also preferable.

The results presented so far are for the case of $a_0 = h$ with various values of E_1/E_2. For values of $a_0 \ll h$, the maximum stresses in the layer are similar to the homogeneous case with modulus E_1. Similarly, for $a_0 \gg h$, the stress field is dominated by the substrate. In the transition zone ($\sim 0.5h < a_0 < 6h$), the stress field depends strongly on the value of E_1/E_2.

2.2 Limits of elastic deformation

As the normal load between the two contacting bodies is applied, they initially deform elastically according to their Young's moduli of elasticity. As the load is increased, one of the two bodies starts to deform plastically. As the normal load is further increased, the plastic zone grows until the whole of the material surrounding the contact has undergone plastic deformation. Metals, alloys and some non-metals and brittle materials deform predominantly by 'plastic shear' or 'slip' in which one plane of atoms slides over the next adjacent plane. The load at which the plastic flow or yield begins is related to the yield point of the softer material in a simple tension or pure shear test through an appropriate yield criterion. Two of the yield criteria most commonly employed for most ductile materials as well as sometimes for brittle materials are Tresca's maximum shear stress criterion and von Mises shear strain energy criterion. The von Mises criterion usually fits the experimental data of metallic specimens better than other theories. However, the difference in the predictions of the two criteria is not large.

According to von Mises yield criterion, the contact pressure, p_0, at which yielding in the layer (in the case of a layered solid) and substrate occurs, is found by equating the square root of the invariant of the stress deviator tensor ($J_2^{1/2}$) to the yield stress in simple shear k, von Mises stress. For static contact, maximum Hertz contact pressure to initiate yield is 2.8 k. Contour plots of $J_2^{1/2}/p_0$ are shown in Fig. 7 for $\mu = 0$, 0.25 and 0.50 for various values of E_1/E_2. For $\mu = 0$, the figure shows symmetrical contours, and the maximum shear stress occurs beneath the surface on the axis of symmetry, the z axis (Fig. 3). As μ is increased, the region of maximum von Mises stress moves from a subsurface location towards the surface and becomes more intense; yield occurs at the surface when μ exceeds about 0.3. With respect to the centre of contact, the maximum von Mises shear stress location moves in the direction of friction force acting on the body (or in the opposite direction of sliding velocity of the body). Contour plots of $E_1/E_2 = 2$ show higher von Mises stresses with significant discontinuities occurring at the interface. The presence of a layer increases the von Mises stress in the layer. For a more compliant layer case of $E_1/E_2 = 0.5$, von Mises stresses are lower than in the case of a non-layered homogeneous body and, further, only a mild discontinuity occurs at the interface. The maximum Hertz pressure at which yield will occur, is presented as a function of coefficient of friction for a homogeneous solid in Fig. 8.

Based on Figs 6 and 7, a schematic diagram of the plastic zone and maximum tensile stress location is shown in Fig. 9. Note that with brittle materials, the appearance of tensile stresses is more important than the value of the maximum shear responsible for yield.

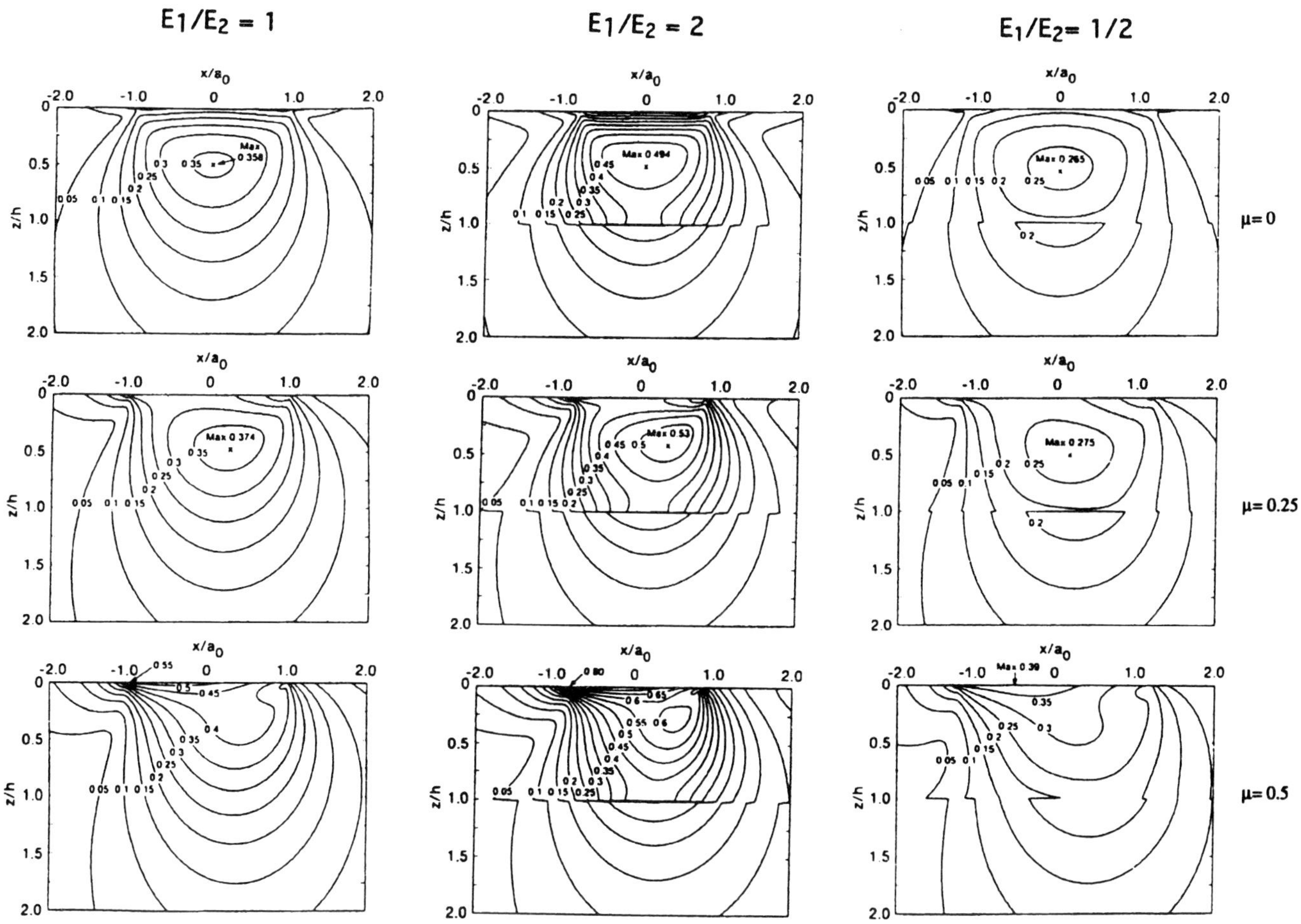

Fig. 7 Contours of constant normalized von Mises stresses $J_2^{1/2}/p_0$, for a rigid spherical indenter acting on a layered elastic half-space at $E_1=E_2$, $E_1=2E_2$, and $E_1=0.5E_2$ when the radius of contact (at E_1/E_2) $a_0=h$ (**61**)

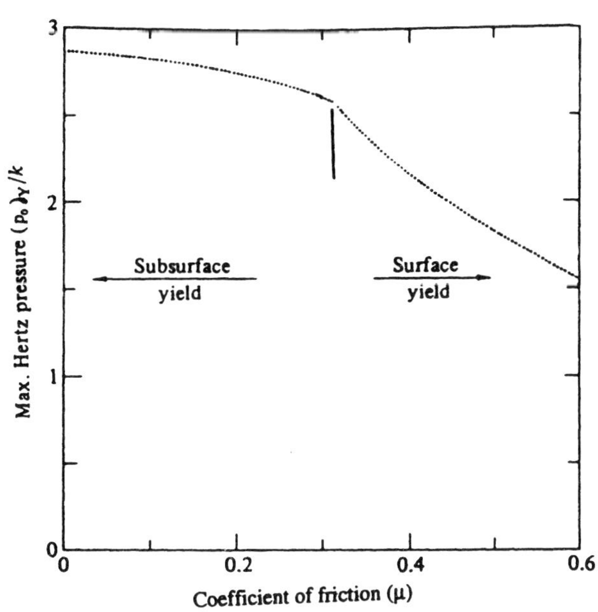

Fig. 8 Effect of coefficient of friction on the maximum Hertz contact pressure for yield (von Mises)

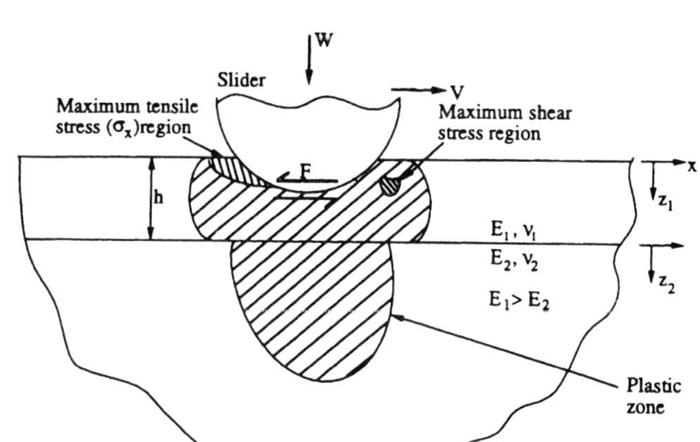

Fig. 9 For the case of a slider in a sliding contact with a layered solid, schematic diagram of plastic zone and maximum tensile and maximum shear stress location in the lower body. Note that stress distributions for the upper body will be that of the lower body with x and z replaced with $-x$ and $-z$, respectively

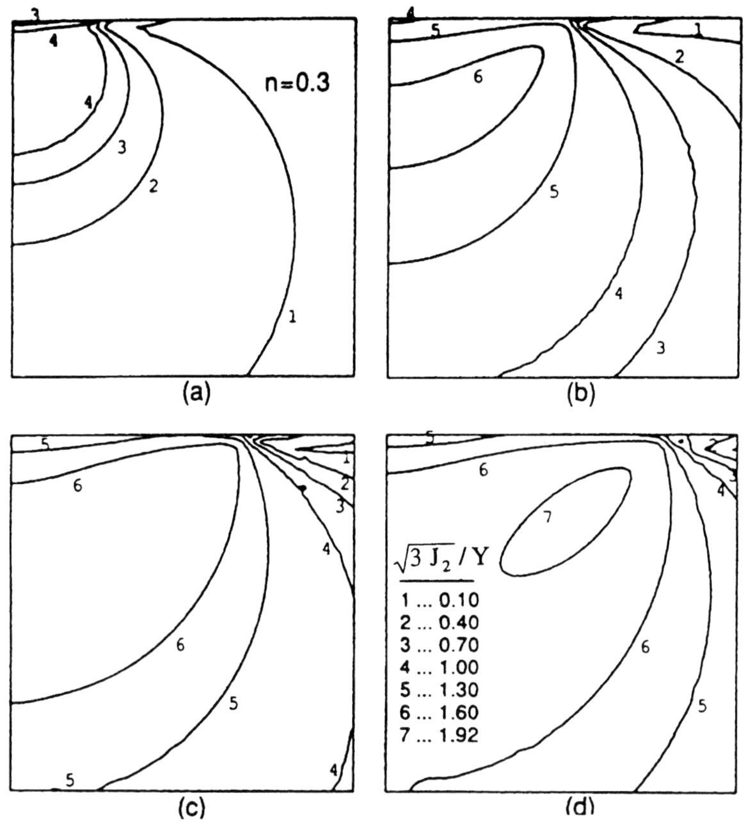

Fig. 10 Contours of normalized von Mises stress in the region $0 \leqslant r/a_y \leqslant 12$, $0 \leqslant z/a_y \leqslant 12$ for loads W/W_y equal to (a) 15.5, (b) 100.6, (c) 200, and (d) 300 for a rigid sphere acting on an elastic–plastic (strain hardening exponent $n=0.3$) half-space. Contour number 4 represents the boundary of plastic zone (**63**)

2.3 Elastic–plastic contact

Kral *et al.* (**63**) carried out a detailed finite element analysis of a rigid sphere in contact with an elastic–plastic strain-hardening half-space. Loads and distances were normalized by the values at the onset of plastic deformation with $E/Y=685$. Figure 10 shows the evolution of normalized von Mises stress ($\sqrt{3J_2}/Y$) contours in the half-space. Y is the initial yield stress in simple tension. As the load increases, the elastic core developed over the subsurface plastic zone diminishes and eventually disappears while the maximum shear stress moves radially outward from the centre and toward the surface. At loads on the order of 300 times the load required to initiate yielding, the material goes through full plasticity. The depth of the plastic zone is about two times the contact radius at the load. The authors reported that the mean contact pressure at full plasticity is about 2.5 times the yield stress with the contact diameter being about 10 times that of the contact diameter at the time yielding is initiated.

Kral *et al.* (**64, 65**) carried out finite element analysis of a rigid sphere against an elastic–plastic layered medium. They reported that a stiffer and harder layer increases the load for inception of yielding for both layer and the substrate. However, as reported earlier, a stiffer layer results in large tensile stresses on the layer surface and shear stresses at the layer–substrate interface, which may result in the formation of cracks and debonding of the interface.

3 MULTIPLE ASPERITY DRY CONTACTS

3.1 Analysis of identical asperities

The first case to consider is the contact between a smooth plane and a nominally flat surface covered with a number of spherical asperities of similar radius and the same height z, relative to the reference plane, Fig. 11. As the surfaces are loaded together, the total displacement (normal approach) δ is equal to $(z-d)$, where d is the current separation of the smooth surface and the reference plane of the rough surface. Each asperity is deformed equally, and carries the same normal load, W_i, so that for N asperities the total load W will be NW_i. For each asperity, the load, W_i, and the area of contact, A_i, are known from the Hertz analysis. Thus, if R is the radius of all

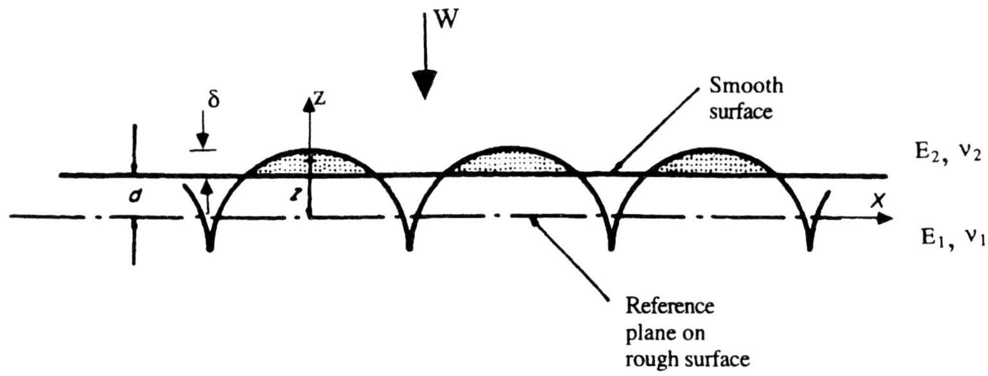

Fig. 11 Schematic of contact of a regular patterned rough surface against a smooth plane surface

identical asperities,

$$W_i = \frac{4E^*}{3} R^{1/2}\delta^{3/2} \tag{1a}$$

and

$$A_i = \pi R\delta \tag{1b}$$

where E^* is the composite or effective modulus. The total load is

$$W = \frac{4}{3\pi^{3/2}} \frac{NE^* A_i^{3/2}}{R} \tag{2}$$

Load is related to the total real area of contact, A $(=NA_i)$ from equations (1) and (2) as

$$W = \frac{4E^* A^{3/2}}{3\pi^{1/2} N^{1/2} R} \tag{3}$$

where

$$\frac{1}{E^*} = \frac{(1-\nu_1^2)}{E_1} + \frac{(1-\nu_2^2)}{E_2}$$

E_1 and E_2 and ν_1 and ν_2 are the Young's moduli of elasticity and Poisson's ratios, respectively, of the mating bodies 1 and 2. Equation (1b) indicates that the surface just outside the footprint is displaced in such a way that the real area of contact is exactly half of the area $2\pi R\delta$ which would be obtained by plastic flattening of the spheres. Equation (3) shows that, for this particular model, the real area of contact is proportional to the two-thirds power of the applied normal load, when the deformation is elastic.

The contact area is circular, having a radius a, and the contact pressure is elliptical with a maximum pressure at the centre of the contact. The mean (p_m) and maximum contact pressures (p_0) are

$$p_m = \frac{2}{3} p_0 = \frac{W_i}{A_i} = \frac{4E^*\delta^{1/2}}{3\pi R^{1/2}} = \left(\frac{16WE^{*2}}{9\pi^3 NR^2}\right)^{1/3} \tag{4}$$

The propensity for yielding is governed by the Tresca's maximum shear stress criterion or von Mises shear strain energy criterion. Yielding is initiated if the maximum shear stress (which occurs at the subsurface in the Hertz problem considered here) exceeds the shear shear strength. For a typical case using both yield criteria, this condition is satisfied,

$$p_m \sim 1.07Y \tag{5a}$$

where Y is the yield strength in simple tension of the softer material (**59**). As the contact pressure is increased further, the plastic zone grows until the whole of the material surrounding the asperity undergoes plastic deformation. The condition for full plasticity for an elastic–perfectly plastic material occurs,

$$\begin{aligned} p_m &\sim 2.8Y, \quad \text{Tabor (66)} \\ &\sim 2.5Y, \quad \text{Kral } et\ al. \text{ (63)} \end{aligned} \tag{5b}$$

Based on a number of numerical analyses and experimental measurements of indentation of half-space of elastic–plastic materials, hardness (H) is related to Y as (**67**)

$$H \sim 2.8Y \tag{6}$$

Therefore, yielding is initiated when

$$p_m \sim H/3 \tag{5c}$$

From equations (4) and (5c), the critical load beyond which plastic deformation occurs is given by

$$\frac{W_{crit}}{N} \sim \frac{\pi^3 R^2}{48E^{*2}}(H^3) \tag{7}$$

It is general practice to introduce a factor of safety to account for the fluctuation of the hardness measurement and all the uncertainties involved in design. The factor includes any dynamic effect during asperity contacts. The value of a factor of safety is normally chosen between 2 and 3 (**68**).

If the load exceeds the critical load, the softer material of the contacting bodies deforms plastically. If the material deforms plastically at the interface, at full plasticity each asperity contact can be thought of going through an indentation process. For elastic–perfectly plastic materials (with no work hardening), the flow pressure under full plasticity is found to be almost independent of the load. In this case,

$$\left(\frac{W_i}{A_i}\right)_p=\left(\frac{W}{A}\right)_p=H \tag{8}$$

thus the real area of contact is proportional to the load.

3.2 Statistical analysis of contacts

If the two rough surfaces, both nominally flat, come in contact until their reference planes (taken to pass through the *mean* of the peak height distribution) are separated by a distance h, then there will be contact at those asperities whose total heights, (z_1+z_2), are greater than h. The contacts can be either elastic or plastic. Greenwood and Williamson (**10**) analysed a rough surface against a smooth surface. They assumed that (a) the rough surface is covered with a large number of asperities, which, at least near their summit, are spherical; (b) asperity summits have a constant radius on each surface, R_{p1} and R_{p2}, and the composite radius (R_p) could be assigned to the rough surface; (c) that their heights vary randomly; and (d) that most engineering surfaces have a Gaussian distribution of peak heights. From Williamson *et al.* (**69**), only surfaces prepared by cumulative processes (such as peening and spark erosion, where the final shape of each small region is the cumulative result of a large number of discrete local events) have a Gaussian height distribution. Single-point processes (such as turning or shaping) and extreme-value processes (such as grinding and milling) lead to only an approximate Gaussian distribution. The assumption of peak radii being constant is clearly not valid.

Greenwood and Tripp (**10**) have treated the contact of two rough surfaces instead of the one rough surface against a flat surface treated by Greenwood and Williamson. For the case of two rough surfaces in contact with the pairs of asperities not aligned and the usual contact between the shoulders of the two hills, they found that, for a Gaussian peak-height distribution, the specification of asperity shape and the locations of asperities on one or both surfaces are unimportant. Therefore, although they have assumed the asperity tips to be spherical for numerical simplicity, this will not affect real area calculations. Also, they showed that the contact of two rough surfaces could be reduced to an equivalent, single, rough surface with a plane. O'Callaghan and Cameron (**71**) and Francis (**18**) also considered a case in which both surfaces can be rough and asperities need not contact at their tops. They concluded that the contact of two rough surfaces is negligibly different from the contact of a smooth and an equivalent rough surface.

The equivalent rough surface is defined as one whose asperity-peak curvature, $1/R_p$, is the sum of the curvatures of two rough surfaces,

$$1/R_p=1/R_{p1}+1/R_{p2} \tag{9}$$

and by elementary statistics, if the peak-height distributions of two rough surfaces follow independent random distribution (not necessarily Gaussian) with standard deviations of σ_{p1} and σ_{p2}, the distribution of the equivalent rough surface has a standard deviation σ_p [e.g. reference (**72**)],

$$\sigma_p=(\sigma_{p1}^2+\sigma_{p2}^2)^{1/2} \tag{10}$$

Equations (9) and (10) are valid when the two surfaces are independent, as is likely when two surfaces are prepared separately. However, when the surfaces have slid together, this assumption may be violated. If so, the expression for σ_p must be modified by a covariance term.

3.2.1 Elastic contact

For elastic contacts under *static conditions* or under *dynamic conditions with no frictional stresses* present at the contact, using Greenwood and Williamson's assumptions, it is possible to calculate the apparent pressure, p_a; mean real pressure, p_r; (elastic) real area of contact, A_{re}; number of contact spots, n; and mean asperity real area of contact as a function of separation, d.

Based on the Greenwood and Williamson analysis, consideration is given to the contact between a plane and a nominally flat surface covered with a large number of spherically tipped asperities of the same radius R_p and with their peak heights described by a probability density function of $p(z)$, Fig. 12.

If the two surfaces come together until their reference planes are separated by a distance d, then there will be contact at any asperity whose height was originally greater than d. Thus, the probability of making contact at any given asperity of height z, is

$$P(z>d)=\int_d^\infty p(z)\,\mathrm{d}z \tag{11a}$$

and if there are N asperities in all, the expected number of contacts will be

$$n=N\int_d^\infty p(z)\,\mathrm{d}z \tag{11b}$$

Also, since $\delta=z-d$, the total real area of contact is

$$A_{re}=\pi N R_p\int_d^\infty (z-d)p(z)\,\mathrm{d}z \tag{11c}$$

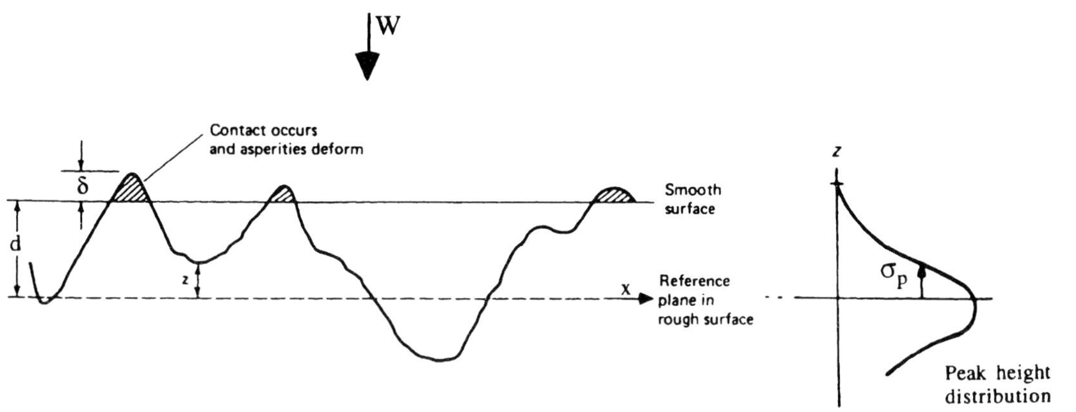

Fig. 12 Schematic representation of the contact between a rough surface and a smooth surface

Similarly, the expected total load is found to be

$$W = p_r A_{re} = p_a A_a = \left(\frac{4}{3}\right) N E^* R_p^{1/2} \int_d^\infty (z-d)^{3/2} p(z)\,dz \tag{11d}$$

It is convenient to introduce non-dimensional variables. The relationships in form of dimensionless variables are presented here,

$$\frac{p_a}{(\eta R_p \sigma_p) E^* (\sigma_p/R_p)^{1/2}} = \left(\frac{4}{3}\right) F_{3/2}(D) \tag{12}$$

$$\frac{p_r}{E^*(\sigma_p/R_p)^{1/2}} = \left(\frac{4}{3\pi}\right) F_{3/2}(D)/F_1(D) \tag{13}$$

$$A_{re} E^* (\sigma_p/R_p)^{1/2}/p_a A_a = \left(\frac{3\pi}{4}\right) F_1(D)/F_{3/2}(D) \tag{14}$$

$$n R_p \sigma_p E^* (\sigma_p/R_p)^{1/2}/p_a A_a = F_0(D) \Big/ \left(\frac{4}{3}\right) F_{3/2}(D) \tag{15}$$

$$(A_{re}/n) R_p \sigma_p = \pi F_1(D)/F_0(D) \tag{16}$$

where D, the dimensionless separation, is d/σ_p; η is the density of asperity summits per unit area (N/A_a) on a surface with smaller density; and $F_m(D)$ is a parabolic cylinder function given by

$$F_m(D) = \int_D^\infty (s-D)^m p^*(s)\,ds \tag{17}$$

where $p^*(s)$ is the *standardized* peak-height-probability density function, in which the height distribution has been scaled to make its standard deviation unity. For the case of peak height distribution following a Gaussian height distribution,

$$F_m(D) = \left[\frac{1}{(2\pi)^{1/2}}\right] \int_D^\infty (s-D)^m \exp(-s^2/2)\,ds$$

$$= \left[\frac{m!}{(2\pi)^{1/2}}\right] [\exp(-D^2/4)] U\left(m+\frac{1}{2}, D\right) \quad \text{for } m \geqslant 0 \tag{18}$$

The values of U are listed by Abramowitz and Stegun (**73**). A short table of functions $F_m(D)$ is also given by Greenwood and Tripp (**70**).

Note that equations (12) to (16) hold for all surface distributions. However, the assumption of two rough surfaces being the same as one equivalent rough surface with a plane is valid only for surfaces having a Gaussian distribution.

A simple relationship exists for an exponential height distribution,

$$p^*(s) = \exp(-s) \tag{19}$$

For this case, functions $F_m(D)$ are just $m! \exp(-D)$, and we have

$$p_r = \left(\frac{4}{3\pi}\right)(1.5!) E^* (\sigma_p/R_p)^{1/2} \tag{20a}$$

$$A_{re} = \left(\frac{3\pi}{4}\right) \frac{1}{(1.5!)} \frac{p_a A_a}{E^*(\sigma_p/R_p)^{1/2}} \tag{20b}$$

$$n = \frac{3}{4} \frac{1}{(1.5!)} \frac{p_a A_a}{R_p \sigma_p E^* (\sigma_p/R_p)^{1/2}} \tag{20c}$$

$$\frac{A_{re}}{n} = \pi R_p \sigma_p \tag{20d}$$

It can be noted that the real area of contact and the number of contacts are both proportional to load, even though the asperities are deforming elastically. The real contact pressure and mean asperity real area of contact are independent of load. For other distributions, such a simple relationship will not apply.

For a Gaussian distribution, D versus p_a is obtained from equation (12). Then, with the help of this relationship and equations (13) to (16), the relationships between p_a and p_r, A_{re}, n, and A_{re}/n are obtained in a dimensionless form. Next, the data are fitted to a power form using a least-squares fit. Approximate direct relationships of D, p_r, A_{re}, n and A_{re}/n with p_a are presented

Table 1 Interplanar separation, mean real pressure, real area of contact, number of contact spots, and mean asperity real area of contact for elastic contacts (**19**)

Interplanar separation	$D=1.40[\log(0.57/P_a)]^{0.65}$
Mean real pressure	$\frac{p_r}{E^*(\sigma_p/R_p)^{1/2}}=0.42P_a^{0.04}\sim 0.32$
Real area of contact	$\frac{A_{re}}{A_a[\eta R_p \sigma_p]}=2.40P_a^{0.96}\sim 3.20P_a$
Number of contact spots	$\frac{n}{\eta A_a}=1.21P_a^{0.88}\sim 2.64P_a=0.5(D\to 0)$
Mean asperity real area of contact	$\frac{A_{re}/n}{R_p \sigma_p}=2.00P_a^{0.08}\sim 1.21$

$P_a=p_a/(\eta R_p \sigma_p)E^*(\sigma_p/R_p)^{1/2}\leqslant 0.57$

in Table 1 (**19**). From equations in Table 1, note that D is a very weak function of p_a; p_r and A_{re}/n are practically independent of p_a; and A_{re} and n are approximately proportional to p_a. An important relationship for the real area of contact in the elastic regime is listed here

$$A_{re}\sim\frac{3.2p_a A_a}{E^*(\sigma_p/R_p)^{1/2}} \tag{21}$$

This model is defined by three parameters: σ_p, R_p and η. Whitehouse and Archard (**11**) regarded the profile of a random surface as a random signal represented by a height distribution and an autocorrelation function. They showed that all features of a surface with Gaussian distribution of heights and an exponential autocorrelation function could be represented by two parameters: σ and β^*. For an exponential autocorrelation function

$$C(\tau)=\exp\left(-\frac{\tau}{\beta^*}\right) \tag{22}$$

where $1/\beta^*$ is the decay rate of function. The correlation length can be defined as $C(\tau)=0.1$ (when $\tau=2.3\ \beta^*$) or $C(\tau)=1/e$ (when $\tau=\beta^*$). Both forms are considered to be a measure of the length at which simple readings become statistically independent of one another. Whitehouse and Archard took β^* as the correlation length given by $C(\tau)=1/e$.

For this surface, Onions and Archard (**12**) expressed the peak heights, curvatures and asperity density in terms of σ and β^*. In their contact model, they did not assume that peak heights follow a Gaussian distribution but follow a distribution derived from the assumed Gaussian distribution of surface heights; second, peak radii are not constant and have a distribution which is dependent upon the height. They reported that the distribution of peak heights is not quite Gaussian and that the peak curvature of the higher peaks tended to have higher values than those at lower levels. Based on their contact model

$$A_{re}\propto\frac{p_a A_a}{E^*(\sigma/\beta^*)} \tag{23}$$

By comparison of contact analyses by Whitehouse and Archard and Greenwood and Williamson, it is observed that σ_p is proportional to σ, R_p is proportional to β^{*2}/σ_p and η is proportional to $1/\beta^{*2}$. Therefore, the three parameters used in Greenwood and Williamson analysis should be related by the equation for a given type of finishing operation

$$\sigma_p R_p \eta=\text{constant} \tag{24}$$

Nayak (**74**) analysed the surface roughness following the pioneering work in statistical geometry of the oceanographer Longuet-Higgins (**75, 76**). He has shown that a random and isotropic surface with a Gaussian height distribution can be completely characterized by the three spectral moments of the profiles: M_0, M_2 and M_4 known as the zeroth, second and fourth spectral moments. M_0, M_2 and M_4 are equivalent to the variance of the distributions of profile heights (σ^2), slopes (σ'^2) and curvature (σ''^2), respectively. He obtained expressions of the distribution of peak height and peak curvature in terms of a parameter α, defined as

$$\alpha=\frac{M_0 M_4}{M_2^2} \tag{25}$$

which defines the width of the power spectrum density of the surface roughness, i.e. range of wave lengths encountered in it. For an isotropic, Gaussian surface, $\alpha\geqslant 1.5$. If the surface is Gaussian, the surface slope in an arbitary direction is essentially a difference between two correlated Gaussian variables and is therefore also Gaussian distributed. Likewise the second derivative of the surface in any general direction is Gaussian distributed.

Given the three spectral moment values M_0, M_2 and M_4 for each of the two rough surfaces, the corresponding values for the equivalent rough surface are computed as respective sums, i.e.

$$(M_0)_e=(M_0)_1+(M_0)_2 \tag{26a}$$

$$(M_2)_e=(M_2)_1+(M_2)_2 \tag{26b}$$

$$(M_4)_e=(M_4)_1+(M_4)_2 \tag{26c}$$

Nayak (**13**) and Bush *et al.* (**14**) carried out the contact analysis by modelling the rough surfaces as isotropic, Gaussian surfaces in terms of spectral moments.

Bush *et al.* (**15**), Gibson (**16**) and McCool (**17**) used Nayak microgeometry assumptions to develop an elastic contact model for anisotropic surfaces. The asperities were represented as elliptical paraboloids with random principal axis orientation and aspect ratio of the grains. They developed expressions for contact area in the terms of five surface parameters—M_0, and M_2 and M_4 along the grains and across the grains. Kotwal and Bhushan (**22**) carried out contact analysis of non-Gaussian surfaces with skewness and kurtosis.

Using Greenwood and Williamson's contact analysis, Fuller and Tabor (**77**) developed an analysis for adhesion of two rough elastic surfaces in contact. They related work of adhesion to a reduction in the free energy of the system resulting from reduction in surface energy as proposed by Johnson *et al.* (**78**). They found that the relative adhesion force between the two contacting solids depends on several parameters such as Young's moduli, surface roughness parameters (σ_p and R_p) and work of adhesion.

3.2.2 Limit of elastic deformation

For a random surface with asperities with Gaussian height distribution and constant radii, the normal approach in terms of p_m from equation (4) is

$$\delta = \frac{9\pi^2}{16} p_m^2 \frac{R_p}{E^{*2}} \tag{27}$$

From equations (5), (6) and (27), the critical value of δ for the asperity necessary to initiate subsurface plastic flow is

$$\delta_p \sim R_p \left(\frac{H}{E^*}\right)^2 \tag{28}$$

Greenwood and Williamson (**10**) defined a plasticity index ψ as the square root of the inverse of δ_p normalized with σ_p as

$$\psi = \left(\frac{\sigma_p}{\delta_p}\right)^{1/2} = \left(\frac{E^*}{H}\right)\left(\frac{\sigma_p}{R_p}\right)^{1/2} \tag{29}$$

This index is indicative of the degree of plasticity. For plastic contact, the real area of contact is given as

$$A_{rp} = \frac{p_a A_a}{H} \tag{30}$$

Assuming $A_{rp}/A_{re} = 0.02$ as the criterion for the onset of a significant degree of plasticity, it was found that if the $\psi < 0.6$, the deformation is largely elastic and if $\psi > 1$, surface deformation is largely plastic. Note that probability of plastic flow is virtually independent of the load and solely a function of the plasticity index as long as the asperities continue to deform independently. The index depends on both the mechanical properties and the surface roughness of the contacting surfaces. Slight modifications to the elastic–plastic models have been presented in several papers [e.g. see references (**7, 18–21**)].

Based on the Greenwood and Williamson approach, Onions and Archard (**12**) also defined a plasticity index based on σ and β^* as

$$\psi = \frac{E^*}{H}\left(\frac{\sigma}{\beta^*}\right) \tag{31}$$

They found that if $\psi > 0.45$, plastic flow occurs even at trivial loads, and if $\psi < 0.25$ plastic flow is most unlikely.

One of the benefits of Onions and Archard's analysis is that plasticity index and contact parameters are expressed in terms of surface parameters, which are most easily measured. Based on Whitehouse and Archard (**11**) and Whitehouse and Phillips (**79**), σ and R_p can be expressed in terms of σ and β^*. All roughness parameters and their relationships are a function of sampling interval (**29, 80**).

It can be noted that in a sliding contact with friction present at the interface, the maximum shear stress would be larger and would occur nearer the surface (Section 2). Therefore, the contacts would become plastic at lower values of ψ. In addition, in a multilayered solid for a fixed value of substrate Young's modulus, the stresses increase for larger values in the Young's modulus of the overcoat (Section 2), implying that the contact becomes plastic at lower values of ψ and vice versa. If the contact radius is much greater than the overcoat thickness, the effect of overcoat Young's modulus would be negligible and the yielding would be dominated by the modulus and hardness of the substrate (**81**). Further, during sliding, polishing of the asperity generally occurs, which results in a smoother surface and an increase in the real area of contact (**3**). During sliding, instantaneous roughness should be used.

For calculations of the real area of contact, *E*, *Y*, and *H* should be measured at a strain rate corresponding to the loading and unloading of the asperity contacts. During sliding, the asperities are loaded and unloaded periodically in a time corresponding to that taken for a moving asperity to traverse its contact diameter. Therefore, the strain rate involved in the loading cycle (which determines the area of contact) can be estimated as the sliding velocity divided by the diameter of an asperity contact.

Finally, in the case of materials that creep to a marked extent such as polymers, the real area of contact will increase with time of application of the load (**3**).

3.2.3 Optimization of mechanical properties and surface roughness parameters

To minimize friction and wear in a machine for given operating conditions, the fraction of real area of contact to apparent area of contact and the real contact pressure should be as low as possible. The real area of contact versus the plasticity indices given from equations (21) or (23) and (30) are plotted in Fig. 13. An examination of Fig. 13 shows that the plastic contact results in a minimum contact area. However, repeated plastic contact would lead to an undesirable permanent deformation and smoothening resulting in elastic contacts (and higher real area of contact). Wear is more probable when asperities touch plastically than in pure elastic contacts. Therefore, it is desirable to design machine components in the elastic contact regime and ψ close to the elastic contact limit ($\psi \sim 0.6$) or $E^*(\sigma_p/R_p)^{1/2}$ to be as high as possible.

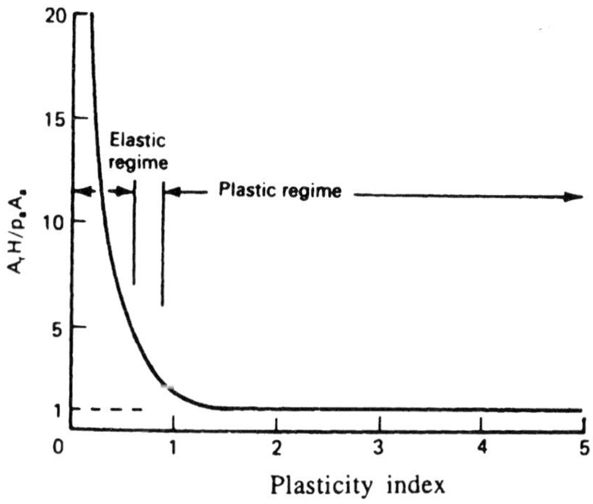

Fig. 13 Influence of plasticity index on the real area of contact (**19**)

Intuitively, it is possible to explain the fact that $E^*(\sigma_p/R_p)^{1/2}$ has to be higher for lower real area of contact by the following observation. The asperities with high E and low R_p produce high contact stresses and result in lower A_{re} for a given load. In addition, high σ_p allows contact with fewer asperities and again produces high contact stresses and results in lower A_{re} for a given load.

In principle, one can produce a well-defined (rather than statistical) roughness with few tall asperities of the same height and a small radius of curvature on a very smooth surface to minimize contact. Selection of number of asperities and radii of curvature of summits should be made such that the mean real pressure is just below the yield strength of the softer material (**68**, **82**). An advantage of this approach is that a well-defined localized roughness is produced and unwanted roughness is eliminated which results in better mechanical durability. However, creation of localized roughness may be an impractical approach because mechanical properties of the asperity tips and load variations in an application may be uncontrolled, making it difficult to select optimal shape and number of asperities.

3.3 Fractal analysis of contacts

Due to the multiscale nature of surfaces, it is found that the surface roughness parameters depend strongly on the resolution of the roughness measuring instrument or any other form of filter, hence not unique for a surface. Therefore, the predictions of the contact models based on conventional roughness parameters may not be unique to a pair of rough surfaces. However, if a rough surface is characterized in a way such that the structural information of roughness at all scales is retained (possible with modern roughness measurement tools such as the atomic force microscope), then it will be more logical to use such a characterization in a contact theory. In

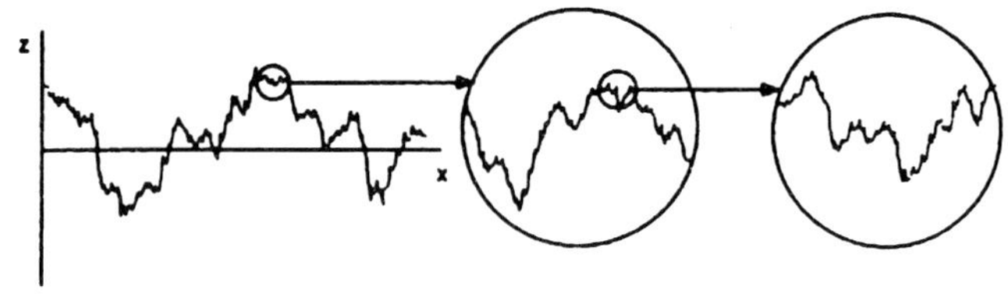

Fig. 14 Qualitative description of statistical self-affinity for a surface profile (**4**)

order to develop such a contact theory, it is first necessary to quantify the multiscale nature of surface roughness.

A unique property of rough surfaces is that if a surface is repeatedly magnified, increasing details of roughness are observed right down to the nanoscale. In addition, the roughness at all magnifications appears quite similar in structure as qualitatively shown in Fig. 14. Such behaviour can be characterized by fractal geometry (**4**–**6**). The main conclusions from these studies are that a fractal characterization of surface roughness is *scale-independent* and provides information of the roughness structure at all the length scales that exhibit the fractal behaviour. Based on this observation, Majumdar and Bhushan (**23**) and Bhushan and Majumdar (**24**) developed a new fractal theory of contact between two rough surfaces.

Structure function and power spectrum of a self-affine fractal surface follow a power law and can be written as [Ganti and Bhushan model (**5**)]

$$S(\tau) = C\eta^{(2D-3)}\tau^{(4-2D)} \tag{32}$$

$$P(\omega) = \frac{c_1 \eta^{(2D-3)}}{\omega^{(5-2D)}} \tag{33a}$$

and

$$c_1 = \frac{\Gamma(5-2D)\sin[\pi(2-D)]}{2\pi} C \tag{33b}$$

Fractal analysis allows the characterization of surface roughness by two parameters D and C, which are instrument independent and unique for each surface. D (generally ranging from 1 to 2) primarily relates to distributions of different frequencies in the surface profile and C to the amplitude of variations at all frequencies. η is the lateral resolution of the measuring instrument, τ is the size of the increment (distance) and ω is the frequency of the roughness. Note that if $S(\tau)$ or $P(\omega)$ are plotted as a function of τ or ω, respectively, on a log–log plot, then the power law behaviour would result in a straight line. The slope of line is related to D and the location of the spectrum along the power axis is related to C.

A fractal model of elastic–plastic contacts (**23**) has been developed to predict whether contacts experience elastic or plastic deformation, and to predict the real area of contact and statistical distribution of contact

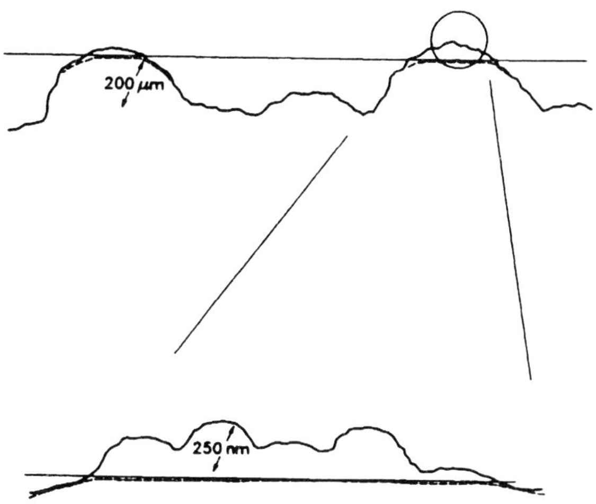

Fig. 15 Schematic of local asperity deformation during contact of a rough surface; upper profile measured by an optical profiler and lower profile measured by AFM. Typical dimensions are shown for a polished, magnetic thin-film rigid disk against a flat magnetic slider surface (**9**)

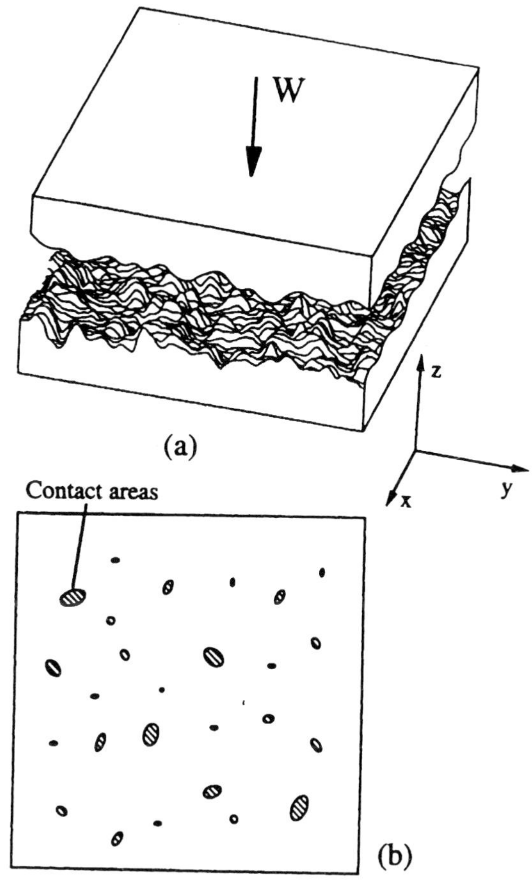

Fig. 16 Schematic of (a) two three-dimensional rough surfaces in contact, and (b) corresponding contact areas

points. Based on the fractal model of elastic–plastic contact, whether contacts undergo elastic or plastic deformation is determined by a critical area that is a function of D, C, hardness, and modulus of elasticity of the mating surfaces. If the contact spot is smaller than the critical area, it undergoes plastic deformation and large spots undergo elastic deformations. The critical contact area of inception of plastic deformation for a magnetic thin-film disk was reported by Majumdar and Bhushan (**23**) to be about 10^{-27} m^2, or so small that all contact spots can be assumed to be elastic at moderate loads.

The question remains as to how large spots become elastic when they must have initially been plastic spots. A possible explanation is shown in Fig. 15 (**9**). As two surfaces touch, the nanoasperities (detected by AFM-type of instruments) first coming into contact have smaller radii of curvature and are therefore plastically deformed instantly, and the contact area increases. When load is increased, nanoasperities in the contact merge, and the load is supported by elastic deformation of the large-scale asperities or microasperities (detected by optical profiler type of instruments).

Majumdar and Bhushan (**23**), Bhushan and Majumdar (**24**), and Bhushan (**6**) have reported relationships for cumulative size distribution of the contact spots, portions of the real area of contact in elastic and plastic deformation modes, and the load–area relationships.

3.4 Numerical three-dimensional contact models and computer simulations

Williamson (**25**), Gupta and Cook (**26**) and Bhushan and Cook (**27**) digitized measured profiles of contacting surfaces and used them for computer simulation. For elastic deformation at individual contacts, the contact pressure and contact area were approximated by the Hertz analysis. For plastic deformation, Gupta and Cook (**26**) analysed the contacts using the plane-strain slip line field solution. In their analysis, they assumed that peaks of asperities were spherical in shape, which is not valid. Finite element methods (**30**) have been used to solve elastic plane-strain contact problems. Komvopoulos and Choi (**31**) used the finite element method (commercial package ABAQUS) to study plane-strain and frictionless elastic contact problems with only few cylindrical asperities with constant radius. For three-dimensional rough surface contact problems with many asperities of arbitary shape, the requirement of a large number of mesh elements makes the finite element approach unfeasible. One of the techniques provides a deterministic solution to stresses and areas for the approach of two three-dimensional real rough surfaces, Fig. 16 (**32**–**38**). This technique takes full account of the interaction of deformation from all contact points and predicts contact geometry of real surfaces under loading. It provides useful information on the contact pressure, number of contacts, their sizes and distributions, and the spacing between contacts. For contacts with a moderate number of contact points, both elastic and elastic–plastic analyses of two rough surfaces can be carried out (**35**, **38**, **39**). These analyses develop the physical model of two rough surfaces in elastic contact. As shown in Fig. 16, the real area of contact between the two bodies occurs at the tips of highest asperities. The contact area is a small fraction of the surface areas of the contacting bodies. Therefore, it can be assumed that the asperity contacts of each body occur on an elastic half-space. Another assumption used in this study is that the area of individual contact is much smaller than the radii of curvature of contacting asperities. This allows the use of the linear theory of elasticity as well as the approximation of plane surface around the real contact area. Sayles and co-workers [e.g. references (**34**, **35**)] have used an iterative process with a conventional matrix inversion technique to solve the contact problem for contact pressure and contact area. The matrix inversion approach starts with the classical approach for finding the relationship between the contact pressure $p(x, y)$ and surface displacement $u_z(x,y)$ due to Boussinesq (**83**) and Cerruti (**84**) who made use of the theory of potential. Their theory is described by Love (**85**). The Boussinesq–Cerruti theory (commonly referred to as Boussinesq theory) gives the normal surface displacement at a general surface point (x, y) due to a concentrated contact (load) at some other point on the surface of a semi-infinite body. If the displacement of a surface point is required due to a number of concentrated contact points over area Ω, the total displacement is

$$u_z(x, y) = \frac{1-\nu^2}{\pi E} \iint_{\Omega} \frac{p(x', y')\,dx'\,dy'}{\sqrt{(x-x')^2+(y-y')^2}} \tag{34}$$

where the Ω is the domain of the contact surface on which surface forces act. For two bodies in contact, the composite surface displacement $\bar{u}_z(x, y)$ is given by equation (34) after replacing $E/(1-\nu^2)$ by the composite Young's modulus E^*.

The contacting interface region is discretized into a mesh of small elements $(2a \times 2b)$ over the domain of entire contact area as shown schematically in Fig. 17 and a piecewise interpolation of contact pressure distribution can be used inside each element. The various Lagrange polynomials can be used for piecewise interpolation of contact pressure inside each element. However, when a mesh element is small enough, the contact pressure at each element can be treated as a constant. In this case, equation (34) reduces to

$$\begin{aligned}(\bar{u}_z)_l &= \frac{1}{\pi E^*} \iint_{\Omega} \frac{p(x', y')\,dx'\,dy'}{\sqrt{(x-x')^2+(y-y')^2}} \\ &= \frac{1}{\pi E^*} \sum_{k=1}^{M} \left\{ \iint_{\Omega_k} \frac{dx'\,dy'}{\sqrt{(x-x')^2+(y-y')^2}} \right\} p_k \\ &= \sum_{k=1}^{M} C_{kl} p_k \end{aligned} \tag{35a}$$

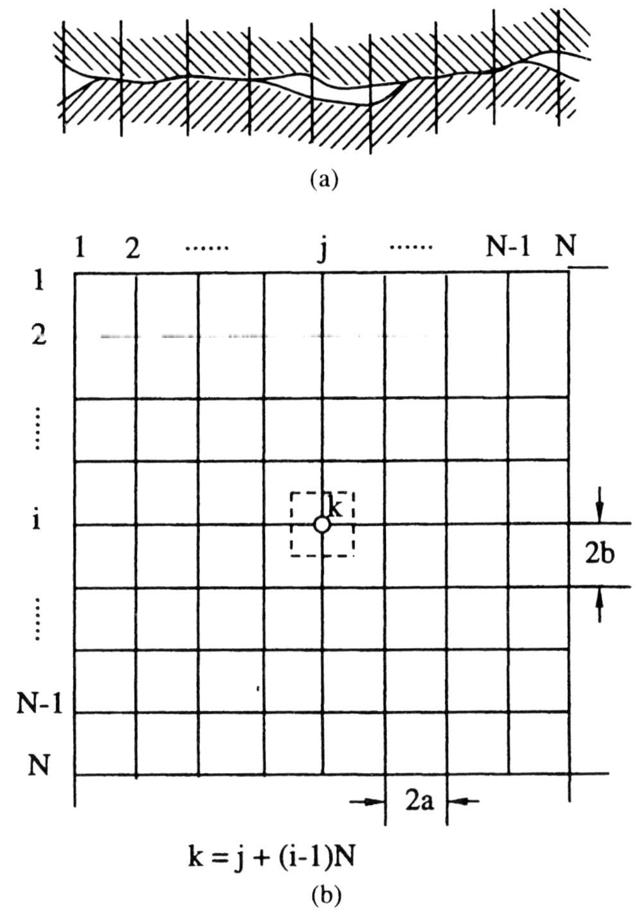

Fig. 17 Discretization of the contacting rough surfaces (a) two-dimensional profile and (b) top view of discretized surface (**38**)

where $(\bar{u}_z)_l$ is the surface displacement at the centre of the rectangular element of $2a \times 2b$, Ω_k is the sub-domain of contact area of a single element, M is the total number of initial contact points (from a total of N^2 elements) where the geometrical interference occurs based on relative rigid body movement of the two rough surfaces, k and l are the two indices representing the contact pressure location and surface displacement location, respectively, and C_{kl} is the element of the so-called 'influence' matrix and based on Love (**85**) is given as

$$\begin{aligned} C_{kl} = \frac{1}{\pi E^*} \Bigg\{ &(x+a) \ln \left[\frac{(y+b)+\sqrt{(y+b)^2+(x+a)^2}}{(y-b)+\sqrt{(y-b)^2+(x+a)^2}} \right] \\ &+ (y+b) \ln \left[\frac{(x+a)+\sqrt{(y+b)^2+(x+a)^2}}{(x-a)+\sqrt{(y+b)^2+(x-a)^2}} \right] \\ &+ (x-a) \ln \left[\frac{(y-b)+\sqrt{(y-b)^2+(x-a)^2}}{(y+b)+\sqrt{(y+b)^2+(x-a)^2}} \right] \\ &+ (y-b) \ln \left[\frac{(x-a)+\sqrt{(y-b)^2+(x-a)^2}}{(x+a)+\sqrt{(y-b)^2+(x+a)^2}} \right] \end{aligned} \tag{36}$$

Equation (35a) can also be expressed in a matrix form as

$$\boldsymbol{u} = \mathbf{C} \cdot \boldsymbol{p} \tag{35b}$$

where **C**, $\boldsymbol{u}$ and $\boldsymbol{p}$ are the influence matrix, displacement vector and contact pressure vector, respectively.

This equation, adjusted by putting the pressure as unity, is used to form the influence coefficient of any contacting point on any other point on the surface. The total deflection at any point can therefore be calculated by summing each element of pressure times its influence coefficient:

Total deflection of a point

$= \sum$ elemental pressures $\times$ influence coefficients

Thus, if the pressures on each contacting element are known, the deflection of any point on the surface and hence deformed geometry can easily be found. However, the elemental pressures are unknown, an iteration process has to be used to solve the equation (35). For a given rigid body normal approach between two rough surfaces, the surface displacements at the contact surface are determined from the geometrical interference of the two bodies. Then, equation (35) is used to solve for the pressure distribution inside the contact region. Those contact elements with negative (tensile) contact pressure are excluded for the next iteration since tensile force is not permitted at the contact surface. The deformations over the entire contact surface are then computed using equation (35) to check that no geometrical interference occurs outside the chosen contact region. The iteration is repeated until convergence occurs to a set of surface elements with positive or zero contact pressure. These surface elements with positive contact pressure define the real area of contact, and the total applied load is the sum of the contact pressure at those surface elements. One major difficulty for contact problems with a large number of contact points is that the storage of influence matrix usually requires a large amount of computer memory since for M contact node points, the total number of influence matrix elements will be M^2. Ren and Lee (**37**) developed a moving grid method to reduce the size of the computer storage for storing a large influence matrix.

For a large number of contact points, the influence matrix may become very large, and possibly ill-posed. The large, ill-posed influence matrix will not only require a great deal of time for the conventional matrix inversion method to complete a large number of iteration cycles, but also, in some cases, cause serious non-converging problems. Instead of using the conventional matrix inversion approach, Tian and Bhushan (**38**) utilized an approach on the basis of the variational principle. The use of the variational principle leads to a standard quadratic mathematical programming problem after an infinite-to-finite dimension transformation. In the variational method, the real area of contact and contact pressure distributions are those which minimize the total complementary potential energy. The variational approach uses a direct quadratic mathematical programming method which guarantees the uniqueness of the solution of rough surface contact problems. It is also able significantly to reduce computing time since it does not need an additional iteration process which is required by the conventional matrix inversion technique, and makes it feasible to solve the three-dimensional rough surface contact problem with a large number of contact points. The variational approach will be described next.

Besides differential equations, the contact problem can also be governed by minimum potential energy theory. The boundary value problem of differential equations for a mechanics system is equivalent to the problem of solving a minimum value of integral equation, i.e. a variational problem. In the variational approach, two minimum energy principles, total elastic strain energy and total complementary potential energy, can be used for solving mechanics problems (**86**). The definition of strain energy and complementary energy is shown in Fig. 18a. The minimum total complementary potential energy principle was used in the analysis developed by Tian and Bhushan (**38**) since it is more convenient to work in terms of unknown contact pressure (**41**) for elastic contact problems. For the minimum total complementary potential energy principle, the variational theory states that of all possible equilibrium stress fields for a solid subjected to prescribed loadings and boundary displacements, the true one corresponding to a compatible strain field, renders the total complementary energy stationary (**86**). For rough surface contact, when the pressure distribution and the real area of contact are unknown, the problem then becomes to find the minimum value of an integral equation that minimizes the total complementary potential energy of the contacting system.

For an elastically deformed body, the total complementary potential energy is given by (**86**)

$$V^* = \int_D U_0^*(\sigma_{ij})\, \mathrm{d}D - \int_\Omega u_i^* T_i\, \mathrm{d}\Omega \tag{37}$$

where V^* is the total complementary potential energy, U_0^* is the internal complementary energy density, D is the domain of deformed body, Ω is the domain of the contact surface on which the surface forces, T_i, act. T_i, σ_{ij} and u_i^* are surface forces, stresses and prescribed displacements, respectively.

For a contact problem of two rough surfaces, the total complementary potential energy is given by

$$V^* = U_E^* - \int_\Omega p(u_{z1}^* + u_{z2}^*)\, \mathrm{d}\Omega$$

$$= U_E^* - \int_\Omega p\bar{u}_z^*\, \mathrm{d}\Omega \tag{38}$$

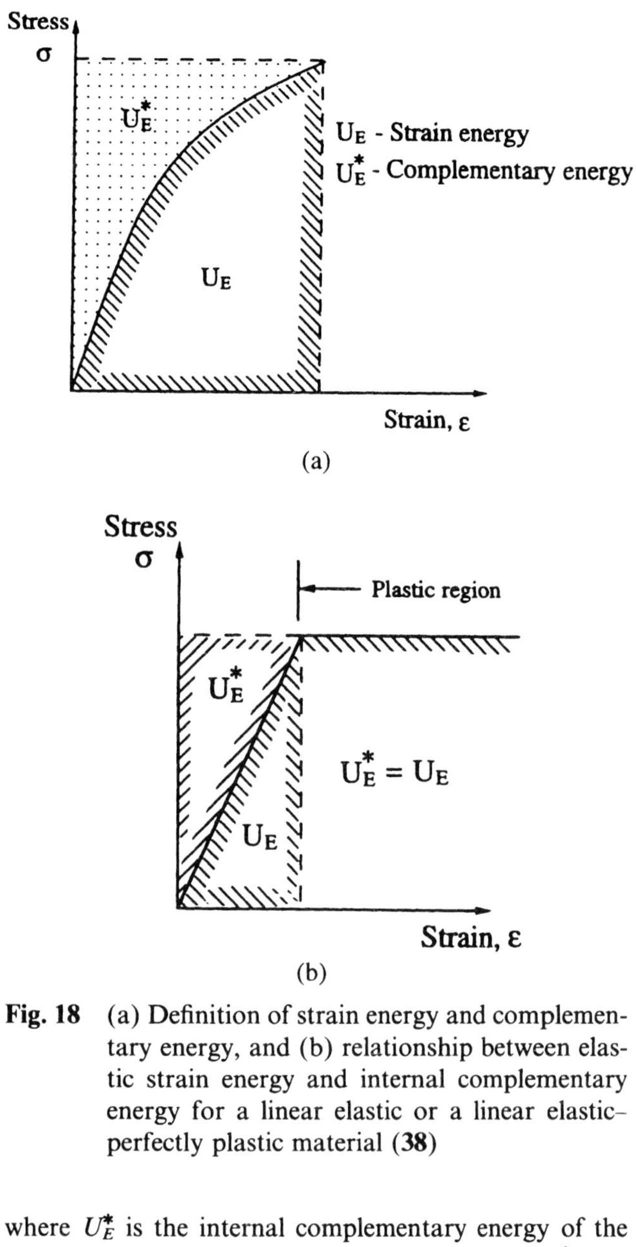

Fig. 18 (a) Definition of strain energy and complementary energy, and (b) relationship between elastic strain energy and internal complementary energy for a linear elastic or a linear elastic–perfectly plastic material (**38**)

where U_E^* is the internal complementary energy of the two stressed bodies, p is the contact pressure, u_{z1}^* and u_{z2}^* are the prescribed displacements of the two contacting bodies identified as 1 and 2, respectively, and $\bar{u}_z^*$ is the total prescribed displacement of the two contacting bodies inside the assumed contact zone based on geometrical interference.

For linear elastic materials, the internal complementary energy U_E^* is numerically equal to the elastic strain energy U_E as shown in Fig. 18b and can be expressed in terms of surface forces and displacements by (**67**)

$$U_E = \frac{1}{2}\int_\Omega p(u_{z1} + u_{z2})\, d\Omega = \frac{1}{2}\int_\Omega p\bar{u}_z\, d\Omega \tag{39}$$

where $\bar{u}_z$ is the composite surface displacement inside the contact zone to be calculated, and equals total contact deformation of two contacting bodies.

Combining equations (38) and (39), the total complementary potential energy for the two contacting bodies becomes

$$V^* = \frac{1}{2}\iint_\Omega p\bar{u}_z\, dx\, dy - \iint_\Omega p\bar{u}_z^*\, dx\, dy \tag{40}$$

Now, to relate $\bar{u}_z$ with p, Boussinesq's solution can be used to relate contact pressure with the surface displacement for normal distributed load on the surface of a semi-infinite body given earlier in equations (35) and (36).

After substituting $\bar{u}_z$ in equation (40) by equation (35), the contact problem of finding a minimum value of the total complementary potential energy now reduces to solving the minimum value problem of the integral equation (40) in terms of contact pressure $p(x, y)$.

In order to obtain an approximate value of minimum complementary potential energy, the integral equation (40) for the total complementary potential energy and equation (35) relating surface displacement to the surface pressure are discretized into a mesh of small elements over the entire contact area (Fig. 17), as described earlier.

After discretization of equations (35) and (40), the total complementary potential energy further reduces to

$$V^* = \frac{1}{2}\sum_{l=1}^{M} p_l\left(\sum_{k=1}^{M} C_{kl}p_k\right) - \sum_{l=1}^{M} p_l\bar{u}_{zl} \tag{41}$$

The total prescribed displacement of the two contacting bodies, $\bar{u}_{zl}^*$, in equation (41) can be determined using the following geometrical interference criterion:

$$\bar{u}_z^* + f(x, y) - \delta = 0 \text{ (within contact area)},$$

$$\bar{u}_z^* + f(x, y) - \delta > 0 \text{ (outside contact area)},$$

where δ is the rigid-body movement under applied load, and $f(x, y)$ is the initial separation of the two contact surfaces and equals

$$f(x, y) = |z_1(x, y) - z_2(x, y)|$$

Here it is assumed that at the initial position, there is no contact between the two surfaces. Thus, within the contact area, the total prescribed displacement of the two contacting bodies, $\bar{u}_z^*$, in equation (35) equals

$$\bar{u}_z^* = \delta - [z_1(x, y) - z_2(x, y)] \tag{42}$$

The total complementary potential energy given in equation (41) is a standard quadratic function of the constant pressure, and can be solved using mathematical programming methods once all possible contact areas are included in equation (41). Since no tension (forces in the direction opposite to motion of contacting bodies) is allowed at the contact interface, the search for a minimum value of the equation (41) is restricted by

$$p_k \geqslant 0, \quad k=1, \ldots, M \tag{43}$$

where M is the total number of initial contact points. It can be noted that the number of final contact points which satisfy the restriction, $p_k \geqslant 0$, would be smaller than M.

One way to find the minimum value of equation (41) under the restriction of equation (43) is to use a standard simplex-type algorithm (**87**). Examples of using the standard simplex-type algorithm for elastic contact problems can be found in some earlier studies (**40, 41**). However, the use of the simplex-type algorithm is limited to an optimization problem which only contains a smaller number of variables. For the search of feasible solution in each step of simplex algorithm is rather tedious and time consuming when the number of variables is large. In the model developed by Tian and Bhushan (**38**) a direct quadratic mathematical programming method is used to obtain the minimum value of complementary potential energy. To reduce the required storage space for the influence matrix, a space-reduced matrix storage was developed to compress the entire influence matrix elements in a space-reduced matrix.

Tian and Bhushan (**38**) developed a computer program to obtain the minimum value of the total complementary potential energy of equation (41) with restrictions of equation (43). In the program, the three-dimensional surface profiles of the two surfaces are initially read in. Contacting surfaces are discretized into small elements corresponding to the different points of surface heights. For a given rigid body approach (or load) between two rough surfaces, the total surface displacement at the surface of real contact is equal to the interference of two contacting bodies. The elements with finite displacement are included in the formulation. An influence matrix is constructed to relate contact pressure to the given displacements dependent on the location of the pressure element and contact points and an expression for the total complementary potential energy involving displacement and pressure is developed. The minimum value of the total complementary potential energy is obtained using a direct quadratic mathematical programming technique. The real area of contact and contact pressure distribution are those which minimize the total complementary potential energy. The analysis solves for contacts with positive pressure. Figure 19 shows the flow chart of the main program. The main feature of this program is that the corresponding contact pressure for a given rigid-body approach can be obtained directly from a single minimization process, thus eliminating the additional iteration cycles in the conventional matrix inversion methods described earlier. By comparison, the elimination of these iteration cycles effectively speeds up the entire process several times as compared to conventional matrix inversion methods.

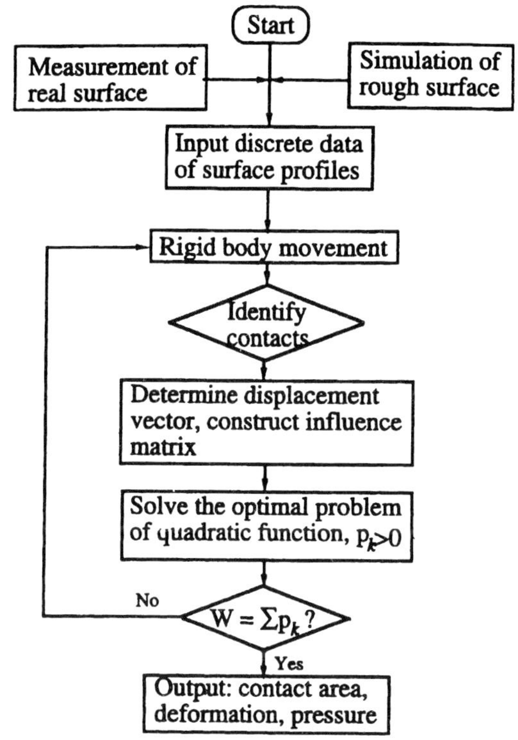

Fig. 19 Flow chart of the computer program for contact analysis of two rough surfaces (**38**)

So far, only elastic contact models have been presented. These models have been extended to elastic–plastic contact analyses for the case in which material is assumed to deform as an elastic–perfectly plastic material (**35, 38, 39, 88**). In these analyses, it is assumed that the regime of plastic deformation is confined to a very small area and does not significantly alter the geometry of the elastically deformed contact surface. Outside the plastic regime, the relationship between contact pressure and elastic deformation [equation (35)] still holds. A contact point deforms plastically once the local contact pressure exceeds the hardness of the softer material. An expression for the total complementary potential energy during plastic deformation is developed by Tian and Bhushan (**38**) and its minimum value is searched to obtain real area of contact and contact pressures.

Subsurface stress fields and the effect of friction on stresses have been analysed by Bailey and Sayles (**42**), Lee and Ren (**89**) and Yu and Bhushan (**43**). For friction effects, the contact pressure and contact locations (contact points) are considered not to be altered by the presence of surface friction. This assumption of no effect on contact pressure is strictly true only when the two bodies are smooth or have the same elastic contacts. In the case of a rough-on-rough contact, asperity interactions may invalidate this assumption, although for the case of a smooth-on-rough contact with approximately equal moduli, it is considered a reasonable assumption. For the case of friction present at the interface, tangential

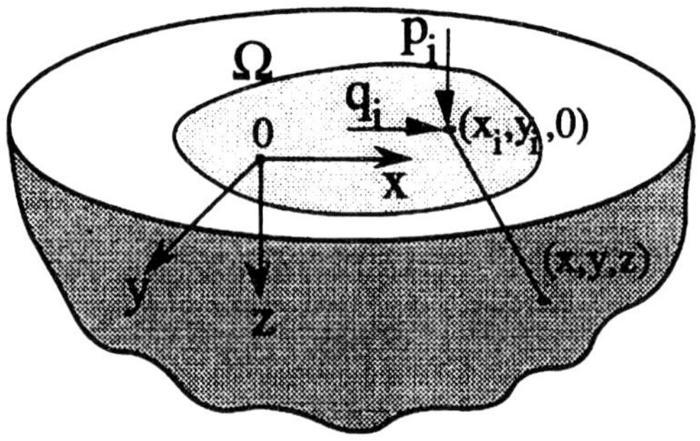

Fig. 20 Schematic of a half-space subjected to both surface normal and tangential forces over an arbitary area Ω. Concentrated normal loading and friction force on the area Ω are shown

force at each contact is equal to its contact pressure multiplied by the coefficient of friction.

Subsurface stress calculations are based on analyses for the stresses produced in a semi-infinite solid by surface normal forces (**85**) and tangential forces (**90**) over a rectangular element or patch. For analytical solutions to stresses and displacements, also see reference (**91**). At each point, the stress is calculated by integrating the influences of surface normal and tangential forces over the entire nominal contact area, see Fig. 20. As presented earlier, the nominal area is divided into rectangular elements (or patches), the stress being simply the sum of the effects of each element. Analytical expressions for stresses at any arbitary point inside the half-space due to a uniformly distributed load on a rectangular element are used to obtain the stress components for all subsurface locations.

Merriman and Kannel (**44**) and Cole and Sayles (**45**) developed a two-dimensional model to analyse the contact of layered elastic bodies (cylinder on a flat body) with real rough surfaces. They presented the contact pressure distributions for layers of varying thicknesses and elastic moduli. They reported that a soft surface layer will reduce the magnitude of peak contact pressures due to roughness.

3.4.1 *Simulation of cylindrical or spherical surface on itself or on a nominally flat surface*

A number of investigators have investigated contact in a two-dimensional (cylinder on itself or on a flat surface) or three-dimensional (sphere on itself or on a flat surface) fashion to simulate contacts in machine elements such as gears, cam–cam follower, and rolling element bearing components (**34**, **37–39**, **42**, **88**, **89**, **92**). Roughness is generally placed on one of the two surfaces and another surface is assumed to be smooth. This paper presents typical results of a spherical surface in sliding contact with a rigid flat surface, based on Lee and Ren (**89**). Figure 21 shows undeformed and deformed surface profiles of a rough sphere and contact pressure profiles of smooth (Hertzian) and rough surfaces. Note that the pressure profile for the rough surface has many pressure peaks with magnitudes far greater than that of the smooth surface contact. The highest pressure peaks are found near the centre of the Hertzian contact. For ductile materials, based on the von Mises shear strain energy criterion, yielding will occur when the von Mises stress $J_2^{1/2}$ reaches the yield stress in simple tension. Figures 22 and 23 show the lines of constant normalized von Mises stresses ($J_2^{1/2}/p_0$) for smooth and rough spheres, with and without interfacial friction. Note that most of the high stresses are concentrated near the top. Further down the surface, the stress field becomes similar to that

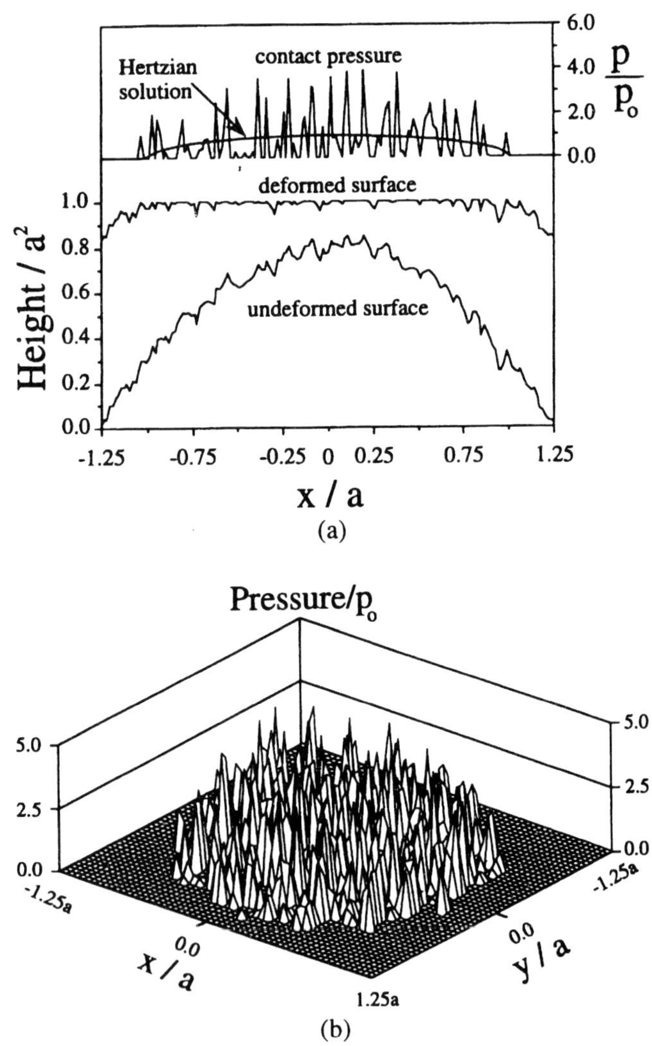

Fig. 21 (a) Undeformed and deformed surface profiles of rough sphere and contact pressures normalized with the maximum Hertzian pressure plotted along $y = z = 0$ line and (b) three-dimensional contact pressures map for a rough sphere in contact with a smooth flat surface or a smooth sphere (Hertzian) (**89**)

of the smooth surface solution. The stresses in the friction case (Fig. 23) are larger than in the frictionless case (Fig. 22). For the friction case, the location of the maximum stress has moved closer to the surface and it is no longer directly below the centre of the Hertzian conjunction. Figure 23c shows the von Mises stresses magnified at the surface for sliding contact of the rough surface. The peak values are much greater than the maximum stress (0.37) of the smooth surface.

Tensile radial stresses make brittle materials susceptible to ring cracks. Therefore, for brittle materials, tensile stress may be more important than maximum shear stress (von Mises stress) at the subsurface. Figure 24 shows three-dimensional plots of the normalized tensile

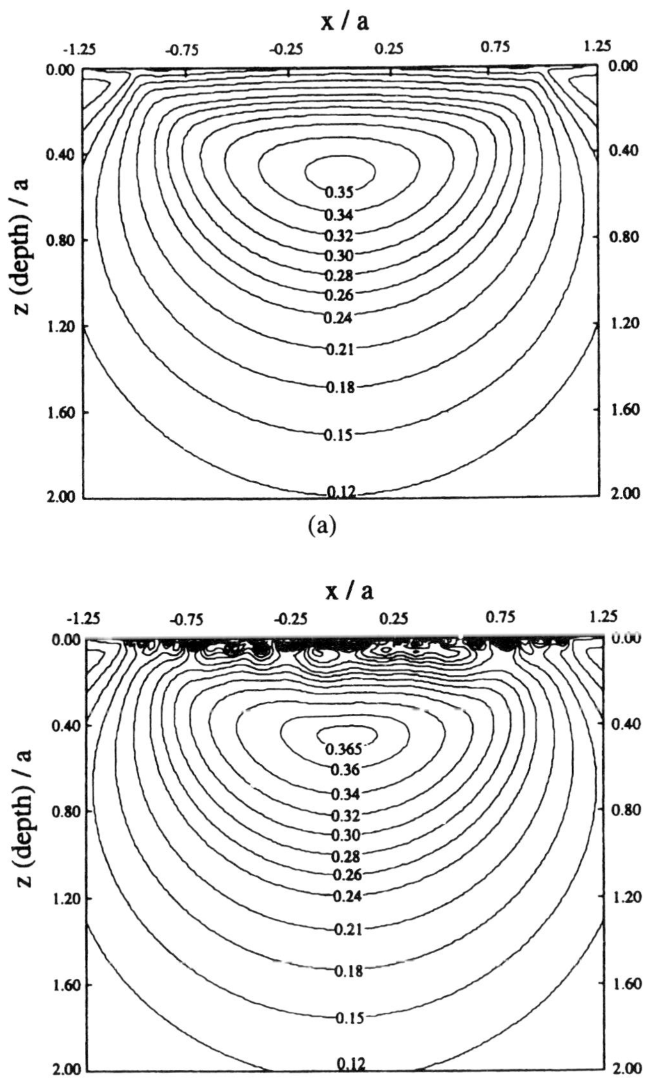

Fig. 22 Lines of constant normalized von Mises stresses ($J_2^{1/2}/p_0$) plotted on the $y=0$ plane for (a) a perfectly smooth sphere in contact with a smooth flat surface and (b) a rough sphere in contact with a smooth flat surface in frictionless contact **(89)**

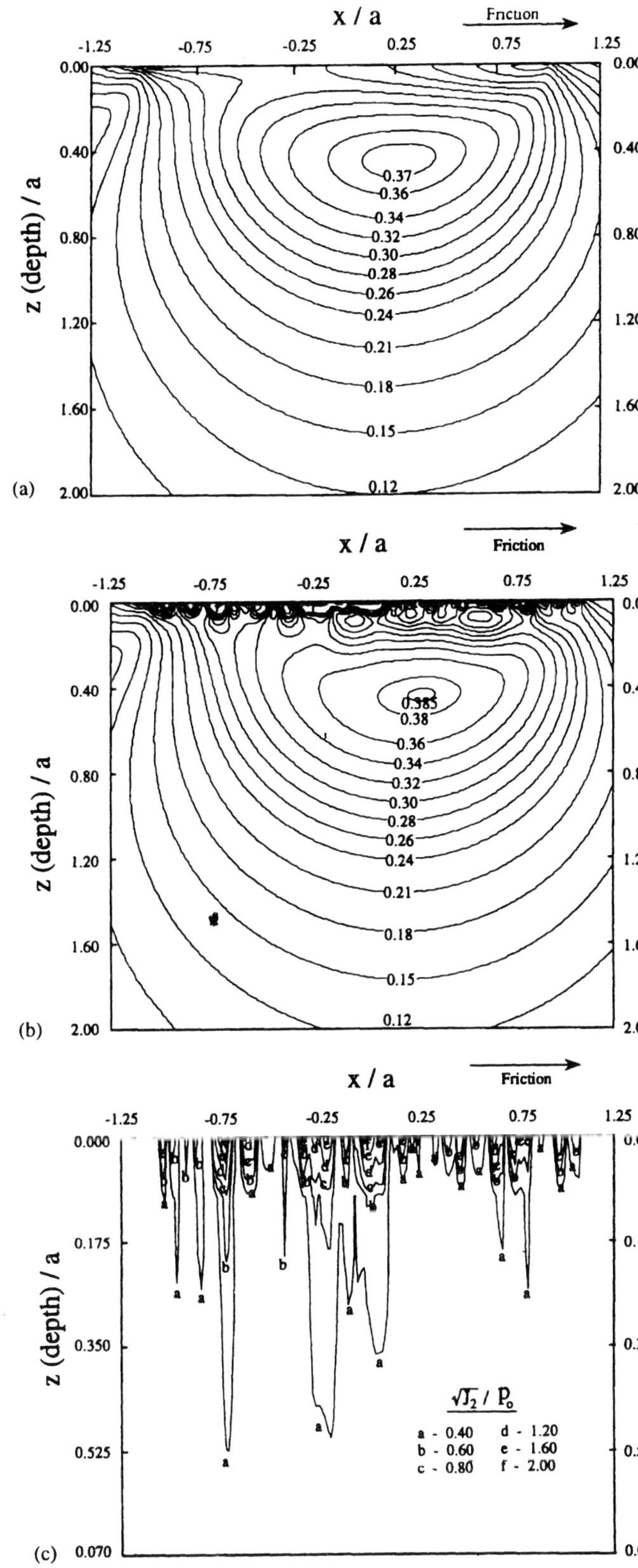

Fig. 23 Lines of constant normalized von Mises stresses ($J_2^{1/2}/p_0$) plotted on the $y=0$ plane for (a) a perfectly smooth sphere in contact with a smooth flat surface, and (b) a rough sphere in contact with a smooth flat surface, and (c) stresses close to the surface in the case (b), for the case of sliding contact with a coefficent of friction of 0.25 **(89)**

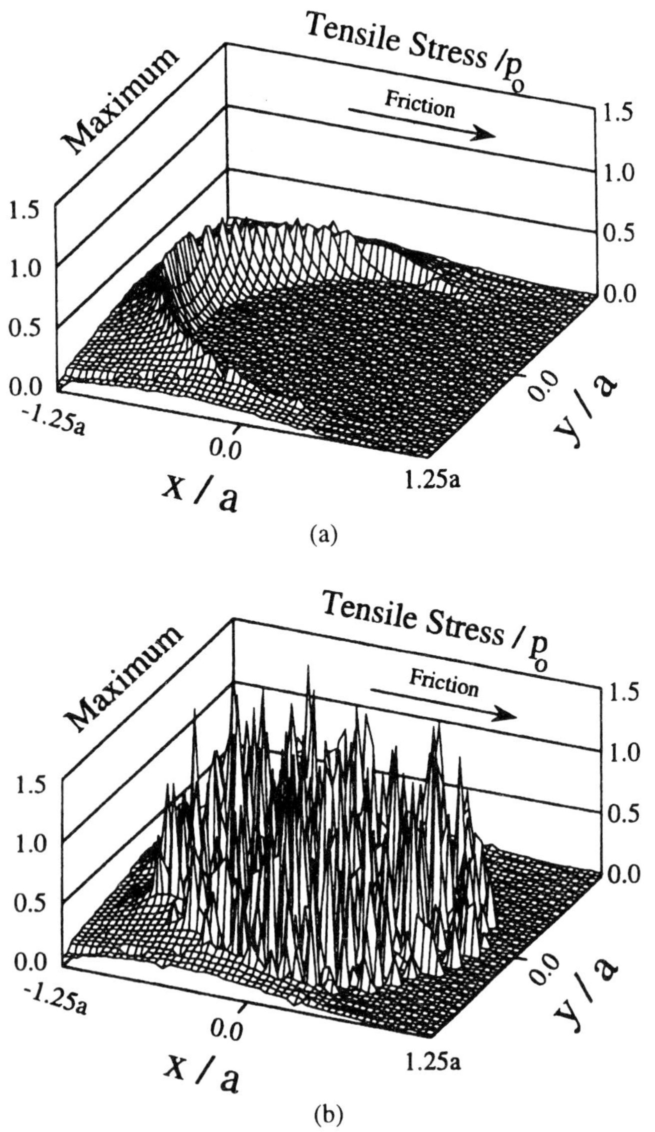

Fig. 24 Normalized tensile principal stress field for (a) a perfectly smooth sphere in contact with a smooth flat surface and (b) a rough sphere in contact with a smooth flat surface, in sliding contact with a coefficient of friction of 0.25 (**89**)

principal stresses for smooth and rough surfaces in sliding contact. For a smooth surface, the maximum value is located at the $x=-a$ and $y=0$. For frictionless contact, the stress field is symmetrical and surrounds the Hertzian conjunction. As the friction increases, the stress field becomes asymmetrical and the maximum grows larger. For the rough surface, there are many peaks which far exceed the maximum value of the smooth surface. The largest tensile stresses occur near the asperity contacts.

Ren and Lee (**93**) showed that the asperity aspect ratio has little effect on the real area of contact and contact pressures.

3.4.2 Simulation of two nominally flat surfaces

Tian and Bhushan (**38**), Bhushan and Chilamakuri (**57**), Poon and Bhushan (**94**, **95**), Yu and Bhushan (**43**) and Chilamakuri and Bhushan (**58**) used their numerical model to predict the real area of contact and contact stresses between a rigid flat and computer generated or measured rough surfaces. A digital filter technique combined with a fast Fourier transform (FFT) developed by Hu and Tonder (**96**) was used to generate three-dimensional Gaussian and non-Gaussian rough surfaces with given standard deviation of surface heights and correlation lengths. A typical computer generated, three-dimensional rough surface map is shown in Fig. 25. The heights of the surface map consist of 256 × 256 data points, and the profile data file has the similar data format of a measured surface map from an atomic force microscope or a non-contact optical profiler. An exponential autocorrelation function was used to generate the rough surface. The standard deviation of surface heights of the rough surface in Fig. 25 is 1 nm, the correlation length at which the autocorrelation value drops to 0.1 of that at origin is 0.5 μm, and the total scan size is 20 × 20 μm for the entire surface. Yu and Bhushan (**43**) calculated surface and subsurface stresses. Figure 26 shows the contact pressure map at a nominal pressure of 32.8 kPa, typical of magnetic head–disk interface. Contact pressure at the asperities is high, compared to the rest of the contact region. Under the applied load, a small number of contact spots (six) are obtained and the deformation is elastic for ceramic to ceramic contacts. Contours of von Mises stresses on the surface and subsurface ($y=11$ μm) on and close to the maximum von Mises stress plane for frictionless and frictional contacts are shown in Fig. 27. Maximum von Mises stress at both frictionless and frictional contacts occurs very close to the surface.

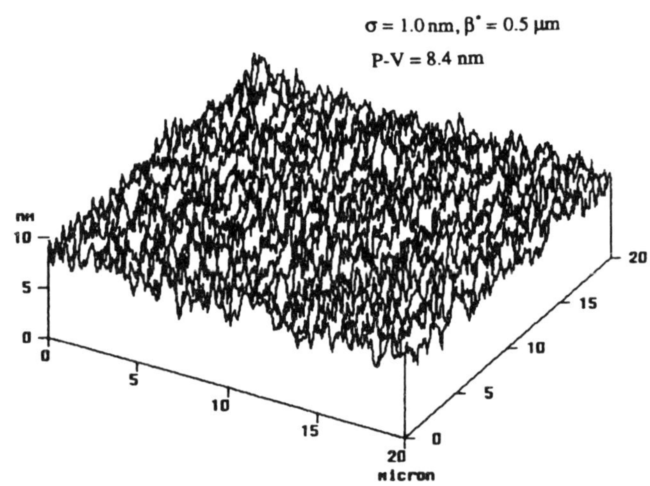

Fig. 25 Computer generated surface profile: standard deviation of surface heights (σ) = 1 nm, correlation length (β^*) = 0.5 μm, and total scan size = 20 × 20 μm^2

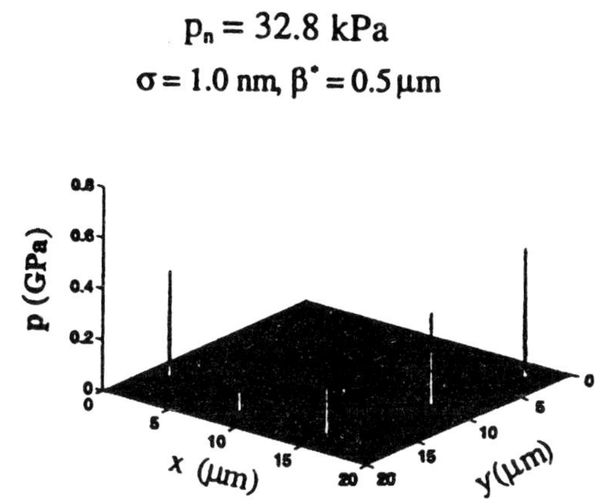

Fig. 26 Contact pressure map of a computer generated rough surface ($\sigma = 1$ nm and $\beta^* = 0.5$ μm) on a rigid smooth flat surface ($E^* = 100$ GPa) at a nominal pressure of 32.8 kPa (**43**)

The tensile principal stress distributions at and near the location of maximum von Mises stress, are shown in Fig. 28. A pulse of high stress appears on the surface ($z = 0$) at the location of maximum $\sqrt{J_2}$, $x = 16.5$ μm and $y = 11$ μm. The effect of friction is to increase the magnitude of the maximum stress but not to change the distribution.

Figure 29 shows the computed results of the real area of contact under various applied nominal pressures. As expected, both the real area of contact and the number of contact points increase with nominal pressure. The bearing area of the contact surface, corresponding to the same geometrical interference, is given by slicing off the surface for a given interference. Tian and Bhushan (**38**) showed that the bearing area greatly overestimates the real area of contact because elastic deformation is not included in the bearing area calculations. The relationship between the maximum contact pressure and the real area of contact and applied nominal pressure (load) for two rough surfaces is shown in Fig. 30. It can be seen that the real area of contact is linearly proportional to the applied pressure when the pressure is relatively small. This relationship between the applied normal pressure and the real area of contact is consistent with what is commonly observed in engineering practice, i.e. the friction force is linearly proportional to the applied pressure. Figure 30 also shows the maximum contact pressure under different applied loads. Contrary to common belief, it can be noted that the contact pressure increases with applied nominal pressure. Figure 31 shows the contact pressure maps and contours of von Mises stresses at the surface and subsurface ($y = 6.5$ μm) for two surfaces with $\sigma = 3$ nm and $\beta^* = 0.1$ μm, at a nominal pressure of 32.8 MPa (1000 × of nominal pressure used in previous figures). The real area of contact is low and the contact pressure and von Mises stresses are high for the surface with $\sigma = 3.0$ nm as compared to other surfaces with $\sigma = 1.0$ nm. For the surface with $\sigma = 3.0$ nm, contact occurs at 1570 contact points, among which contact pressure at the surface and von Mises stress very close to the surface are very high at six points and plastic deformation occurs. Since the maximum von Mises stress occurs very close to the surface, plastic deformation and consequently wear will occur on and close to the surface.

Poon and Bhushan (**97**) developed computer generated Gaussian surfaces with various values at σ and β^*. For these surfaces, in the elastic contact regime, they calculated real area of contact and contact pressure. They found that contact area is inversely proportional to, and the contact pressure is proportional to σ/β^*, as proposed by Onions and Archard (**12**). Therefore, plastic deformation can be avoided by decreasing σ or increasing β^* to promote better interface durability. Yu and Bhushan (**43**) studied the effect of σ, β^*, and friction on subsurface stresses. Increasing σ and friction, and decreasing β^* generally increased the magnitude of stresses.

Poon and Bhushan (**94, 95**) studied the effect of high-frequency roughness (sampling resolution of roughness measurement) on contact pressure and contact area. They measured rough surfaces of Ni–P coated Al–Mg (Ni–P) and glass-ceramic (GC) substrates used for construction of magnetic disks, by an AFM using different scan sizes $L \times L$ with $L = 64$, 32, 16, 8, 4, 1, 0.5, 0.25 and 0.1 μm, each containing 256 × 256 points. Therefore, the sampling interval τ decreases linearly with L. The variation of σ and β^* with scan size for the Ni–P and GC surfaces is shown in Fig. 32. From the curves of σ with L, σ of each surface initially increases with L and approaches a constant value when L reaches about 16 μm. This indicates that both surfaces contain a long-wavelength limit of about 16 μm. Therefore β^*, which represents the wavelength structure of a random surface, should approach a constant value as L reaches 16 μm. But Fig. 32 shows that β^* increases with L. The increase is due to increasing τ as L increases. As τ increases, the high frequencies in the surface will appear as a lower frequency in the digitized signal (**7**) and error in sampling occurs. The surfaces being studied contain predominantly submicron details and increasing τ will result in a loss of high-frequency information. As τ increases, high-frequency details of the original profile gradually disappear resulting in higher β^*. σ is the vertical parameter and is not sensitive to τ but generally increases with L. β^* is the spatial parameter affected by both τ and L.

Next, the effect of sampling interval on contact area and contact pressure is shown in Fig. 33. For both surfaces, contact area decreases and contact pressure increases as the sampling interval decreases. The contact pressure of the GC surface is always higher than that of the Ni–P surface. As the sampling interval becomes smaller, the maximum contact pressure can approach the hardness value. The subsurface stress with low interfacial friction is always higher than the surface stress and plastic deformation is expected to occur before surface stress reaches the hardness.

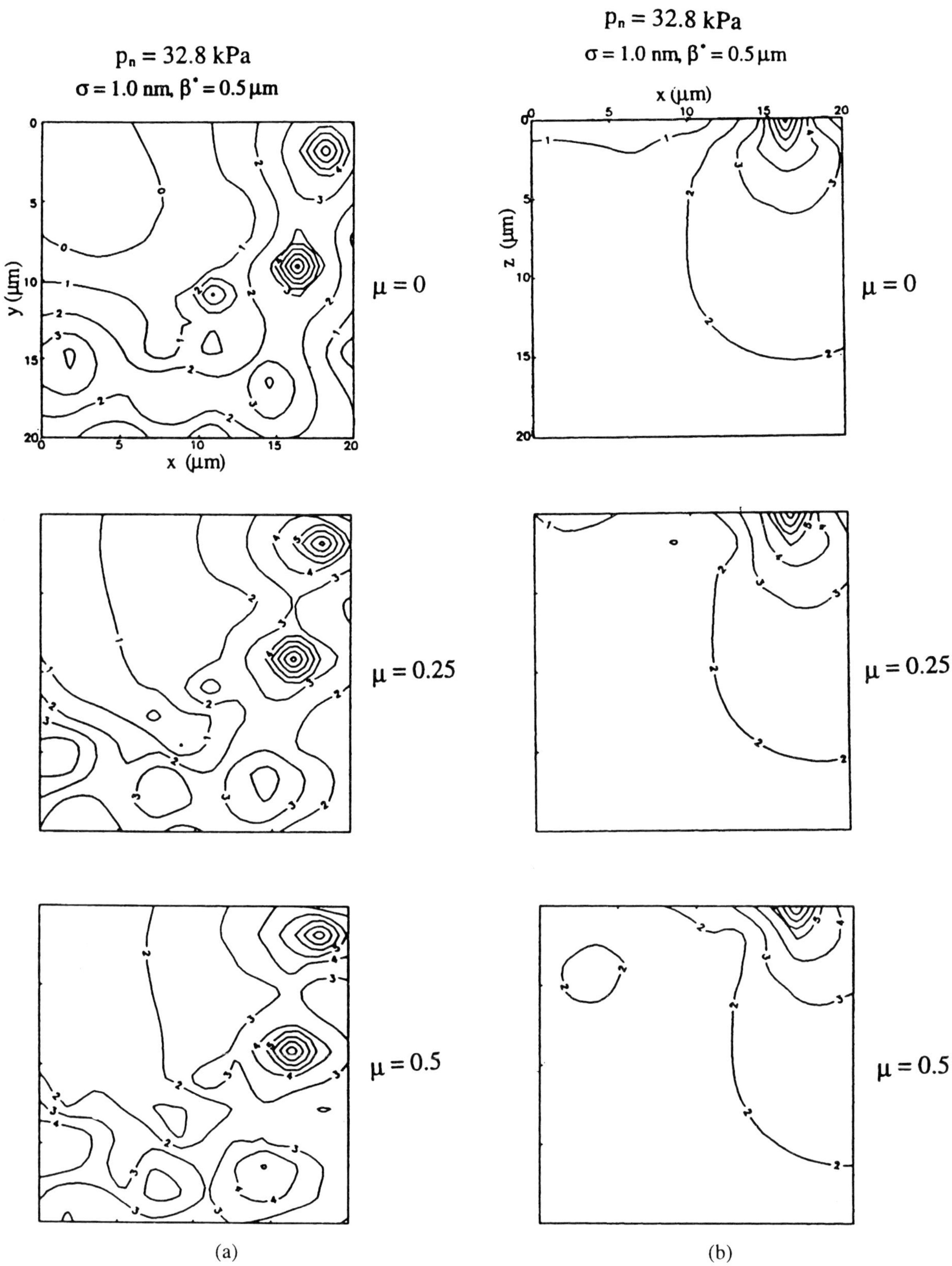

Fig. 27 Contours of von Mises stresses (a) on the surface and (b) and (c) in the subsurface (y = 11 μm) at $\mu = 0$, 0.25 and 0.50, for the case of a rough surface ($\sigma = 1$ nm, $\beta^* = 0.5$ μm) on a rigid smooth flat surface ($E^* = 100$ GPa) at a nominal pressure of 32.8 kPa. The contour levels are natural log values of the calculated stresses expressed in kPa (**43**)

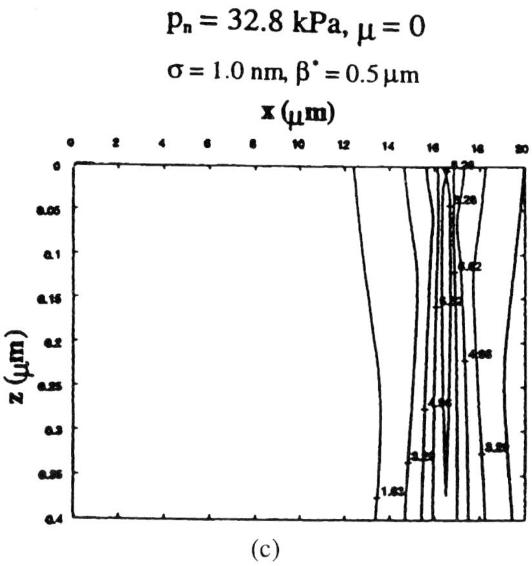

(c)

Fig. 27 (Continued)

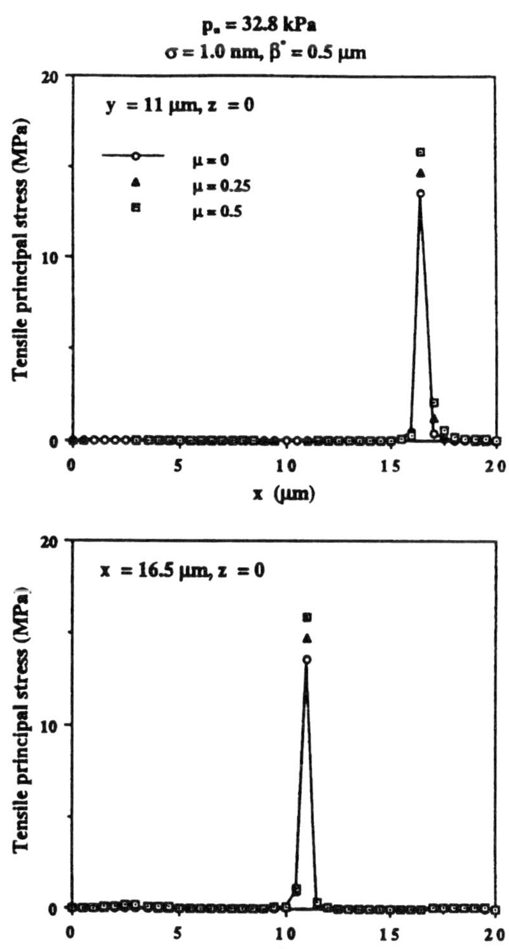

Fig. 28 Distributions of tensile principal stresses on the surface along x and y axes at and near the location of maximum von Mises stress ($x = 16.5$ μm, $y = 11$ μm) for the case of a rough surface ($\sigma = 1$ nm, $\beta^* = 0.5$ μm) on a rigid smooth flat surface ($E^* = 100$ GPa) at a nominal pressure of 32.8 kPa (**43**)

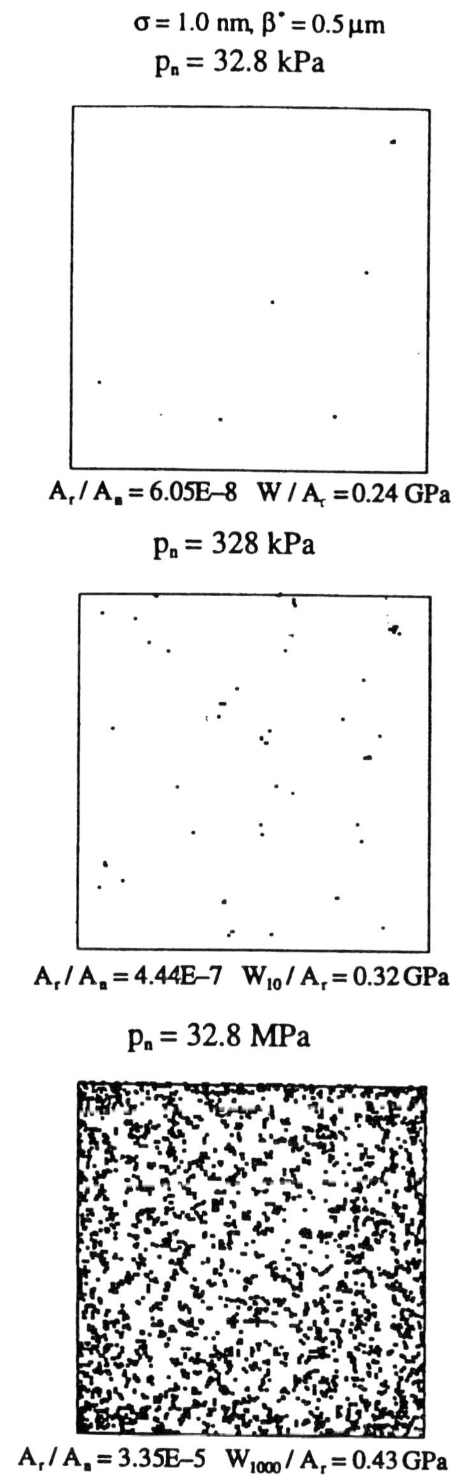

Fig. 29 Two-dimensional images of real area of contact between a computer generated rough surface ($\sigma = 1$ nm, $\beta^* = 0.5$ μm) on a rigid smooth flat surface ($E^* = 100$ GPa) at three different nominal pressures (**43**)

From Fig. 33, the asperity contact pressure of both surfaces reaches the hardness of the softer material as τ decreases. This implies that the asperity contact of real surfaces starts from plastic deformation, the scale of

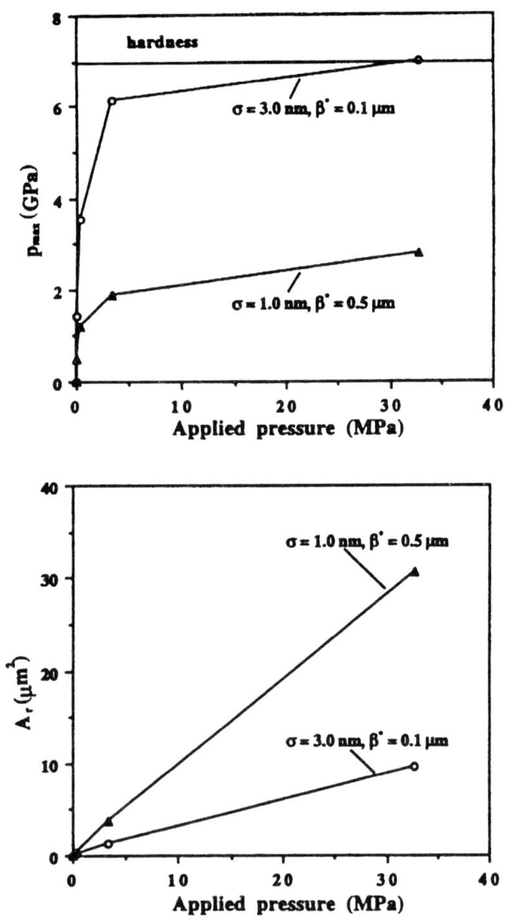

Fig. 30 Predicted maximum contact pressure and real area of contact as a function of applied nominal pressure for the case of two rough surfaces on a rigid smooth flat surface ($E^* = 100$ GPa) (**43**)

which depends on different surface topographical structures. The GC surface has a larger scale of plastic deformation because it has higher frequency structures, compared to the Ni–P surface. Thus, roughness distribution plays a key role in plastic deformation on asperity contacts.

3.5 Dynamic friction modelling

In dynamic friction models, a constant value of coefficient of friction is assumed and perturbations under dynamic conditions are investigated (**46**, **47**, **98**, **99**). Oden and Martins have presented continuum models for solving a large class of dynamic frictional phenomena and computation methods for their analyses. Their particular interests were adequately to predict stick–slip motion, frictional damping in structural dynamics, and sliding resistance. Through experimental observations and theoretical analyses, they identified the key features that the interface models should exhibit and developed interface models to reflect these features. Their considerations of the normal deformability of the interface asperities, non-linear boundary effect, and generalized friction law are different from the classical approaches of the statistical description of the interface geometry and normal contact, non-penetration unilateral contact condition, and Coulomb's friction law. In their interface models, as shown in Fig. 34, a continuous material body β is considered to be in contact with another material β_1 over a contact surface $\Gamma_c \subset \partial\beta$. $\boldsymbol{u}$ is denoted the displacement of particles $X \in \beta$ relative to a fixed reference configuration which is located at the undeformed body. It is assumed that the contact surface Γ_c has an exterior normal vector $\boldsymbol{n}$ and that the actual interface (asperities, oxide film, gas, work-hardened material, etc.) is initially of thickness t_0. The initial gap between Γ_c and the other contact surface is defined as the distance and denoted as g, which is measured along a line normal to Γ_c between the highest asperities of these surfaces in the reference configuration. The interface thickness after deformation is denoted as t and the actual displacement of Γ_c in the direction of $\boldsymbol{n}$ is $u_n = \boldsymbol{u} \cdot \boldsymbol{n}$. Thus, the approach of the contact surfaces is

$$\delta = t_0 - t = (u_n - g)_+ \tag{44}$$

This definition is valid for a rigid body β_1 bounded by an ideally flat surface. Applications of this model to cases of deformable and rough surfaces have been carried out by Kikuchi and Oden (**100**). Let σ_{ij} denote the Cartesian components of the Cauchy stress tensor at a particle X in β, then the normal stress developed in Γ_c is

$$\sigma_n = \sigma_{ij} n_i n_j \tag{45}$$

The relationship between σ_n and δ is obtained by plotting the variation in the normal force versus the penetrating approach as obtained from experimental results, which appears to be a power-law relation as given by

$$-\sigma_n = c_n (u_n - g)_+^{m_n} \tag{46}$$

where c_n and m_n are material parameters determined from interface compression experiments. If there is friction present on the contact interfaces, the following friction law applies, as σ_T denotes the tangential stress on the contact surface:

$$\begin{aligned} & u_n \leqslant g \Rightarrow \sigma_T = 0, \\ & u_n > g \Rightarrow |\sigma_T| \leqslant c_T (u_n - g)_+^{m_T}, \quad \text{or} \\ & \qquad |\sigma_T| < c_T (u_n - g)_+^{m_T} \quad \text{if } \dot{u}_T = U, \text{ or} \\ & \qquad |\sigma_T| = c_T (u_n - g)_+^{m_T} \\ & \qquad\qquad \text{if } \exists \lambda \geqslant 0,\ \dot{u}_T - U = -\lambda \sigma_T, \end{aligned} \tag{47}$$

where c_T and m_T are material constants depending on interface properties, $\dot{u}_T$ is the tangential velocity of material particles on Γ_c, and U is the prescribed tangential

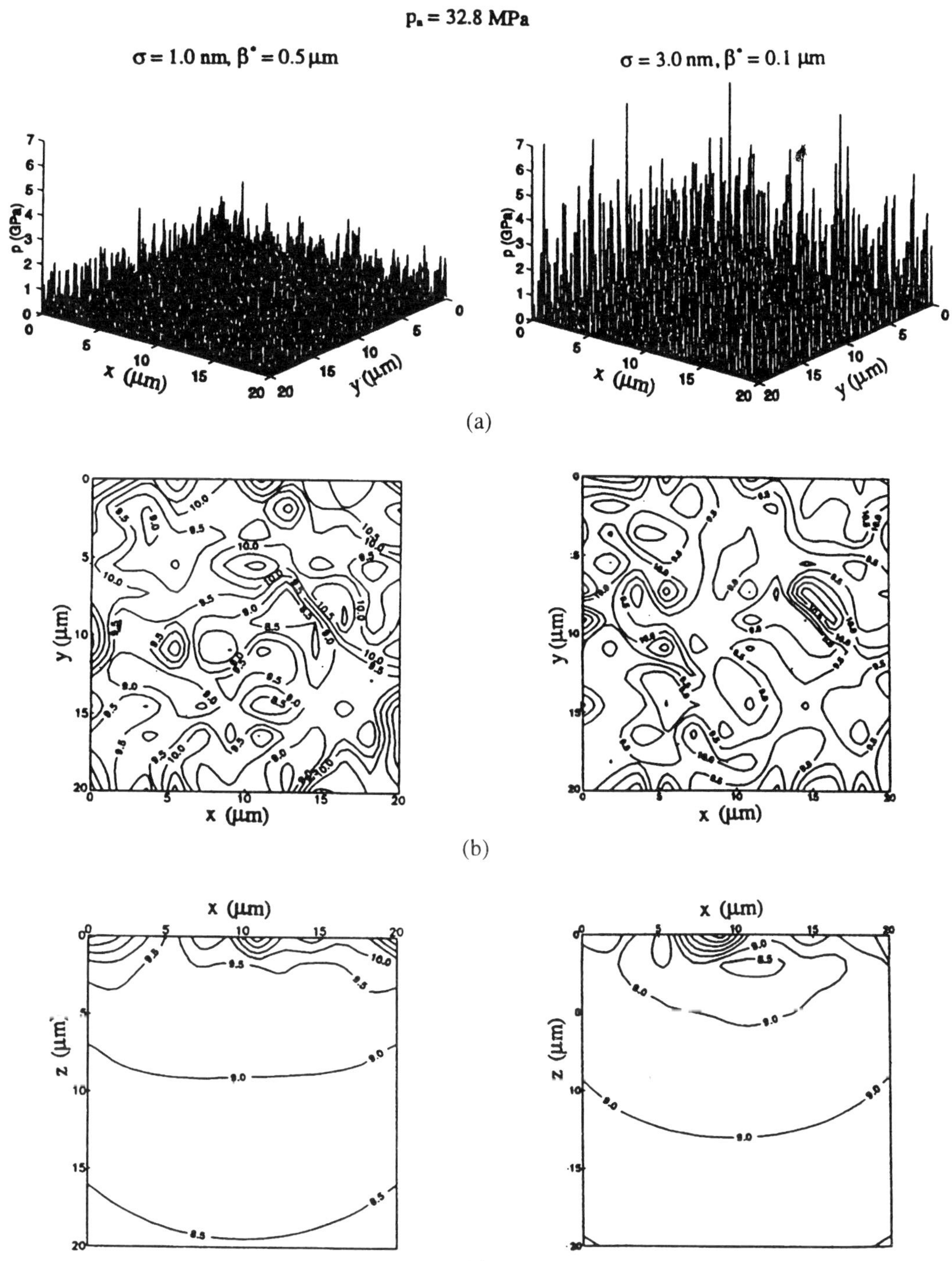

Fig. 31 (a) Contact pressure maps and (b) contours of von Mises stresses on the surface and (c) in the subsurface at $y = 6.5$ μm at a nominal pressure of 32.8 MPa for two rough surfaces on a rigid smooth flat surface ($E^* = 100$ GPa) (**43**)

velocity of β_1. This friction law, equation (47), is a generalization of Coulomb's friction law and allows for possible deviations from Amontons' laws, i.e. a possible dependence of the friction coefficient μ on the normal stress according to

$$\mu = C|\sigma_n|^{\alpha} \tag{48}$$

if the normal dissipate effects are negligible (denoted as $b_n \approx 0$), where $\alpha = m_T/m_n - 1$ and $C = c_T/c_n^{m_T/m_n}$. If $m_T = m_n$ (and $b_n \approx 0$) the usual Coulomb's law of friction is recovered with $\mu = c_T/c_n$.

Rabier *et al.* (**101**) further expressed Oden and Martins' (**46**) phenomenological laws of contact interfaces in the following forms for elastostatic problems:

$$\sigma_n = -c_n \delta^{m_n} \tag{49}$$

$$\sigma_T = c_T \delta^m \acute{A} \quad \text{(if friction is present)} \tag{50}$$

where $\acute{A}$ is a unit tangential vector field parallel to the

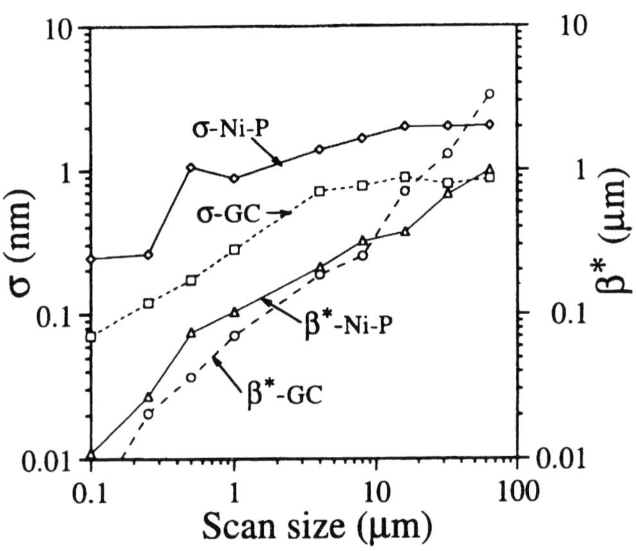

Fig. 32 Variation of standard deviation of surface heights and correlation length with scan size (**93**)

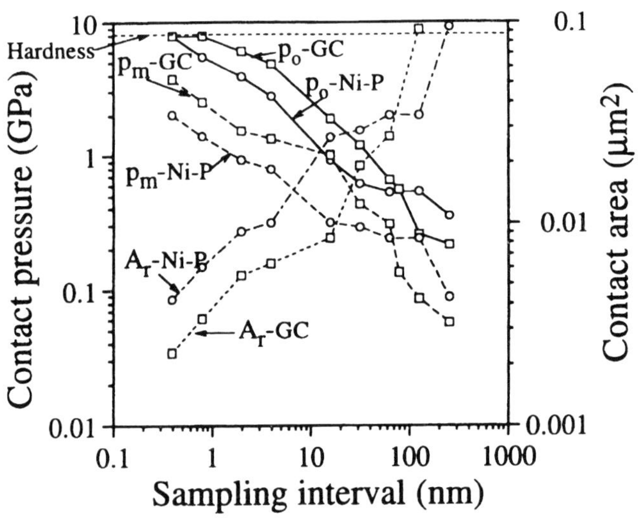

Fig. 33 Variation of mean contact pressure (p_m), maximum contact pressure (p_0) and contact area with sampling interval for Ni–P and GC surfaces (**93**)

sliding velocity. They formulated a class of contact problems in elastostatics according to the governing laws of equations (49) and (50), and showed Oden and Martins' models did provide a mathematical framework that admits the establishment of existence and uniqueness results under appropriate hypotheses on the regularity and smallness of the data, particularly the smallness of the coefficient of friction for prescribed external loads.

Martins *et al.* (**47**) continued Oden and Martins' work (**46**) to study small speed sliding phenomena. They reviewed the relationship between static and kinetic friction, examined the role played by the dynamic properties of experimental apparatus, and studied the importance of normal degree-of-freedom under frictional sliding. The time and dissipate effects are included in the constitutive equation and frictional law as given respectively in the following:

$$\sigma_n = -c_n \delta^{m_n} + b_n \delta^{ln} \delta_{,t} \tag{51}$$

$$-\sigma_T \in c_T \delta^{m_T} \operatorname{sgn}(v) \tag{52}$$

where the subscript t denotes differentiation with respect to time and sgn (v) denotes the multivalued relation

$$\zeta \in R \rightarrow \operatorname{sgn}(\zeta) = \begin{cases} 1 & \text{if } \zeta > 0 \\ [-1, 1] & \text{if } \zeta = 0 \\ -1 & \text{if } \zeta < 0 \end{cases} \tag{53}$$

They found that contrary to classical interpretations, no distinction between the static and kinetic coefficient of friction is required for a body to have an accelerated motion. Their other conclusion is that the experimentally observed dependence of static coefficient of friction on the time of stationary contact, distinctions between the static and kinetic coefficient of friction, and dependence of the latter coefficient on sliding speed at small speeds are not necessarily intrinsic properties of dry metallic interfaces. Physically small external perturbations and geometric and dynamic properties of an experimental apparatus may lead to these experimental observations even when the coefficient of friction is constant and independent of sliding speed or time of contact.

4 MULTIPLE ASPERITY WET CONTACTS

With the presence of a thin liquid film with a small contact angle (wetting characteristics) such as a lubricant or an adsorbed water layer at the contact interface, curved menisci form around contacting and non-contacting asperities due to surface energy effects. The attractive meniscus force arises from the negative Laplace pressure inside the curved meniscus as a result of surface tension (**48**–**50**). The product of this pressure difference and the immersed surface area of the asperity is an attractive (adhesive) force, referred to as the meniscus force. This intrinsic attractive force may result in high static friction (stiction), kinetic friction and wear. This problem is particularly important in the computer data storage industry (**3**, **51**).

In order to calculate the meniscus force, different meniscus geometries and assumptions have been used by various authors. Figure 35 presents the existing models for calculation of the meniscus force between a sphere and a flat, either with a liquid droplet (cases A and B) or with a uniform liquid film (case C). In the figure, f_m is the force due to a single meniscus along the normal direction. For multiple menisci, the total meniscus force is determined by summing the meniscus forces over all

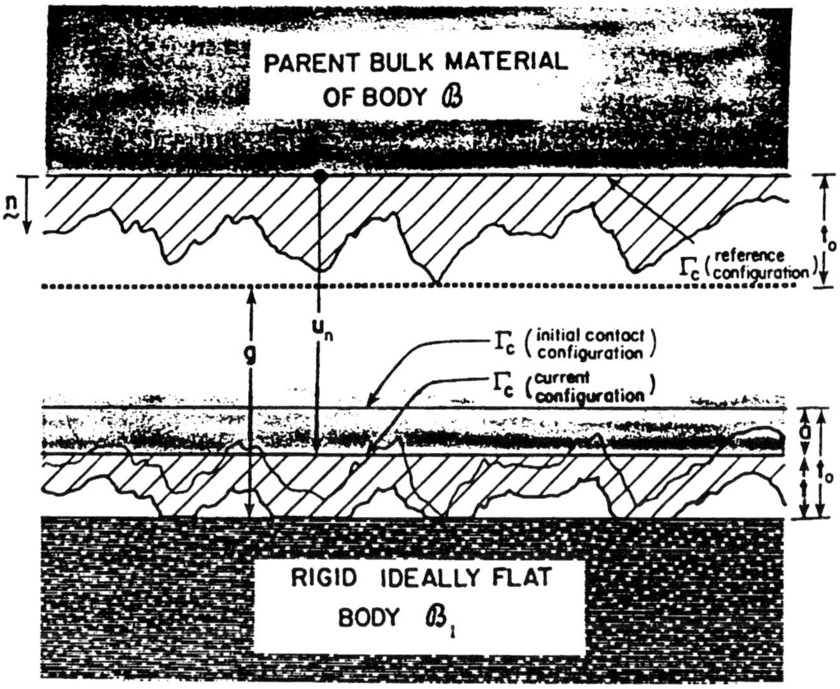

Fig. 34 Initial gap, normal displacement and penetrating approach at the contact surface Γ_c **(46)**

Case	Meniscus Geometry	Meniscus Force f_m and Major Assumptions	Sources
A	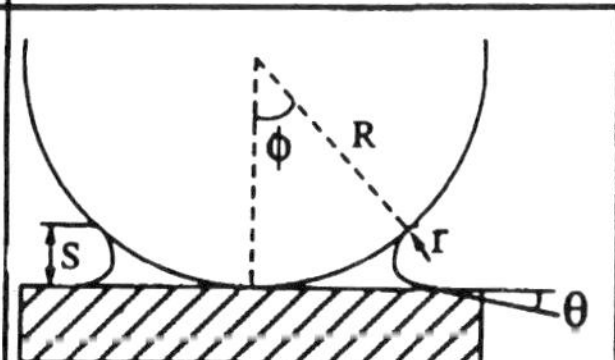	$f_m = 4\pi R\gamma\cos\theta$ (small ϕ, uniform r, $s = 2r\cos\theta$)	Bowden and Tabor (1950)
B	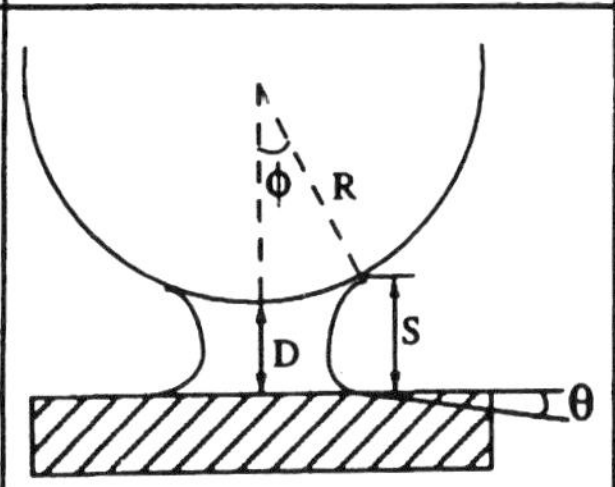	$f_m = \frac{4\pi R\gamma\cos\theta}{1+\frac{D}{s}}$ (small ϕ, uniform r, constant meniscus volume)	Israelachvili (1992)
C	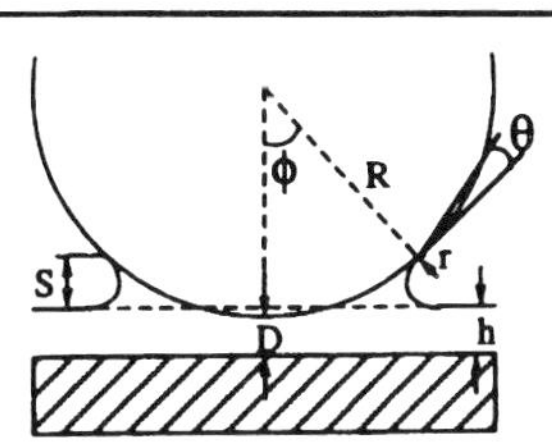	$f_m = 2\pi R\gamma(1+\cos\theta)$ (small ϕ, uniform s and h)	Gao et al. (1995)

Fig. 35 Different meniscus geometries and formulas for calculation of the meniscus force for a single asperity in contact with liquid of surface tension γ. In cases A and B, the contact angle θ is assumed to be the same for the liquid in contact with either surface; if the contact angles are different, $\cos\theta$ should be replaced by $(\cos\theta_1 + \cos\theta_2)/2$. In case C, θ is the contact angle of the liquid with the sphere. Note that all meniscus forces are attractive and directed normal to the meniscus contact

menisci formed at the interface. However, the contact statistics need to be determined.

The total normal force on the wet interface is the externally applied normal force plus the meniscus force. Therefore, during sliding, the force required to inititiate or sustain sliding is equal to the sum of the intrinsic (true) friction force, F_i, and the stiction force, F_s, the latter being a combination of the friction force due to the meniscus and the viscous effects (**3**)

$$F = F_i + F_s = \mu_r(W + F_m) + F_v \tag{54}$$

where μ_r is the true coefficient of friction in the absence of meniscus, and smaller than the measured value of $\mu = F/W$. The sum of W and F_m is the total normal load. F_m is the meniscus force in the normal direction, and F_v is the viscous force in the sliding direction. The friction force ($\mu_r W$) depends on the material properties and surface topography, while F_m depends on the roughness parameters as well as the type of liquid and its film thickness.

For static friction calculations, the viscous effect can be neglected; therefore, the measured coefficient of static friction μ_s including the effect of the meniscus force is given by

$$\mu_s = \frac{F}{W} = \mu_r\left(1 + \frac{F_m}{W}\right) \tag{55}$$

where $\mu_r F_m$ is the friction force due to liquid-mediated adhesion.

4.1 Statistical analysis of contacts

Several models have been developed to predict meniscus forces developed at a wet interface using a statistical approach (**52**–**55**, **22**). A schematic of a rough surface in contact with a smooth surface with a continuous liquid film on the smooth surface, is shown in Fig. 36. Note that both contacting and near-contacting asperities wetted by the liquid film contribute to the total meniscus force. The Greenwood and Williamson (**10**) approach described earlier, is used to model the contact. The peak heights are assumed to follow a Gaussian distribution function and peak radii are assumed to be constant. In general, given the peak-height distribution function $p(z)$, the mean peak radius (R_p), the thickness of liquid film (h), the liquid surface tension (γ), and the contact angle for the liquid in contact with the rough surface (θ), the total meniscus force (F_m) at the sliding interface is obtained by summing up the meniscus forces from all individual contacting and non-contacting asperities that form menisci over the nominal contact area, shown in Fig. 36a,

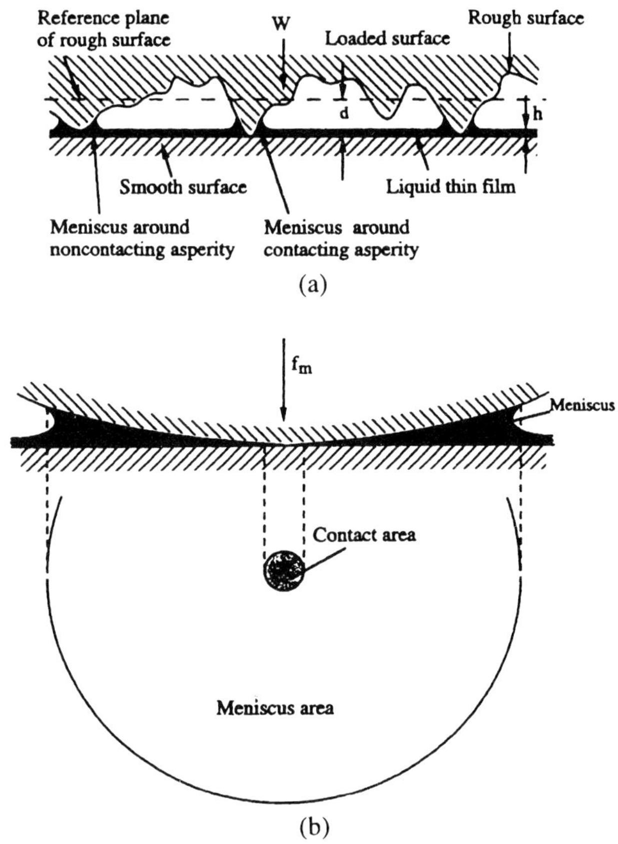

Fig. 36 (a) Schematic for a rough surface in contact with a flat surface with a liquid film, and (b) schematic of contact area and meniscus area in a contacting asperity

$$F_m = \int_{d-h}^{\infty} f_m N_0 p(z)\,dz = 2\pi R_p \gamma(1+\cos\theta) N_0 \int_{d-h}^{\infty} p(z)\,dz \tag{56}$$

where N_0 is the total number of peaks in the nominal contact area. The interplanar separation, d, is determined from

$$W + F_m = \frac{4}{3} E^* R_p^{1/2} N_0 \int_{d}^{\infty} (z-d)^{3/2} p(z)\,dz \tag{57}$$

An iterative numerical approach is used to solve equations (56) and (57).

It is evident that the maximum meniscus force can be obtained by setting h very large so that the integral in equation (56) approaches its maximum value of unity. Therefore the maximum possible meniscus force is

$$F_{max} = 2\pi R_p \gamma(1+\cos\theta) N_0 \tag{58}$$

regardless of the distribution function of peak heights. Conversely, when the film thickness h is very small, i.e.

less than a molecular layer thick, F_m is zero since no meniscus can be formed and the problem reduces to dry contact.

Gao *et al.* (**54**) showed that the meniscus force increases as a function of liquid film thickness (h). For a given film thickness, the meniscus force decreases with an increase in standard deviation of peak heights (σ_p) and it increases with an increase of radii of peaks (R_p) and number of peaks (N_0), Fig. 37. Gao *et al.* (**54**) and Chilamakuri and Bhushan (**82**) produced the design curves to provide a relationship between critical σ_p and R_p for minimum stiction and no plastic deformation.

Meniscus force analyses discussed so far consider only what happens at equilibrium. When a body first comes to rest on another body, in the presence of a liquid film, the interface is not in equilibrium. In the early stages of meniscus formation, meniscus force increases as a result of flow of liquid into the low-pressure region created inside the curved liquid–air interface causing them to grow until the Laplace pressure is sufficiently reduced to match the disjoining pressure of the liquid remaining on the surface. For the entire interface to reach equilibrium, liquid from the wet body surrounding the other body has to flow into the interface until the capillary pressure equals the disjoining pressure of the liquid film in the rest of the wet body. Bhushan *et al.* (**55**) developed a kinetic meniscus model to predict the time dependence of static friction. In this analysis, the liquid flow at the contacting and near contacting asperities is calculated based on Laplace pressure. The liquid meniscus at equilibrium is obtained. The meniscus growth with time is used in this analysis to calculate the time dependent static friction. They found that the relative meniscus force reached a maximum at a rest time called the 'equilibrium' time. The equilibrium time is found to increase with an increase in liquid film thickness, liquid viscosity and average asperity radius. Growth rate of the relative meniscus force is inversely proportional to the liquid viscosity. This model explains the time dependent stiction encountered in magnetic disk drives (**3**).

Kotwal and Bhushan (**22**) developed a statistical model for non-Gaussian surfaces with skewness and kurtosis. They used a Pearson system of frequency curves, based on the method of moments, to obtain analytical expressions for the probability density functions for distributions with various skewness and kurtosis values. These probability density functions were then used to calculate the real area of contact, contact pressure and meniscus forces. They reported that non-Gaussian surfaces with a range of positive skewness (between 0.3 and 0.7) and a high kurtosis (greater than 5) significantly lower the real area of contact and meniscus forces and these surfaces are somewhat insensitive to the liquid film as far as magnitude of meniscus force is concerned. Further discussion will be presented in the next section.

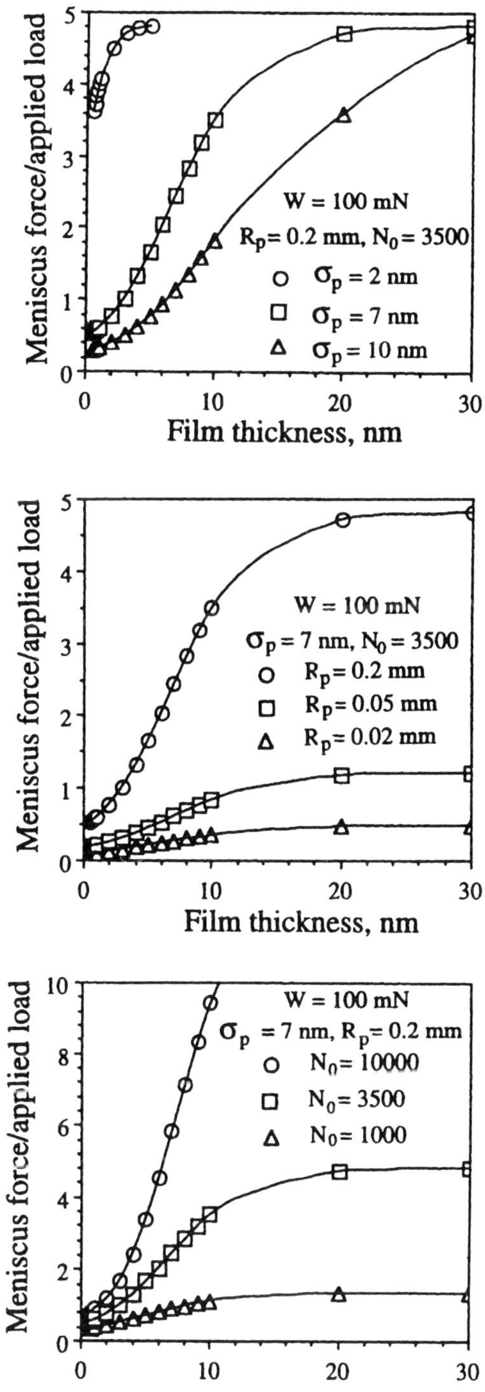

Fig. 37 Ratio of the meniscus force to applied load (F_m/W) as a function of water film thickness at different σ_p, R_p and N for magnetic head–disk interface (**54**)

4.2 Numerical three-dimensional contact models

A first numerical meniscus model was developed by Tian and Bhushan (**56**). They were first to modify the classical meniscus theory of a single asperity (sphere on a flat) contact presented earlier, to include the effect of multi-asperity contacts with a pre-existing ultra-thin liquid film during contact of two rough surfaces. They modified the numerical contact model based on the variational approach presented earlier, to take into account the effect of meniscus forces.

For a meniscus formed around a pair of contacting asperities, the pressure difference across any meniscus surface, independent of asperity sizes and shapes, can be determined by the Laplace equation (**49**)

$$\Delta p = \gamma\left(\frac{1}{r_1} + \frac{1}{r_2}\right) \tag{59}$$

where Δp is the pressure difference between the pressures across the liquid and air interface and is commonly referred to as the Laplace pressure (or capillary pressure) in the liquid; and $1/r_1$ and $1/r_2$ are the meniscus curvatures along the mutually orthogonal planes. If $|r_1| \ll |r_2|$ (note that this condition is always satisfied for contacting asperities at the contact interface of rough surfaces where asperity height is several orders of magnitude smaller than asperity radius), then equation (59) becomes

$$\Delta p = \gamma / r_1 \tag{60}$$

The meniscus force (normal force) due to the pressure difference is then given by

$$F_m = \iint_\Omega \Delta p(x, y)\, d\Omega = \gamma \iint_\Omega \frac{1}{r_1}\, d\Omega \tag{61}$$

where Ω is the projected area of meniscus enclave which intersects the upper contacting asperity at meniscus height, s. For multiple isolated menisci scattered over the whole contact interface, Ω should be the sum of the projected area of each meniscus enclave.

To solve equation (61), it is necessary to know both the meniscus radius at different locations and the projected area of the meniscus enclave. These two parameters are in turn functions of the shape (which determines meniscus radius) and the volume of the meniscus. The determination of the exact shape and volume of the meniscus in the thin film situation shown in Fig. 36 is not tractable. Tian and Bhushan (**56**) adopted a relatively simplified approach as follows. If the effective section of a meniscus is approximated as a section of arc (which is true if both gravitation and van der Waals forces are negligible), then the relationship between the effective radius of the meniscus and the meniscus height is given by

$$\bar{r} \approx \frac{s}{1 + \cos\theta} \tag{62}$$

where $\bar{r}$ is the effective radius of the meniscus. If it is further assumed that the variation of meniscus height around an irregular asperity is small (this is probably the case), then the meniscus force in equation (61) is reduced to

$$F_m = \gamma \iint_\Omega \frac{1}{\bar{r}}\, d\Omega \approx \gamma(1 + \cos\theta) \iint_\Omega \frac{1}{s}\, d\Omega$$
$$\approx \gamma(1 + \cos\theta)\frac{\Omega}{\bar{s}} \tag{63}$$

where $\bar{s}$ is the mean meniscus height for an irregular asperity. It can be noted that, for rough surfaces, Ω is a function of mean meniscus height. For a typical spherical contacting asperity, a simple relation exists between Ω and s ($\Omega \approx 2\pi Rs$) (**50**), and equation (62) reduces to the classical expression. However, for randomly distributed menisci around irregular contacting asperities, no relationship between Ω and s is available.

Figure 38 shows some concepts for selecting the projected meniscus area. (In Fig. 38a, for convenience, the thin film of liquid is put on the smooth surface instead of on the rough surface. This is because the two situations are numerically equivalent.) Menisci will form around the contacting and the non-contacting asperities which touch the liquid. The areas where asperities touch the liquid are referred to as wetted areas (as shown in Fig. 38a). To determine the projected meniscus area, $\Omega(s)$, it is necessary first to obtain the cross cut area of the rough surface at a given mean meniscus height, $\bar{s}$. The cross cut area at the mean meniscus height is then

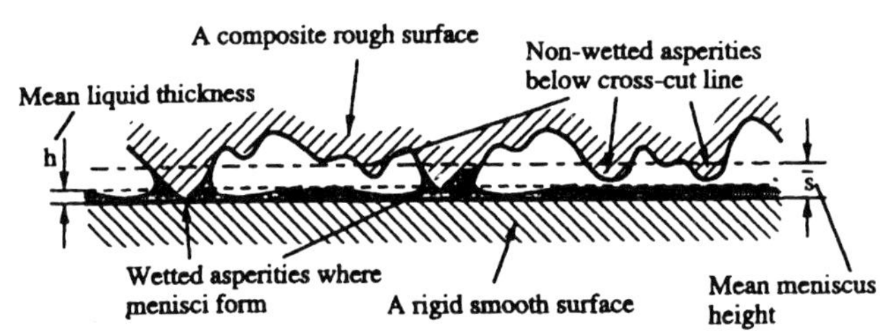

(a)

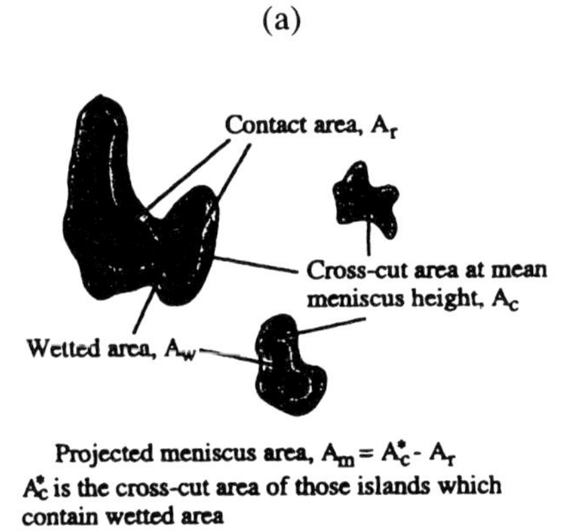

(b)

Fig. 38 (a) A schematic diagram of a rough surface in contact with a smooth surface with a liquid film, and (b) a schematic diagram of contact area, wetted area and cross cut area at mean meniscus height for meniscus bridges around contacting and non-contacting asperities (**56**) grouped into individual area islands. There is no geometrical link among individual islands. Those islands which overlap the wetted area are selected as the projected meniscus area, $\Omega(\bar{s})_i$, and those islands which do not overlap the wetted area are discarded. The aim of this step is to eliminate those areas which do not touch the liquid; therefore, no meniscus forms around non-wetted asperities. The total projected meniscus area is equal to

$$A_m = \sum_i \Omega(\bar{s})_i = \sum_i (A_c^{*i} - A_r^i) \tag{64}$$

where A_m is the total projected meniscus area, A_c^{*i} is the area of the ith island which contains wetted asperities from the cross cut area at the mean meniscus height, A_r^i is the real area of contact located inside the ith island. Finally, the meniscus force is given by equation (63) replacing Ω by A_m,

$$Fm \approx \frac{\gamma}{\bar{s}}(1 + \cos\theta) \sum_i (A_c^{*i} - A_r^i) \tag{65}$$

Tian and Bhushan (**56**) developed a computer program to select the individual meniscus island and to compute the total projected meniscus area according to the method just described. Two parameters, meniscus height and liquid film thickness, are required for the analyses, which can be obtained for a given relative humidity. Water film thickness for a given relative humidity can be measured. For capillary condensation from a humid environment, the meniscus radius at equilibrium, the so-called Kelvin radius r_K, is related to relative humidity (relative vapour pressure), by the Kelvin equation (**49**)

$$r_K = \frac{\gamma V}{RT \log RH} \tag{66}$$

where RH is the relative humidity in fraction, V is the molar volume, R is the gas constant and T is the absolute temperature. $\gamma V/(RT) = 0.54$ nm for water at 20°C. For RH = 1, $r_K = \infty$. For a given meniscus radius, equation (62) is then used to calculate the meniscus height. The adsorbed water film thickness for a given relative humidity can be obtained from an empirical relation (**102**),

$$h = h_1(\mathrm{RH}) + h_2 \exp[\alpha(\mathrm{RH} - 1)] \tag{67}$$

where $h_1 = 0.3$ nm, $h_2 = 0.5$ nm, $\alpha = 20$ and RH is relative humidity fraction ranging from 0 to 1 for a carbon-coated disk surface. For a given liquid film thickness, meniscus height is a function of liquid film thickness and needs to be estimated.

In the numerical wet model developed by Tian and Bhushan (**56**), elastic–plastic dry contact of rough surfaces is first analysed. In the next step, a liquid film of known mean thickness is introduced over the deformed rough surfaces. Wetted areas are determined by selecting the areas where asperities of both contacting surfaces touch the liquid. The total projected meniscus area is

$p_a = 32.8$ kPa

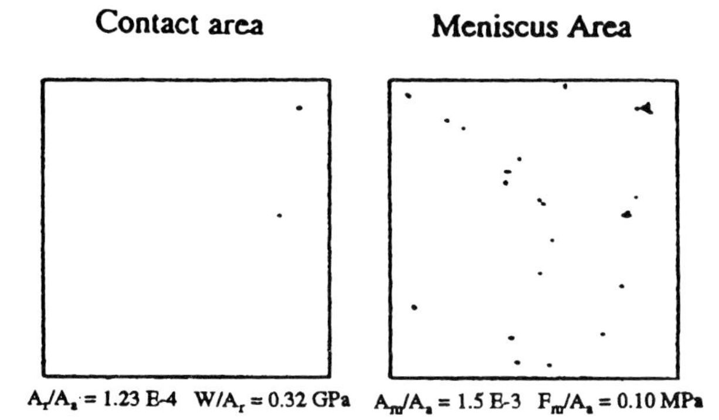

Fig. 39 Contact area and meniscus area for the case of computer generated rough surface ($\sigma = 1$ nm, $\beta^* = 0.5$ µm) in contact with a smooth, flat surface with a composite elastic modulus of 100 GPa and with a water film ($\gamma = 73$ dynes/cm, $\theta = 60°$) thickness of 1 nm and meniscus height of 1 nm (**97**)

determined by selecting those area islands of cross cut area at a given mean meniscus height that overlap the wetted area. The meniscus force is then calculated using equation (65). Figure 39 shows representative contact area and meniscus area maps for a computer generated rough surface in contact with a smooth surface in the presence of water film. As expected, the meniscus area is larger than the contact area and meniscus force is three times that of the normal force.

Tian and Bhushan (**56**) and Poon and Bhushan (**97**) used the model to predict the effects of relative humidity and liquid film thickness on the meniscus force [also see Bhushan (**51**)]. Effect of relative humidity on an interface is shown in Fig. 40a. Effect of liquid film thickness and interface roughness on computer generated rough surfaces in contact with a smooth surface is shown in Fig. 40b. An increase in either relative humidity or liquid film thickness increases the liquid present at the interface. The thicker a liquid film, the more asperities touch the liquid surface and the menisci form on a larger number of asperities. In addition, with a thicker film, a larger volume of liquid is present around the asperities resulting in greater amount of meniscus volume accumulated at the contact interface and greater meniscus height. These effects lead to larger meniscus forces. There is a critical film thickness for a surface with given roughness, above which the meniscus force increases rapidly. The critical film thickness is on the order of 3/4 of the liquid film thickness.

Bhushan and Chilamakuri (**57, 58**) used the numerical model to study contact between a flat rigid surface and a non-Gaussian rough surface with skewness and kurtosis. Non-Gaussian surfaces may exhibit skewness (Sk) and kurtosis (K). Rough surfaces with different skewness and kurtosis were generated using the two-dimensional digital filter technique, referenced earlier. Figure 41 shows

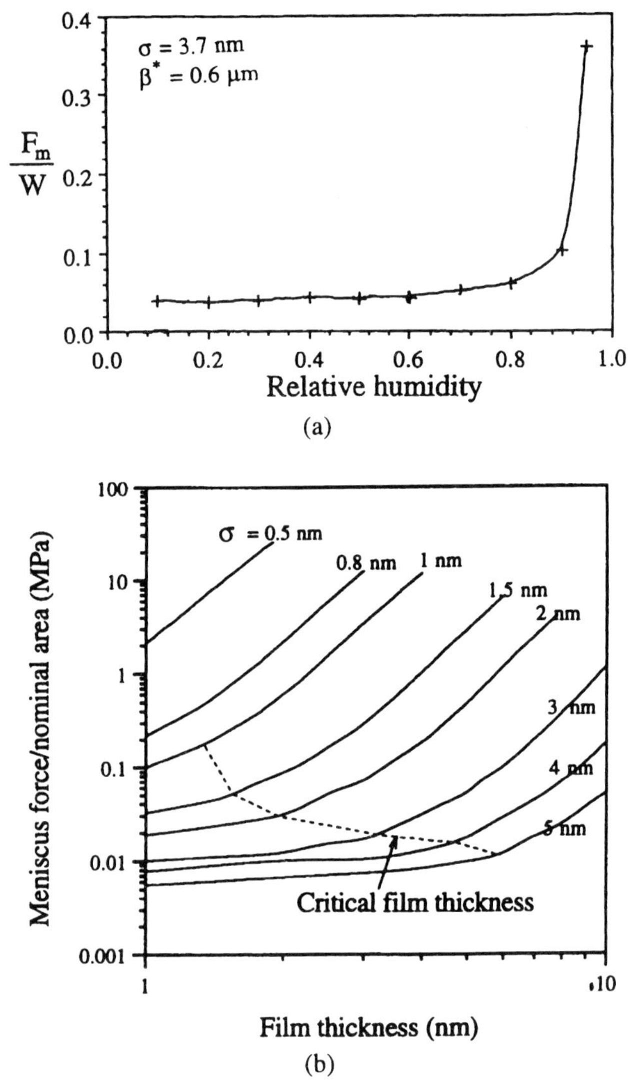

Fig. 40 (a) The effect of relative humidity on the relative meniscus force for a glass-ceramic disk substrate in contact with a smooth surface (**56**). (b) The effect of water film thickness and surface roughness on the relative meniscus force for computer generated Gaussian surfaces (correlation distance $\beta^* = 0.5$ μm) in contact with a smooth surface. The dotted line defines the critical film thickness for different σ (**97**)

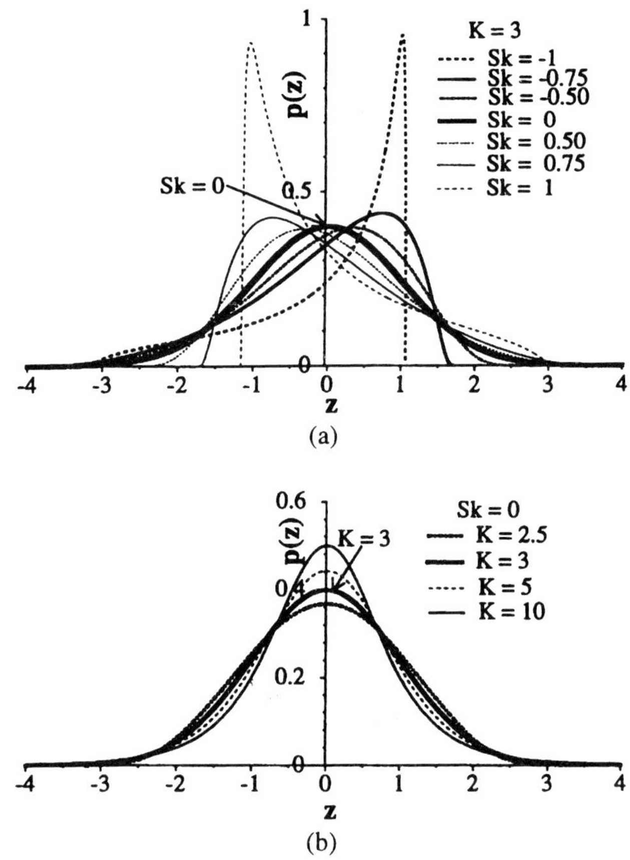

Fig. 41 Probability density functions for surfaces with (a) different skewness and (b) different kurtosis values (**57**)

the probability density functions of surfaces with various skewness and kurtosis values. Figure 42a shows the effect of skewness and kurtosis on the fractional real area of contact (A_r/A_a, where A_a is the apparent area) and relative meniscus force (F_m/W) at different nominal pressures. A positive skewness between 0 and 0.2 at low pressure and about 0.2 at higher pressures results in lowest real area of contact and meniscus force. Contact area and meniscus force decrease with an increase in the kurtosis. Fewer peaks present on a surface with positive skewness or high kurtosis can explain the trends. Figure 42b shows the variation of relative meniscus force with h/σ ratio for different skewness and kurtosis values. Note that the sensitivity of h/σ to the meniscus force decreases at a range of positive skewness of 0 to 0.2, and kurtosis of about five or larger is optimum.

5 CONCLUDING REMARKS

When two nominally flat surfaces are placed in contact, surface roughness causes contact to occur at discrete contact spots. The sum of the areas of all the contact spots constitutes the real (true) area of contact. Deformation occurs in the region of the contact spots, establishing stresses that oppose the applied load. Relative sliding introduces a tangential force (referred to as friction force) at the contact interface. The mode of surface deformation is either elastic or elastic–plastic. Operation of the interface at high relative humidities introduces a water film at the interface. In addition, a liquid film may be deliberately applied at the interface to reduce friction and wear. A liquid film at the interface results in attractive meniscus forces that can also increase friction and wear.

A rough surface is a random system and stochastic models are used to characterize and model the contact

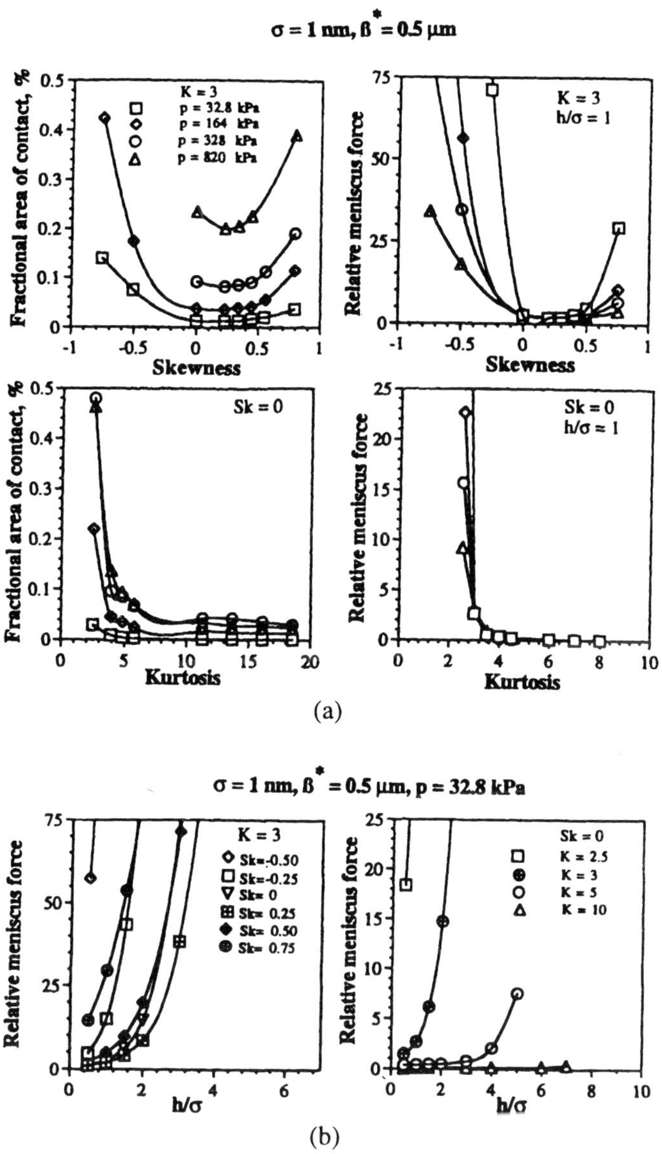

Fig. 42 (a) Fractional real area of contact and relative meniscus force as a function of skewness and kurtosis at various nominal pressures, and (b) relative meniscus force as a function of h/σ for different skewness and kurtosis values, for an interface in the presence of perfluoropolyether film ($\gamma = 25$ dynes/cm, $\theta = 10°$) (**57**)

of two rough surfaces. The classical statistical model for a combination of elastic and elastic–plastic contacts between rough surfaces is that of Greenwood and Williamson (**10**) which uses significant and unrealistic assumptions. Nevertheless the model allows one to identify important roughness parameters and mechanical properties that govern contact mechanics. The basic model has been the subject of much research in the last three decades and many refinements (rather insignificant) have been proposed. The statistical models have been extended to take into account the effect of a liquid film present at the interface in wet contacts (**3**).

With the advent of computer technology, numerical models have been developed, particularly in the past decade, that can analyse the contact of two rough surfaces with measured three-dimensional roughness maps and mechanical properties. An important numerical model was developed by Sayles and co-workers [e.g. reference (**35**)], which uses an iterative process with a conventional matrix inversion technique to solve the contact problem for contact pressure and contact area. To handle a large number of contact points, Tian and Bhushan (**38**) developed a numerical model based on the variational principle. This model is very powerful as it can handle a large number of contact points (tens of thousands). A first numerical contact model for a wet interface has been developed by Tian and Bhushan (**56**). The effect of friction can rather easily be included in these models. Multilayered structures in three-dimensional models have not yet been analysed. Sliding and failure criteria have not been used much in the numerical models to predict interface failure.

These models have been used to study the effect of roughness distribution and mechanical properties on the real area of contact, surface and subsurface stresses and meniscus forces (in the case of wet contacts). The numerical models are now extensively used to design texture for magnetic disks and sliders in the computer data storage industry (**3**, **51**).

REFERENCES

1 **Bowden, F. P.** and **Tabor, D.** *The Friction and Lubrication of Solids, Part I*, 1950 (Clarendon Press, Oxford).

2 **Bowden, F. P.** and **Tabor, D.** *The Friction and Lubrication of Solids, Part II*, 1964 (Clarendon Press, Oxford).

3 **Bhushan, B.** *Tribology and Mechanics of Magnetic Storage Devices*, 2nd edition, 1996 (Springer, New York).

4 **Majumdar, A.** and **Bhushan, B.** Role of fractal geometry in roughness characterization and contact mechanics of rough surfaces. *Trans. ASME, J. Tribol.*, 1990, **112**, 205–216.

5 **Ganti, S.** and **Bhushan, B.** Generalized fractal analysis and its applications to engineering surfaces. *Wear*, 1995, **180**, 17–34.

6 **Bhushan, B.** *Handbook of Micro/Nanotribology*, 1995 (CRC Press, Boca Raton, Florida).

7 **Thomas, T. R.** *Rough Surfaces*, 1982 (Longman, London).

8 **Whitehouse, D. J.** *Handbook of Surface Metrology*, 1994 (Institute of Physics Publishing, Bristol).

9 **Bhushan, B.** and **Blackman, G. S.** Atomic force microscopy of magnetic rigid disks and sliders and its applications to tribology. *Trans. ASME, J. Tribol.*, 1991, **113**, 452–457.

10 **Greenwood, J. A.** and **Williamson, J. B. P.** Contact of nominally flat surfaces. *Proc. R. Soc. Lond.*, 1966, **A 295**, 300–319.

11 **Whitehouse, D. J.** and **Archard, J. F.** The properties of random surface of significance in their contact. *Proc. R. Soc. Lond.*, 1970, **A 316**, 97–121.

12 **Onions, R. A.** and **Archard, J. F.** The contact of surfaces having a random structure. *J. Phys. D.: Appl. Phys.*, 1973, **6**, 289–304.

13 **Nayak, P. R.** Random process model of rough surfaces in plastic contact. *Wear*, 1973, **26**, 305–333.

14 **Bush, A. W., Gibson, R. D.** and **Thomas, T. R.** The elastic contact of a rough surface, *Wear*, 1975, **35**, 87–111.

15 **Bush, A. W., Gibson, R. D.** and **Keogh, G. P.** Strongly anisotropic rough surfaces. *Trans. ASME, J. Lubric. Technol.*, 1979, **101**, 15–20.

16 **Gibson, R. D.** The surface as a random process. In *Rough Surfaces* (Ed. T. R. Thomas), 1982 (Longman, London).

17 **McCool, J. I.** Predicting microfracture in ceramics via a microcontact model. *Trans. ASME, J. Tribol.*, 1986, **108**, 380–386.

18 **Francis, H. A.** Application of spherical indentation mechanics to reversible and irreversible contact between rough surfaces. *Wear*, 1977, **45**, 221–269.

19 **Bhushan, B.** Analysis of the real area of contact between a polymeric magnetic medium and a rigid surface. *Trans. ASME, J. Tribol.*, 1984, **106**, 26–34.

20 **McCool, J. I.** Comparison of models for the contact of rough surfaces. *Wear*, 1986, **107**, 37–60.

21 **Chang, W. R., Etsion, I.** and **Bogy, D. B.** An elastic–plastic model for the contact of rough surfaces. *Trans. ASME, J. Tribol.*, 1987, **109**, 257–263.

22 **Kotwal, C. A.** and **Bhushan, B.** Contact analysis of non-Gaussian surfaces for minimum static and kinetic friction and wear. *Tribol. Trans.*, 1996, **39**, 890–898.

23 **Majumdar, A.** and **Bhushan, B.** Fractal model of elastic–plastic contact between rough surfaces. *Trans. ASME, J. Tribol.*, 1991, **113**, 1–11.

24 **Bhushan, B.** and **Majumdar, A.** Elastic–plastic contact model of bifractal surfaces. *Wear*, 1992, **153**, 53–64.

25 **Williamson, J. B. P.** The microtopography of surfaces. *Proc. Instn. Mech. Engrs*, 1967/68, **182** (Part 3k), 21–30.

26 **Gupta, P. K.** and **Cook, N. H.** Junction deformation models for asperities in sliding interaction. *Wear*, 1972, **20**, 73–108.

27 **Bhushan, B.** and **Cook, N. H.** On the correlation between friction coefficients and adhesion stresses. *Trans. ASME, J. Engng Mater. Technol.*, 1975, **97**, 285–287.

28 **Sayles, R. S.** and **Thomas, T. R.** Computer simulation of the contacting rough surfaces. *Wear*, 1978, **49**, 273–296.

29 **Poon, C. Y.** and **Bhushan, B.** Comparison of surface roughness measurements by stylus profiler, AFM and non-contact optical profiler. *Wear*, 1995, **190**, 76–88.

30 **Akyuz, F. A.** and **Merwin, M.** Solution of nonlinear problems of elastoplasticity by finite element method. *AIAA J.*, 1968, **6**, 1825–1831.

31 **Komvopoulos, K.** and **Choi, D. H.** Elastic finite element analysis of multi-asperity contact, *Trans. ASME, J. Tribol.*, 1992, **114**, 823–831.

32 **Francis, H. A.** The accuracy of plane-strain models for the elastic contact of three-dimensional rough surfaces. *Wear*, 1983, **85**, 239–256.

33 **Lai, W. T.** and **Cheng, H. S.** Computer simulation of elastic rough contacts, *ASLE Trans.*, 1985, **28**, 172–180.

34 **Webster, M. N.** and **Sayles, R. S.** A numerical model for the elastic frictionless contact of real rough surfaces. *Trans. ASME, J. Tribol.*, 1986, **108**, 314–320.

35 **West, M. A.** and **Sayles, R. S.** A 3-dimensional method of studying 3-body contact geometry and stress on real rough surfaces. *Proc. 14th Leeds-Lyon Symposium on Tribology, Tribology Series 12,* 1987, pp. 195–200, (Elsevier, Amsterdam).

36 **Lubrecht, A. A.** and **Ioannides, E.** A fast solution of the dry contact problem and associated subsurface stress field using multilevel techniques. *Trans. ASME, J. Tribol.*, 1991, **113**, 128–133.

37 **Ren, N.** and **Lee, S. C.** Contact simulation of three-dimensional rough surfaces using moving grid method. *Trans. ASME, J. Tribol.*, 1993, **115**, 597–601.

38 **Tian, X.** and **Bhushan, B.** A numerical three-dimensional model for the contact of rough surfaces by variational principle. *Trans. ASME, J. Tribol.*, 1996, **118**, 33–42.

39 **Poon, C. Y.** and **Sayles, R. S.** Numerical contact model of a smooth ball on an anisotropic rough surface. *Trans. ASME, J. Tribol.*, 1994, **116**, 194–201.

40 **Conry, T. F.** and **Seireg, A.** A mathematical programming method for design of elastic bodies in contact. *Trans. ASME, J. Appl. Mechanics*, 1971, **38**, 387–392.

41 **Kalker, J. J.** and **van Randen, Y.** A minimum principle for frictionless elastic contact with application to non-Hertzian half-space contact problems. *J. Engng Math.*, 1972, **6**, 193–206.

42 **Bailey, D. M.** and **Sayles, R. S.** Effect of roughness and sliding friction on contact stresses. *Trans. ASME, J. Tribol.*, 1991, **113**, 729–738.

43 **Yu, M. H.** and **Bhushan, B.** Contact analysis of three-dimensional rough surfâces under frictionless and frictional contact. *Wear*, 1996, **200**, 265–280.

44 **Merriman, T.** and **Kannel, J.** Analyses of the role of surface roughness on contact stresses between elastic cylinders with and without soft surface coating. *Trans. ASME, J. Tribol.*, 1989, **111**, 87–94.

45 **Cole, S. J.** and **Sayles, R. S.** A numerical model for the contact of layered elastic bodies with real rough surfaces. *Trans. ASME, J. Tribol.*, 1992, **114**, 334–340.

46 **Oden, J. T.** and **Martins, J. A. C.** Models and computational methods for dynamic friction phenomena. *Computer Methods in Appl. Mechanics and Engng*, 1985, **52**, 527–634.

47 **Martins, J. A. C., Oden, J. T.** and **Simoes, F. M. F.** A study of static and kinetic friction. *Int. J. Engng Sci.*, 1990, **28**, 29–92.

48 **Derjaguin, B. V., Chuarev, N. V.** and **Muller, V. M.** *Surfaces Forces*, 1987 (Consultants Bureau, New York).

49 **Adamson, A. W.** *Physical Chemistry of Surfaces*, 1990 (Wiley, New York).

50 **Israelachvili, J. N.** *Intermolecular and Surface Forces*, 2nd edition, 1992 (Academic Press, San Diego, California).

51 **Bhushan, B.** Methodology for roughness measurement and contact analysis for optimization of interface roughness. *IEEE Trans. Magn.*, 1996, **32**, 1819–1825.

52 **Li, Y.** and **Talke, F. E.** A model for the effect of humidity on stiction of the head/disk interface. In *Tribology and Mechanics of Magnetic Storage Systems*. (Ed. B. Bhushan), 1990, pp. 79–84, (SP-29, STLE, Park Ridge, Illinois).

53 **Tian, H.** and **Matsudaira, T.** The role of relative humidity, surface roughness and liquid build-up on static friction behavior of the head/disk interface. *Trans. ASME, J. Tribol.*, 1993, **115**, 28–35.

54 **Gao, C., Tian, X.** and **Bhushan, B.** A meniscus model for optimization of texturing and liquid lubrication of magnetic thin film rigid disks. *Tribol. Trans.*, 1995, **38**, 201–212.

55 **Bhushan, B., Kotwal, C. A.** and **Chilamakuri, S. K.** Kinetic meniscus model for prediction of rest stiction. *Trans. ASME, J. Tribol.*, 1997, **119** (in press).

56 **Tian, X.** and **Bhushan, B.** The micro-meniscus effect of a thin liquid film on the static friction of rough surface contact. *J. Phys. D: Appl. Phys.*, 1996, **29**, 163–178.

57 **Bhushan, B.** and **Chilamakuri, S.** Non-Gaussian surface roughness distribution of magnetic media for minimum friction/stiction. *J. Appl. Phys.*, 1996, **79**, 5794–5996.

58 **Chilamakuri, S. K.** and **Bhushan, B.** Contact analysis of non-Gaussian random surfaces. *Proc. Instn Mech. Engrs, Part J, J. Engng Tribol.*, 1998, **212** (J1), 19–32.

59 **Bhushan, B.** Contact mechanics of rough surfaces in tribology: single asperity contact. *Appl. Mech. Rev.*, 1996, **49**, 275–298.

60 **Hamilton, G. M.** and **Goodman, L. E.** Stress field created by a circular sliding contact. *Trans. ASME, J. Appl. Mechanics*, 1996, **33**, 371–376.

61 **O'Sullivan, T. C.** and **King, R. B.** Sliding contact stress field due to a spherical indenter on a layered elastic half-space. *Trans. ASME, J. Tribol.*, 1998, **110**, 235–240.

62 **Lawn, B.** *Fracture of Brittle Solids*, 1993 (Cambridge University Press, Cambridge).

63 **Kral, E. R., Komvopoulos, K.** and **Bogy, D. B.** Elastic–plastic finite element analysis of repeated indentation of a half-space by a rigid sphere. *Trans. ASME, J. Appl. Mechanics*, 1993, **60**, 829–841.

64 **Kral, E. R., Komvopoulos, K.** and **Bogy, D. B.** Finite element analysis of repeated indentation of an elastic–plastic layered medium by a rigid sphere. Part I: surface results. *Trans. ASME, J. Appl. Mechanics*, 1995, **62**, 20–28.

65 **Kral, E. R., Komvopoulos, K.** and **Bogy, D. B.** Finite element analysis of repeated indentation of an elastic–plastic layered medium by a rigid sphere. Part II: subsurface results. *Trans. ASME, J. Appl. Mechanics*, 1995, **62**, 29–42.

66 **Tabor, D.** *The Hardness of Metals*, 1951 (Clarendon Press, Oxford).

67 **Johnson, K. L.** *Contact Mechanics*, 1985 (Cambridge University Press, Cambridge).

68 **Bhushan, B.** and **Tian, X.** Contact analysis of regular patterned rough surfaces in magnetic recording. *Trans. ASME, J. Electronic Packaging*, 1995, **117**, 26–33.

69 **Williamson, J. B. P., Pullen, J.** and **Hunt, R. T.** The shape of solid surfaces. *Surface Mechanics*, 1970, p. 24 (ASME, New York).

70 **Greenwood, J. A.** and **Tripp, J. H.** The contact of two nominally flat rough surfaces. *Proc. Instn Mech. Engrs*, 1970–71, **185**, 625–633.

71 **O'Callaghan, M.** and **Cameron, M. A.** Static contact under load between nominally flat surfaces in which deformation is purely elastic. *Wear*, 1976, **36**, 76–97.

72 **Gibra, I. N.** *Probability and Statistical Inference for Scientists and Engineers*, 1973 (Prentice-Hall, Englewood Cliffs, New Jersey).

73 **Abramowitz, M.** and **Stegun, I. A.** *Handbook of Mathematical Functions*, 1965, pp. 685–720 (Dover, New York).

74 **Nayak, P. R.** Random process model of rough surfaces. *Trans. ASME, J. Lubric. Technol.*, 1971, **93**, 398–407.

75 **Longuet-Higgins, M. S.** The statistical analysis of a random, moving surface. *Phil. Trans. R. Soc. Lond.*, 1957, **A 249**, 321–387.

76 **Longuet-Higgins, M. S.** Statistical properties of an isotropic random surface. *Phil. Mag. R. Soc. Lond.*, 1957, **A 250**, 157–174.

77 **Fuller, K. N. G.** and **Tabor, D.** The effect of surface roughness on the adhesion of elastic solids. *Proc. R. Soc. Lond.*, 1975, **A 345**, 327–342.

78 **Johnson, K. L., Kendall, K.** and **Roberts, A. D.** Surface energy and the contact of elastic solids. *Proc. R. Soc. Lond.*, 1971, **A 324**, 301–313.

79 **Whitehouse, D. J.** and **Phillips, M. J.** Discrete properties of random surfaces. *Proc. R. Soc. Lond.*, 1978, **A 290**, 267–298.

80 **Poon, C. Y.** and **Bhushan, B.** Surface roughness analysis of glass-ceramic substrates and finished magnetic disks, and Ni–P coated Al–Mg and glass substrates. *Wear*, 1995, **190**, 89–109.

81 **Bhushan, B.** and **Doerner, M. F.** Role of mechanical properties and surface texture in the real area of contact of magnetic rigid disks. *Trans. ASME, J. Tribol.*, 1989, **111**, 452–458.

82 **Chilamakuri, S. K.** and **Bhushan, B.** Optimization of asperities for laser-textured magnetic disk surfaces. *Tribol. Trans.*, 1997, **40**, 303–311.

83 **Boussinesq, J.** *Application des Potentials a l'Etude de l'Equilibre et du Mouvement des Solides Elastiques*, 1885 (Gauthier-Villars, Paris, France).

84 **Cerruti, V.** *Roma Acc. Lincei, Mem. fis. mat.*, 1882.

85 **Love, A. E. H.** The stress produced in a semi-infinite solid by pressure on part of the boundary. *Proc. R. Soc. Lond.*, 1929, **A 228**, 377–420.

86 **Richards, T. H.** *Energy Methods in Stress Analysis*, 1977, (Ellis Horwood, Chichester).

87 **Wolfe, P.** The simplex method for quadratic programming. *Econometrica*, 1959, **27**, 382–395.

88 **Lee, S. C.** and **Ren, N.** Behavior of elastic–plastic rough surface contacts as affected by surface topography, load and material hardness. *Tribol. Trans.*, 1996, **39**, 67–74.

89 **Lee, S. C.** and **Ren, N.** The subsurface stress field created by three-dimensionally rough bodies in contact with traction. *Tribol. Trans.*, 1994, **37**, 615–621.

90 **Ahmadi, N., Keer, L. M., Mura, T.** and **Vithoontien, V.** The interior stress field caused by tangential loading of a rectangular patch on an elastic half space. *Trans. ASME, J. Tribol.*, 1987, **109**, 627–629.

91 **Kalker, J. J.** Numerical calculation of the elastic field in a half-space. *Communications in Appl. Numer. Methods*, 1986, **2**, 401–410.

92 **Poon, C. Y.** and **Sayles, R. S.** Contact analysis of a smooth ball on an anisotropic rough surface. *Trans. ASME, J. Tribol.*, 1994, **116**, 850–859.

93 **Ren, N.** and **Lee, S. C.** The effects of surface roughness and topography on the contact behavior of elastic bodies. *Trans. ASME, J. Tribol.*, 1994, **116**, 804–810.

94 **Poon, C. Y.** and **Bhushan, B.** Rough surface contact analysis and its relation to plastic deformation at the head disk interface. *J. Appl. Phys.*, 1996, **79**, 5799–5801.

95 **Poon, C. Y.** and **Bhushan, B.** Nano-asperity contact analysis and surface optimization for magnetic head slider/disk contact. *Wear*, 1996, **202**, 83–98.

96 **Hu, Y. Z.** and **Tonder, K.** Simulation of 3-D random surface by 2-D digital filter and Fourier analysis. *Int. J. Mach. Tool Manufact.*, 1992, **32**, 82–90.

97 **Poon, C. Y.** and **Bhushan, B.** Numerical contact and stiction analyses of Gaussian isotropic surfaces for magnetic head slider/disk contact. *Wear*, 1996, **202**, 68–82.

98 **Hess, D. P.** and **Soom, A.** Normal and angular motions at a rough planar contact during sliding with friction. *Trans. ASME, J. Tribol.*, 1992, **114**, 567–578.

99 **Polycarpou, A. A.** and **Soom, A.** A two-dimensional mixed friction model for a lubricated line contact. *Trans. ASME, J. Tribol.*, 1996, **118**, 183–189.

100 **Kikuchi, N.** and **Oden, J. T.** *Contact Problems in Elasticity*, 1985, (SIAM, Philadelphia).

101 **Rabier, P., Martins, J. A. C.** and **Oden, J. T.** Existence and local uniqueness of solutions to contact problems in elasticity with nonlinear friction laws. *Int. J. Engng Sci.*, 1986, **24**, 1755–1768.

102 **Bhushan, B.** and **Zhao, Z.** Friction/stiction and wear studies of magnetic thin-film disks with two perfluoropolyether lubricants. *IEEE Trans. Magn.*, 1997, **31**, 918–925.

Contact analysis of non-Gaussian random surfaces

S K Chilamakuri and **B Bhushan**
Computer Microtribology and Contamination Laboratory, Department of Mechanical Engineering, The Ohio State University, Columbus, Ohio, USA

Abstract: Surfaces produced by machining processes such as grinding, shaping and turning and some magnetic disk texturing processes such as sputtering and laser processing are frequently non-Gaussian. Most of the contact models developed in the past three decades assume Gaussian distribution of surface heights in their analyses. In the present paper, contact analysis of non-Gaussian surfaces has been conducted. A computer program was developed to generate non-Gaussian surfaces with specified standard deviation, autocorrelation length, skewness and kurtosis. Contact area, maximum contact pressure and relative meniscus force as a function of skewness and kurtosis were studied at different values of σ and β^*. In these studies, it is observed that a surface with slight positive skewness and kurtosis in the range of 4–7.5 results in an optimum surface with a minimum contact area and meniscus force. Further, stiction of a surface with high kurtosis is somewhat insensitive to liquid film thickness. A surface with negative skewness and/or kurtosis of less than 3 will lead to severe friction/stiction problems and a surface with very high kurtosis and σ may lead to plastic deformation. Non-Gaussian surfaces offer promise for the design of interface roughness to provide low friction/stiction and wear.

Keywords: contact area, meniscus force, surface roughness, skewness, kurtosis

NOTATION

$A(\omega_x, \omega_y)$	Fourier transform of $\eta(I, J)$
$h(k, l)$	filter function
$H(\omega_x, \omega_y)$	frequency response or transfer function of the system $Z(\omega_x, \omega_y) = H(\omega_x, \omega_y) \times A(\omega_x, \omega_y)$
K	kurtosis
K_z, K_η	kurtosis of output (z) and input (η) sequences respectively
L	scan length
m	mean
$p(z)$	probability density function
$R(\tau)$	autocorrelation function
$R_z(k, l)$	autocorrelation function in discrete form
S_B, S_L, S_U	three main types of frequency curves used in Johnson's translator system
S_z, S_η	power spectral density function of z and η respectively
Sk	skewness
Sk_z, Sk_η	skewness of output (z) and input (η) sequences respectively
x, y, z	coordinates
$Z(\omega_x, \omega_y)$	Fourier transform of $z(I, J)$
β^*	correlation length for an isotropic surface
β^*_x, β^*_y	correlation lengths in the x and y directions respectively
γ, λ, ξ, δ	unknown constants to be determined by the method of moments
η	input sequence
η'	standardized normal variable sequence
η''	variable sequence with known skewness and kurtosis
σ	standard deviation
τ	sampling distance

The MS was received on 8 August 1996 and was accepted for publication on 29 August 1997.

1 INTRODUCTION

Contact of two rough surfaces occurs at a small fraction of the nominal area. The applied normal load is carried by the highest asperities which results in extremely high normal pressures at the contacting tips. The real area of contact plays a very important role in the friction and wear (**1**). In the presence of a thin film of liquid lubricant or adsorbed water layer at the interface, menisci form around the contacting asperities and non-contacting asperities. The meniscus force will pull the contacting bodies even further and results in a higher contact load and in turn higher friction force. The meniscus contribution to the friction can be very large to a point where the surfaces may get stuck to each other. This phenomenon in the magnetic disk drive industry is known as stiction (**1**). The contact behaviour of rough surfaces depends on the surface roughness distribution, liquid film, mechanical properties of contacting bodies and operating environment. In the past three decades, many analytical and numerical models have been developed for contact simulation of rough surfaces (**2**–**7**). In most models, a Gaussian surface height distribution is assumed. However, assumption of most surface heights having a Gaussian distribution is not strictly true. Most of the common machining processes produce surfaces with non-Gaussian distributions (**7**–**10**). Turning, shaping and electrodischarge machining produce a peaked surface with positive skewness. Grinding, honing, milling and abrasion processes produce grooved surfaces with negative skewness but high kurtosis values. Laser polishing produces surfaces with high kurtosis. Mechanical texturing of disks with free abrasives or abrasive cloth result in a surface with deep valleys with negative skewness. Disk surfaces produced by sputter texturing (**11**) have mainly peaks on the disk surfaces which are expected to have positive skewness. Laser texturing also produces non-Gaussian surfaces (**12**, **13**).

Recently Kotwal and Bhushan (**14**) developed an analytical model for contact analyses of non-Gaussian surfaces. In this study they extended the Greenwood and Williamson model (**2**) to non-Gaussian surfaces by considering the probability density functions of these surfaces. In this model they have assumed a constant asperity density. However, for non-Gaussian surfaces, asperity density changes with skewness and kurtosis. They found that surfaces with positive skewness and high kurtosis exhibit a low contact area and meniscus force. In this study Kotwal and Bhushan (**14**) neglected the effect of skewness and kurtosis on contact pressure.

In the present study a method to produce non-Gaussian rough surfaces on the computer is developed and contact analyses are performed on these surfaces. Variations of contact area, maximum contact pressure and relative meniscus force as functions of skewness and kurtosis at different values of standard deviation of surface heights and correlation length are studied.

2 GENERATION OF NON-GAUSSIAN RANDOM SURFACES ON THE COMPUTER

A Gaussian random surface with an exponential autocorrelation function can be represented by two surface roughness parameters—the standard deviation of surface heights, σ (or r.m.s value or R_q), and the correlation length, β^* (β^*_x, β^*_y in the case of an anisotropic surface); σ and R_q are vertical roughness parameters, whereas β^* is a spatial parameter giving information about how surface heights are distributed in space. The knowledge of spatial variation is essential, as it affects the shape and density of the asperities. For a two-dimensional profile, the autocorrelation function is defined as

$$R(\tau) = \frac{1}{L}\int_L z(x)z(x+\tau)\,\mathrm{d}\tau \tag{1a}$$

or

$$R(\tau) = \frac{1}{L}\sum_{i=0}^{N-1} z_i z_{i+\tau}\,\Delta\tau \tag{1b}$$

where τ is the sampling interval, N is the number of points and $\Delta\tau = L/N$.

In the present study, the autocorrelation function is assumed to be exponential and is given by

$$R(\tau) = \sigma^2 \exp\left(-\frac{\tau}{\beta}\right) \tag{2}$$

where $1/\beta$ is the decay rate of the function. The correlation length, β^*, can be defined as the distance when $R(\tau)/\sigma^2 = 0.1$ (when $\tau = 2.3\beta$) (**15**) or $R(\tau)/\sigma^2 = 1/e$ (when $\tau = \beta$) (**16**). Both forms are considered to be a measure of the length at which single height readings become statistically independent of one another. In the present paper β^* is given by $R(\tau)/\sigma^2 = 0.1$.

A non-Gaussian surface may also exhibit skewness (Sk) and kurtosis (K). The probability density function for surface heights gives the probability of locating a point at a height z and is denoted by $p(z)$. The shape of the probability function offers useful information on the behaviour of the distribution. This shape can be described in terms of moments of the function about the mean, referred to as central moments. The first moment is m, which is taken to be zero and the second moment is the variance σ^2, the square of the standard deviation,

$$\sigma^2 = \int_{-\infty}^{\infty} z^2 p(z)\,\mathrm{d}z \tag{3}$$

The third moment is the skewness, which represents asymmetric spread of the height distribution and is given by

$$\mathrm{Sk} = \frac{1}{\sigma^3}\int_{-\infty}^{\infty} z^3 p(z)\,\mathrm{d}z \tag{4}$$

where $p(z)$ is the probability density function of surface heights, z. A Gaussian surface has zero skewness with an equal number of peaks and valleys at certain heights. Profiles with peaks removed have negative skewness whereas profiles with valleys filled or high peaks have positive skewness. The fourth moment is termed kurtosis, which represents the peakedness of the distribution and is given by

$$K = \frac{1}{\sigma^4}\int_{-\infty}^{\infty} z^4 p(z)\,dz \tag{5}$$

A Gaussian surface has a kurtosis of 3. If $K < 3$, the distribution has relatively few high peaks and low valleys. If $K > 3$, the distribution has relatively many high peaks and low valleys.

Non-Gaussian surfaces with various Sk and K values for given σ and β^* are generated on the computer using a two-dimensional digital filter technique suggested by Hu and Tonder (**17**). A digital filter is a system that transforms an input sequence $\eta(I, J)$ into an output sequence $z(I, J)$. First an input sequence of independent random numbers having a Gaussian distribution of a known standard deviation, $\eta(I, J)$, is generated using a random number generator. To simulate real rough surfaces, however, surfaces having an expected autocorrelation function and height distribution need to be generated. The two-dimensional digital filter technique is employed to generate an output sequence $z(I, J)$ of a known autocorrelation by a linear transformation system defined as

$$z(I, J) = \sum_{k=0}^{n-1}\sum_{l=0}^{m-1} h(k, l)\eta(I - k, J - l) \tag{6}$$

where $I = 0, 1, \ldots, N - 1$, $J = 0, 1, \ldots, M - 1$, $n = N/2$, $m = M/2$ and h is a filter function. The Fourier transform of equation (6) is given by

$$Z(\omega_x, \omega_y) = H(\omega_x, \omega_y)A(\omega_x, \omega_y) \tag{7}$$

where A and Z are Fourier transformations of the input η and output sequences z respectively, and H is the transfer function of the system given by

$$H(\omega_x, \omega_y) = \sum_{k=0}^{n-1}\sum_{l=0}^{m-1} h(k, l)\,e^{-jk\omega_x - jl\omega_y} \tag{8}$$

The autocorrelation function (ACF) of a random surface in the discrete form is defined as

$$R_z(k, l) = \frac{1}{NM}\sum_{I=0}^{N-1}\sum_{J=0}^{M-1} z(I, J)z(I + k, J + l) \tag{9}$$

The power spectral density (PSD) function of the surface is the Fourier transform of R_z, given by

$$S_z(\omega_x, \omega_y) = \frac{1}{nm}\sum_{k=-(n/2)+1}^{(n/2)-1}\ \sum_{l=-(m/2)+1}^{(m/2)-1} R_z(k, l)\,e^{-jk\omega_x - jl\omega_y} \tag{10}$$

If $S_\eta(\omega_x, \omega_y)$ denotes the PSD function of the input sequence, the relation between S_η and S_z for a linear system has the form

$$S_z(\omega_x, \omega_y) = |H(\omega_x, \omega_y)|^2 S_\eta(\omega_x, \omega_y) \tag{11}$$

Since η is a sequence composed of independent random numbers, its PSD function is a constant. The inverse Fourier transform of $H(\omega_x, \omega_y)$ obtained from equation (11) gives the filter function h:

$$h(k, l) = \frac{1}{nm}\sum_{\omega_x=-(n/2)+1}^{(n/2)-1}\ \sum_{\omega_y=-(m/2)-1}^{(m/2)-1} H(\omega_x, \omega_y)\,e^{-jk\omega_x - jl\omega_y} \tag{12}$$

Now $h(k, l)$ from equation (12) is used in equation (6) to obtain $z(I, J)$.

So far, a technique has been described to generate Gaussian random surfaces. For the generation of non-Gaussian random surfaces, the Gaussian input sequence is transformed to an input sequence with appropriate skewness, Sk_η, and kurtosis, K_η, by using the Johnson translator system of distribution (**18, 19**). However, during later transformation using the two-dimensional filter technique to obtain an output sequence, the distribution of the output sequence will not in most cases be of the same form as that of the input sequence. In this case, the skewness and kurtosis of the output sequence are related to those of the input sequence as (**17**)

$$(Sk_z)^{1/2} = \frac{\sum_{i=0}^{q}\theta_i^3}{(\sum_{i=0}^{q}\theta_i^2)^{3/2}}(Sk_\eta)^{1/2} \tag{13}$$

$$(K_z)^{1/2} = \frac{\sum_{i=0}^{q}\theta_i^4 K_\eta + \sum_{i=0}^{q-1}\sum_{j=i+1}^{q}\theta_i^2\theta_j^2}{(\sum_{i=0}^{q}\theta_i^2)^2} \tag{14}$$

where Sk_z and K_z are the required skewness and kurtosis, and Sk_η and K_η are the input skewness and kurtosis for Johnson's translator system, and

$$\begin{aligned}\theta_i &= h(k, l), \qquad k = 0, \ldots, n - 1,\\ l &= 0, \ldots, m - 1, \qquad i = km + 1\end{aligned} \tag{15}$$

The non-Gaussian input sequence with modified skewness and kurtosis obtained from equations (13) and (14) is generated using Johnson's translator system. Any good translator system should be simple, monotonic and easy to calculate. Moreover, it should be able to handle most distributions. The Johnson system of frequency curves based on the method of moments provides a family of curves that can be used to generate an equation for the distribution for which the first four moments (mean, σ, Sk and K) are known. This system of equations is simpler than the Pearson system of frequency curves (**18, 19**). In this system there are three main types of frequency curves, S_U, S_L and S_B, defined as:

$$S_U: \quad \eta' = \gamma + \delta \sinh^{-1}\left(\frac{\eta'' - \xi}{\lambda}\right) \tag{16}$$

$$S_L: \quad \eta' = \gamma + \delta \log\left(\frac{\eta'' - \xi}{\lambda}\right) \quad (\eta'' > \xi) \tag{17}$$

$$S_B: \quad \eta' = \gamma + \delta \log\left(\frac{\eta'' - \xi}{\xi + \lambda - \eta''}\right) \quad (\xi < \eta'' < \xi + \lambda) \tag{18}$$

where η' is a standardized normal variable sequence, η'' is the variable sequence with given skewness and kurtosis, and γ, δ, ξ and λ are constants to be determined for the first four given moments by using method of moments. Finally, steps of the two-dimensional digital filter technique outlined earlier [equations (6) to (12)] are used to transform the non-Gaussian input sequence η'' to z.

To summarize, non-Gaussian surfaces are generated using the following steps:

1. Generate a Gaussian input sequence $\eta'(I, J)$ using a random number generator.
2. Determine the skewness and kurtosis of the input sequence for desired skewness and kurtosis of an output sequence, using equations (13) and (14).
3. Transform the Gaussian input sequence to the input sequence $\eta''(I, J)$, which has a distribution with appropriate skewness, Sk_η, and kurtosis, K_η, by using the Johnson translator system of distributions.
4. Assume that the ACF for the simulated surface has the form

$$R_z(k, l) = \sigma^2 \exp\left\{-2.3\left[\left(\frac{k}{\beta_x^*}\right)^2 + \left(\frac{l}{\beta_y^*}\right)^2\right]^{1/2}\right\} \tag{19}$$

   where σ is the standard deviation of surface heights and β_x^*, β_y^* are the correlation lengths of surface profiles in the x and y directions. An isotropic surface is considered, i.e. $\beta_x^* = \beta_y^* = \beta^*$.
5. Calculate $S_z(\omega_x, \omega_y)$ using equations (19) and (10).
6. For the known values of S_η and S_z, calculate $H(\omega_x, \omega_y)$ using equation (11).
7. Obtain $h(k, l)$ by the inverse Fourier transform of $H(\omega_x, \omega_y)$ using equation (12).
8. Obtain $z(I, J)$ by performing the filtering operation using equation (6).

Using this method, non-Gaussian surfaces are generated at different values of σ, β^*, Sk and K. To ensure that the autocorrelation function of the generated surface is exponential, as assumed in equation (19), an average autocorrelation function of 256 profiles for a surface was calculated using equation (1b) and it was found to be closely exponential. A typical plot of calculated and expected ACF is shown in Fig. 1. Further, to check the statistical parameters of the generated surfaces, a statistical program was developed to calculate the σ, β^*, Sk and K values of the generated surface. This program calculates the σ, Sk and K values of the surface from the discretized data. The correlation length β^* value is calculated by finding the distance at which the value of the function is 0.1 of the value at $\tau = 0$ and then the average β^* for the surface is obtained by taking the average of β^* of 256 profiles. The calculated values of σ, β^*, Sk and K are found to be within 5 per cent of the input values for the surface generation program, which are shown in Table 1. This guarantees the accuracy of this program.

Probability density functions of the generated surfaces are constructed by plotting the fraction of surface heights lying between two specific heights as a function of height, as shown in Fig. 2. Figure 3 shows two- and three-dimensional surface maps of non-Gaussian surfaces. Figure 4 shows two-dimensional surface profiles at different β^*, Sk and K values. A surface with very high negative skewness and kurtosis values less than 3 is almost flat with some deep valleys, whereas a surface with high positive skewness and kurtosis values greater than 3 has fewer

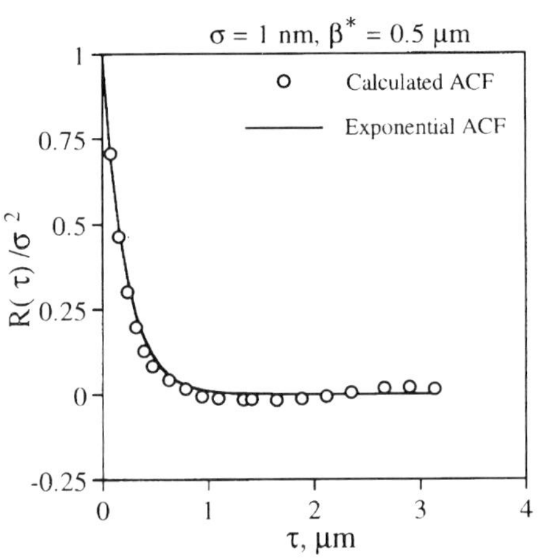

Fig. 1 Autocorrelation function of a typical computer generated profile

Table 1 Comparison of input and calculated statistical parameters of generated non-Gaussian surfaces

Case	σ (nm)	β^* (μm)	Sk	K
Input to the surface generation program	1.0	0.5	0.5	5.0
Calculated values from the generated surface	1.001	0.51	0.49	4.913

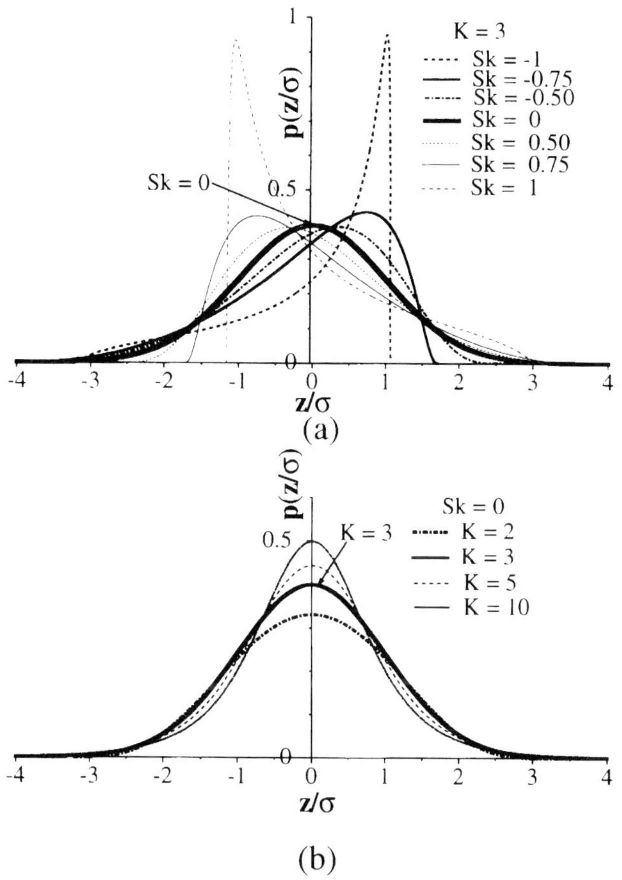

Fig. 2 Probability density functions for surfaces with different (a) skewness (Sk) and (b) kurtosis (K) values

peaks. Figure 5 schematically shows differences between negatively and positively skewed surfaces and surfaces with high and low kurtosis. Next, the asperity density, peak-to-mean height and peak radius of the generated surfaces are calculated and the effects of Sk and K on these parameters are studied. Figure 6 shows the effect of skewness and kurtosis on the asperity density, peak-to-mean height (the distance between the highest peak and the mean plane) and peak radius. The asperity density and peak-to-mean height increases with skewness up to Sk = 0.5 and then starts decreasing with an increase in Sk, whereas skewness has a negligible effect on the peak radius. The asperity density and peak radius decrease with an increase in kurtosis whereas the peak-to-mean height increases with kurtosis.

Finally, the effect of σ and β^* on asperity statistics is studied. To study this effect, asperity statistics, asperity density, peak-to-mean height and peak radius of surfaces having different σ and β^* were calculated using the previously mentioned statistical program. In this program, peak density was calculated using a threshold of 10 per cent of σ (**1**). The peak radius was obtained by calculating the second derivative of the profile peaks. These results are shown in Table 2. It is noted that the asperity density decreases with an increase of β^*, whereas the peak radius increases with an increase of β^* and the peak-to-mean height has little effect. It is also observed that the peak radius decreases and the peak-to-mean height increases with an increase in σ.

3 CONTACT MODEL

A three-dimensional contact model developed by Tian and Bhushan (**5**) was used to perform contact analysis of a computer-generated non-Gaussian flat surface. In an earlier study Kotwal and Bhushan (**14**) showed that two non-Gaussian surfaces in contact were equivalent to a flat surface in contact with a composite rough surface. The model (**5**) is based on the variational principle in which the real area of contact and contact pressure is that which minimizes the total complementary potential energy. The contact model can be used to simulate contact between any two bodies with digitized surface heights representing the real surfaces or a computer-generated surface. For wet contact analysis a meniscus model developed by Tian and Bhushan (**20**) was used. In this stiction model a uniform film thickness is assumed everywhere on the rough surface and meniscii form at the contacting and non-contacting asperities. Then the relative meniscus force (the ratio of meniscus force and normal load) is calculated. For the contact analysis, surfaces having different correlation lengths β^* and σ with different values of skewness and kurtosis were generated using a scan size (profile length in the x and y directions) of 20 μm × 20 μm. Analyses were carried out for an Al–Mg/Ni–P disk with elastic modulus = 110 GPa, Poisson's ratio = 0.36 and hardness = 7 GPa against an Al_2O_3–TiC slider with elastic modulus = 450 GPa, Poisson's ratio = 0.23 and hardness = 22 GPa. The composite modulus of elasticity of this interface is about 100 GPa and the hardness of the softer material is 7 GPa. For most cases, a liquid film of perfluoropolyether lubricant with surface tension = 0.025 N/m and contact angle = 10° was analysed. A 50 per cent nanoslider (a conventional magnetic slider with a 50 per cent recording rail) with a nominal area of 0.915 mm^2 (with a contact area between the slider and the disk) and a normal load of 3 g (0.03 N) is considered here for which the nominal pressure is 32.8 kPa.

4 RESULTS AND DISCUSSION

The effects of skewness and kurtosis on the fractional contact area (the ratio between the real area of contact and the nominal area), maximum contact pressure and relative meniscus force (the ratio between the meniscus force and the normal load) are shown in Fig. 7. A surface with

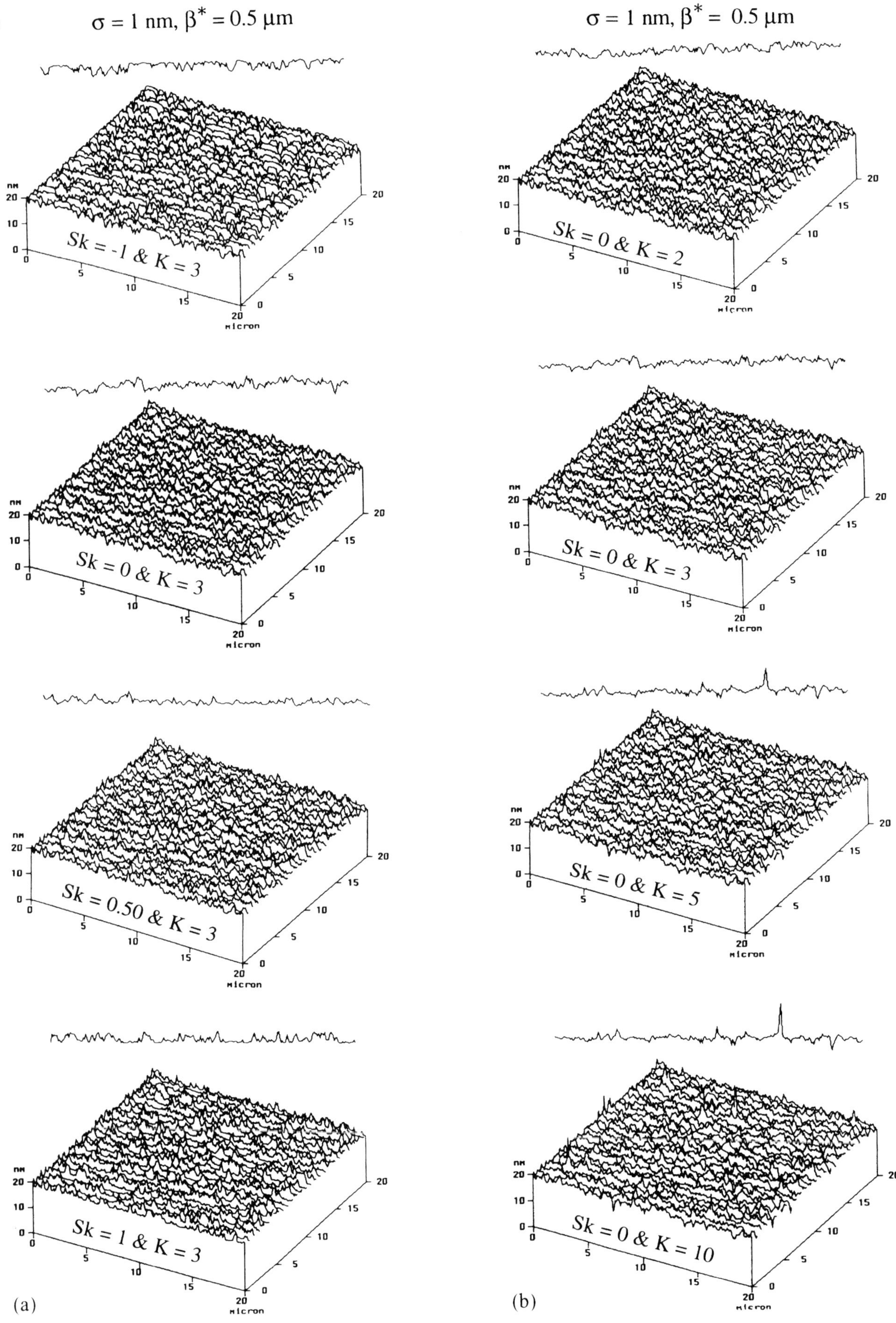

Fig. 3 Two- and three-dimensional surface maps of computer-generated non-Gaussian surfaces with different (a) skewness (Sk) and (b) kurtosis (K) values

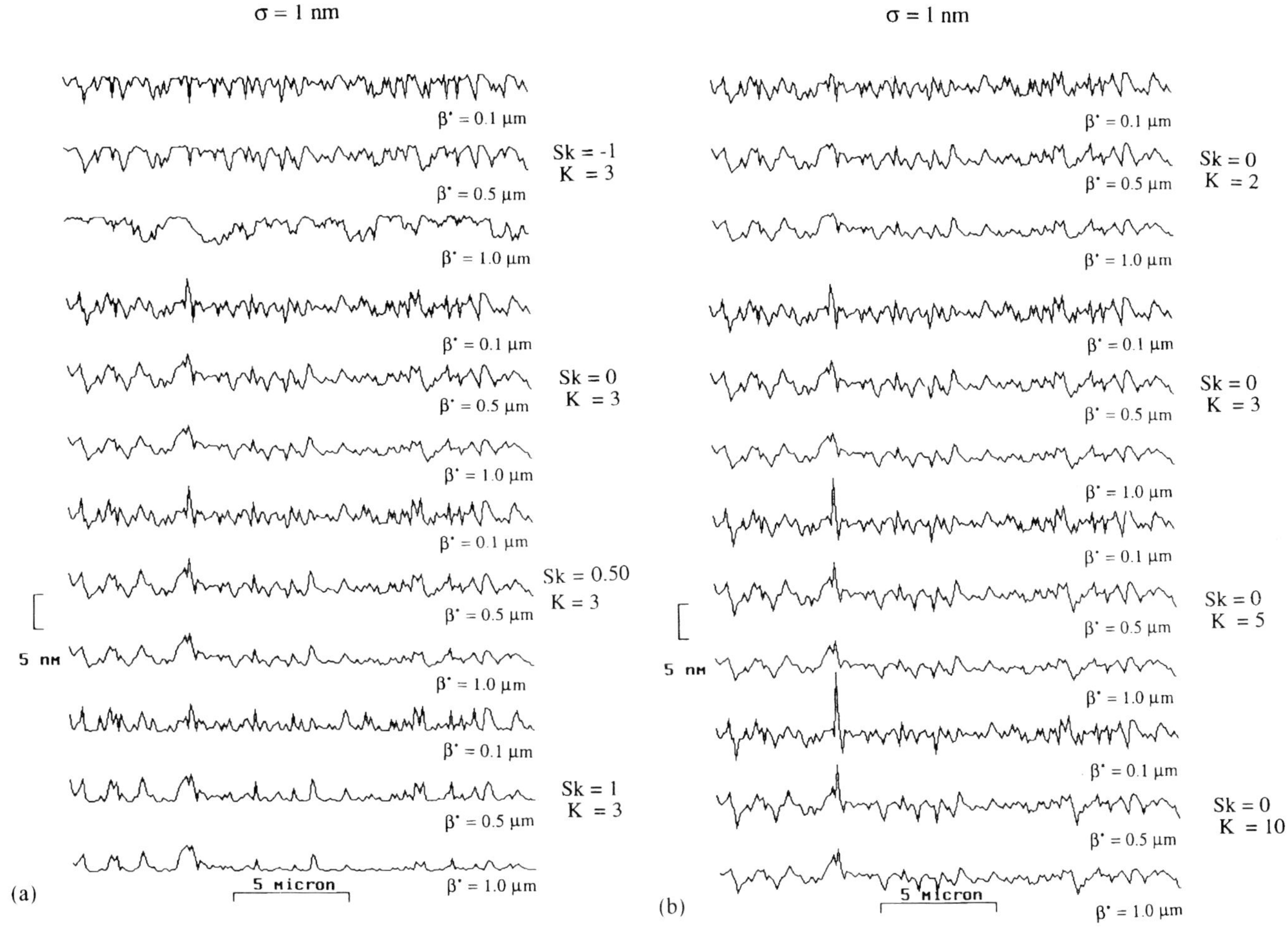

Fig. 4 Two-dimensional surface profiles of non-Gaussian surfaces at different β^* and (a) skewness (Sk) and (b) kurtosis (K) values

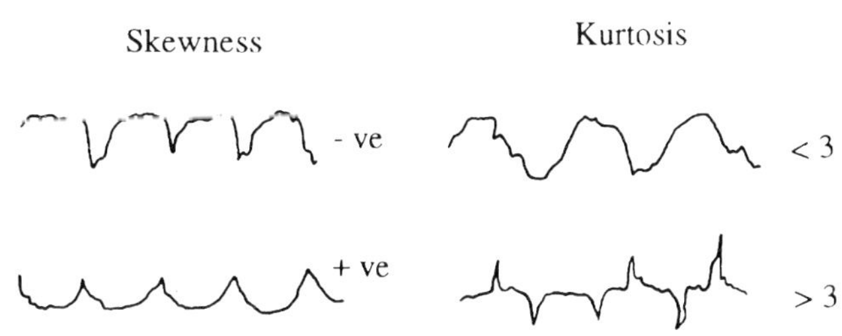

Fig. 5 Schematic of two-dimensional profiles of non-Gaussian surfaces

slightly positive skewness between 0 and 0.25 gives a minimum fractional contact area, meniscus force and maximum contact pressure. This is not very clear at lower normal loads because there are only a few contact points at this load. However, at higher loads a clear minimum can be seen between 0.25 and 0.5. The fractional contact area and relative meniscus force increase rapidly as skewness decreases from −0.75 to −1. A surface with a slightly positive skewness consists of the optimum number of peaks to support the applied load which results in an optimum contact area, whereas a surface with a high negative skewness has a large number of peaks at a certain height with deep valleys that give a high contact area, high meniscus force and low contact pressure. An increase in the number of peaks with positive skewness results in an increase in the contact area and meniscus force with

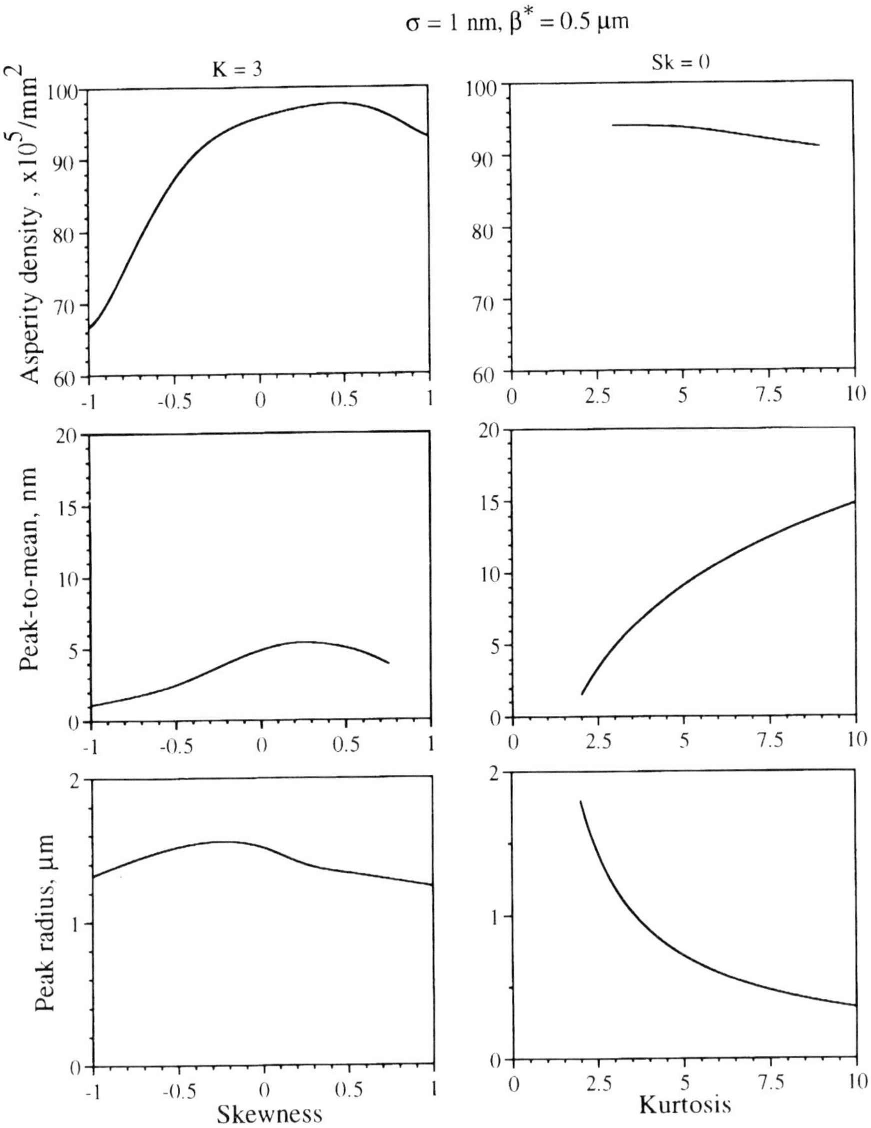

Fig. 6 Asperity density, peak-to-mean height and peak radius as functions of skewness and kurtosis

Table 2 Asperity density, peak-to-mean height and peak radius data at different β^* and σ of Gaussian surfaces

(a) $\sigma = 1$ nm

β^* (μm)	Asperity density ($\times 10^5/\text{mm}^2$)	Peak-to-mean height (nm)	Peak radius (μm)
0.1	249	4.49	0.714
0.5	131	4.40	1.514
1.0	93	4.51	2.262

(b) $\beta^* = 0.5$ μm

σ (nm)	Asperity density ($\times 10^5/\text{mm}^2$)	Peak-to-mean height (nm)	Peak radius (μm)
1	131	4.40	1.514
5	131	22.0	0.303

skewness above a value of 0.5. As shown in Fig. 7, the fractional contact area and relative meniscus force decrease with an increase in the kurtosis value. However, as expected, the maximum pressure increases with kurtosis due to a decrease in the contact area with kurtosis. An increase in the peak-to-mean distance and a decrease in the peak density with an increase in kurtosis (Fig. 6) result in a lower contact area and a higher contact pressure at high kurtosis values.

Figure 8 shows three-dimensional pressure profiles at different skewness and kurtosis values. A surface with a skewness of -1 has a greater number of contact points which lowers contact pressure whereas a surface with slight positive skewness (0 or 0.5) has higher pressure spikes, as

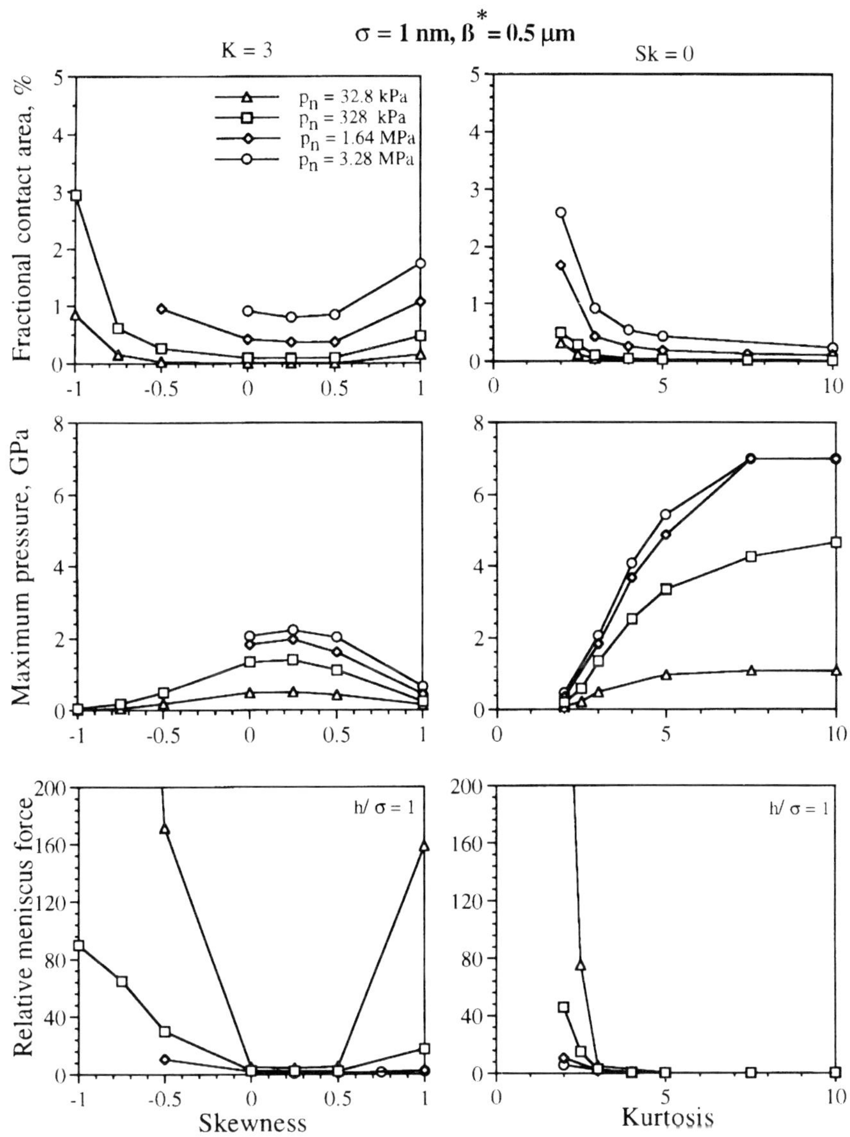

Fig. 7 Variations of fractional contact area, maximum contact pressure and relative meniscus force as functions of skewness (Sk) and kurtosis (K) at different nominal normal pressures

shown in the figure. The number of contacts decrease and the peaks become higher and sharper with an increase in kurtosis, which results in higher pressure spikes at higher kurtosis values. Higher contact pressures may lead to a significant amount of plastic deformation at nano and microscales. Defects in the material and fluctuations in the applied load may increase normal load or decrease effective hardness by a factor of 2 or 3 (**1**). Therefore the effective maximum pressure may be higher by a factor of 2 or more than that shown in Figs 7 and 8. As explained in the following subsections, contact pressure also increases with an increase in σ and a decrease in β^*. Thus surfaces with high kurtosis and σ may fail due to plastic deformation. Failure due to plastic deformation can be avoided by using a surface with a high β^* and/or a low σ, to be discussed later.

4.1 Effect of autocorrelation length (β^*)

As mentioned earlier, β^* gives the spatial distribution of asperity peaks. Figure 9a shows the variation of the fractional contact area, maximum pressure and relative meniscus force at different values of β^*. As shown in these figures, β^* has a negligible effect on the fractional contact area and relative meniscus force. However, the maximum pressure decreases with an increase in the β^* value. In an earlier study, Poon and Bhushan (**21**) studied the effect of σ and β^* on the contact area and contact pressure for computer-generated Gaussian surfaces. In this study they showed that contact pressure is inversely proportional to β^*, which agrees well with the present results. The authors believe that the increase in the peak radius results in a

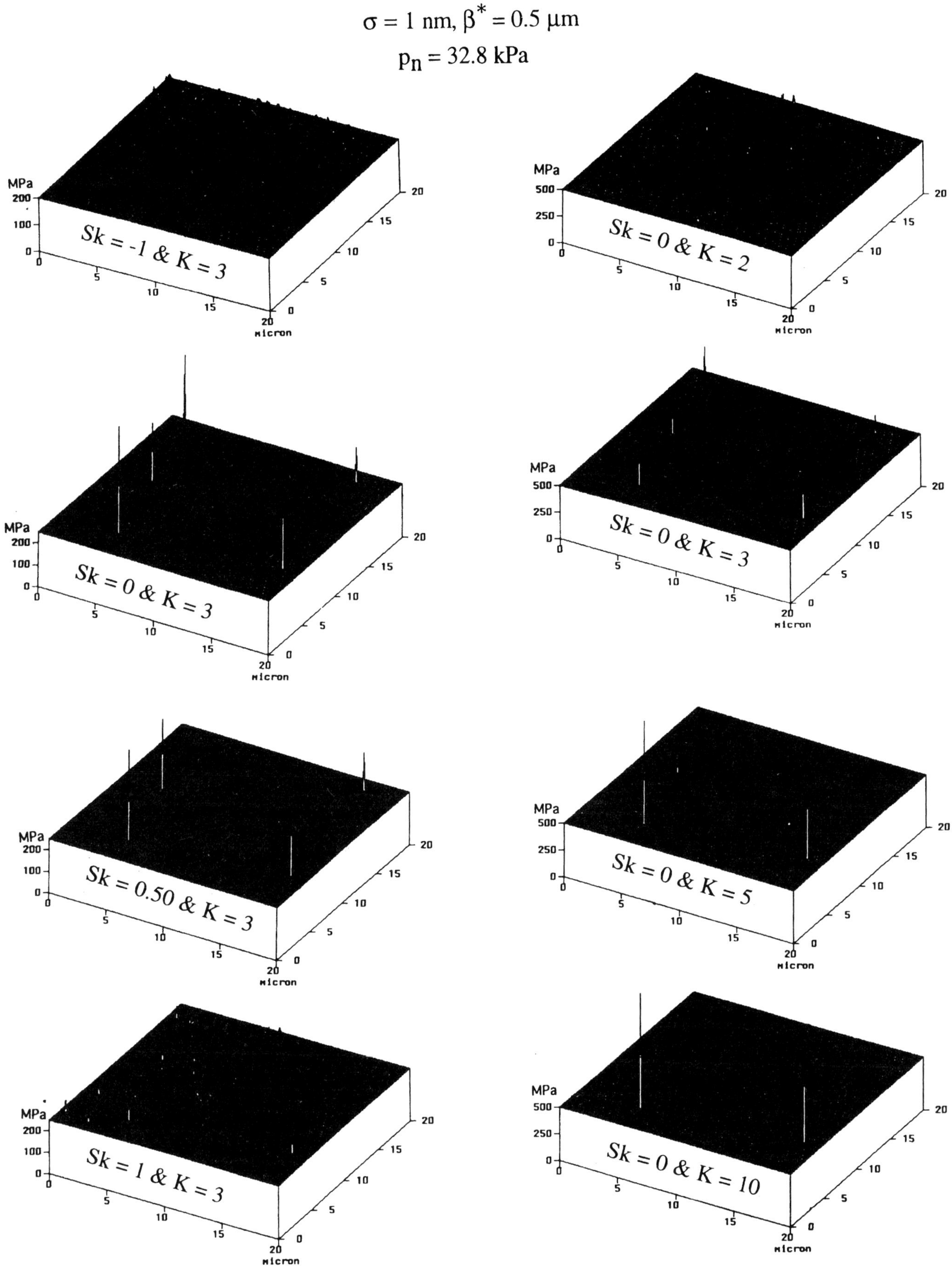

Fig. 8 Contact pressure profiles at different skewness (Sk) and kurtosis (K) values

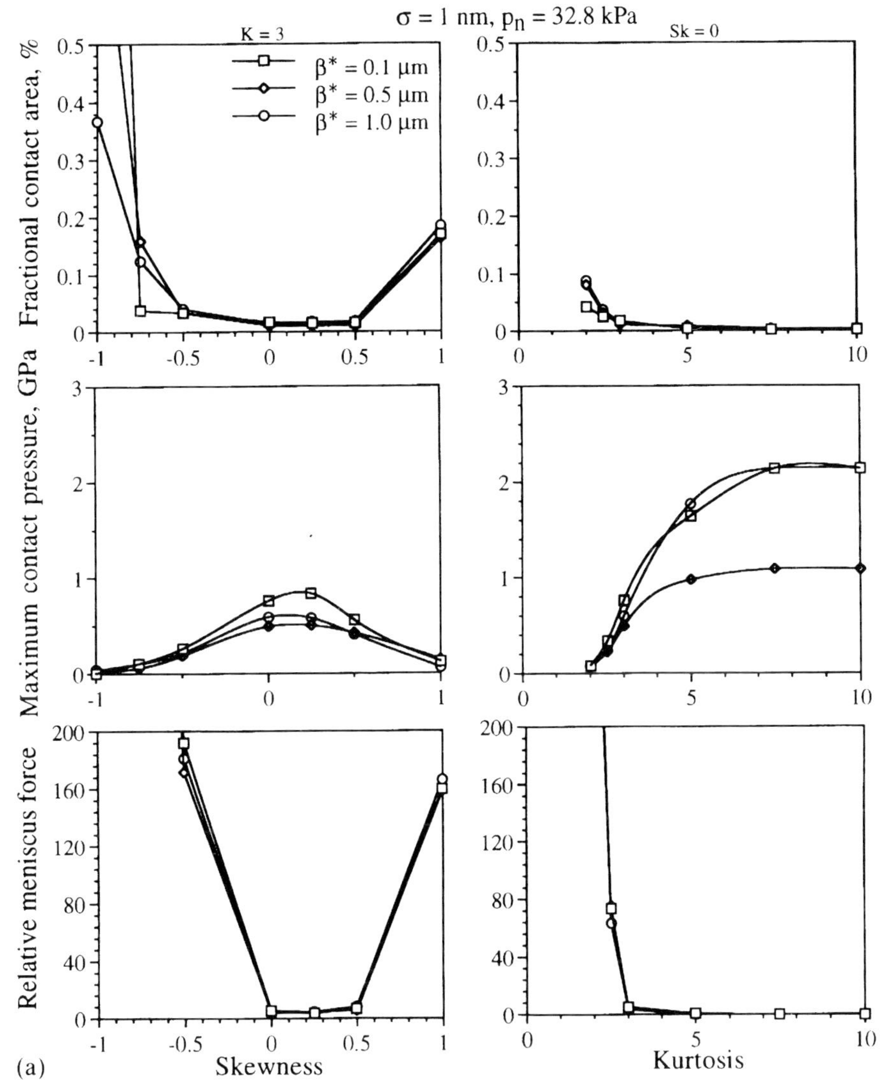

Fig. 9a Variations of fractional contact area, maximum contact pressure and relative meniscus force as functions of skewness and kurtosis at three β^* values

decrease in the maximum pressure, with an increase in β^* (see Table 2). It is clear from Fig. 9a that contact pressure exceeds the effective hardness value of the material at higher kurtosis values. Thus high contact pressures which may lead to severe plastic deformation can be avoided by using a surface with high β^*.

4.2 Effect of standard deviation of surface heights (σ)

The fractional contact area, maximum pressure and relative meniscus force are plotted as a function of skewness and kurtosis and are shown in Fig. 9b. The fractional contact area and relative meniscus force decrease with an increase of σ. The decrease in the peak radius with σ gives a lower contact area and a higher maximum pressure at higher σ, as shown in Fig. 9b. At higher σ the critical range of skewness (0–0.5) and kurtosis (> 5) at which the contact area and relative meniscus force are optimum changes to Sk = 0–0.5 and $K > 3$. However, high σ may lead to very high contact pressures, which may in turn lead to the failure of the material due to plastic deformation (especially at high kurtosis values). The effect of skewness is not very clear in Figs 9a and b because of the very few contact points. This effect can be seen more clearly at higher loads.

Figure 10 shows the variation of relative meniscus force with film thickness at different skewness and kurtosis values at $\sigma = 1$ and 5 nm. It is clear from this figure that surfaces become insensitive to film thickness at $\sigma = 5$ nm or higher. The critical film thickness at which relative meniscus force increases rapidly moves to the left with an increase in kurtosis.

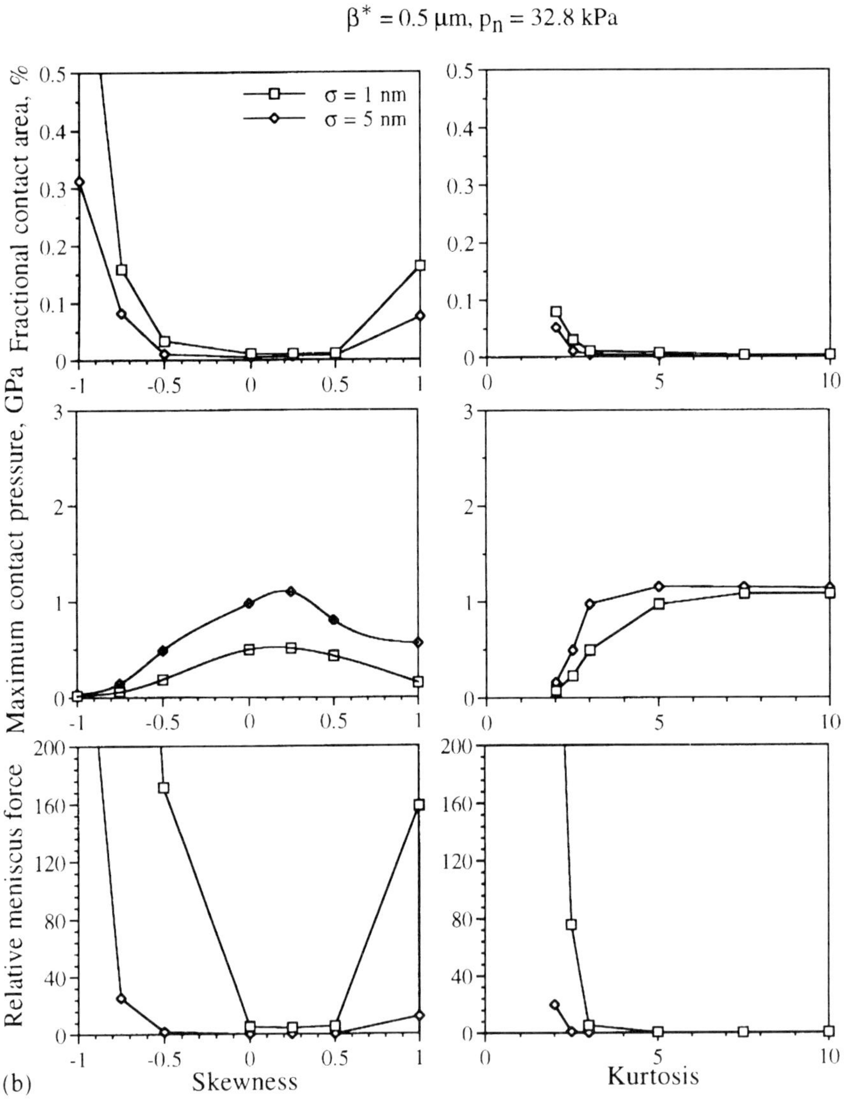

Fig. 9b Variations of fractional contact area, maximum contact pressure and relative meniscus force as functions of skewness and kurtosis at two σ values

4.3 Interplay of Sk and *K*

Figure 11 shows the combined effect of skewness and kurtosis on the contact area, maximum contact pressure and relative meniscus force. As shown in this figure, at higher kurtosis ($K \geqslant 5$) skewness has little effect on these parameters. At $\mathrm{Sk} \leqslant -0.5$ and $\mathrm{Sk} \geqslant 1$, kurtosis values greater than 5 result in a minimum contact area and relative meniscus force. However, between Sk = 0 and 0.5 a kurtosis value greater than 3 gives a minimum contact area and meniscus force.

5 CONCLUSIONS

A computer program was developed to generate non-Gaussian random surfaces with given values of skewness and kurtosis. A surface with positive skewness between 0 and 0.25 gives a minimum contact area and relative meniscus force; however, contact pressure is maximum for this surface. A surface with high negative skewness results in a high contact area and high meniscus force. The contact area and relative meniscus force decrease whereas maximum pressure increases with an increase in kurtosis of the surface. The correlation length has a minor effect on the contact area and relative meniscus force but contact pressure decreases with an increase in correlation length. For a surface with high σ and high kurtosis the contact pressure may reach the value of hardness, which may result in plastic deformation. Therefore, surfaces with high σ and kurtosis should be avoided at higher loads. The contact pressure can be slightly reduced by increasing the β^* value. The critical film thickness at which the meniscus force increases rapidly with film thickness increases with an increase in kurtosis and σ.

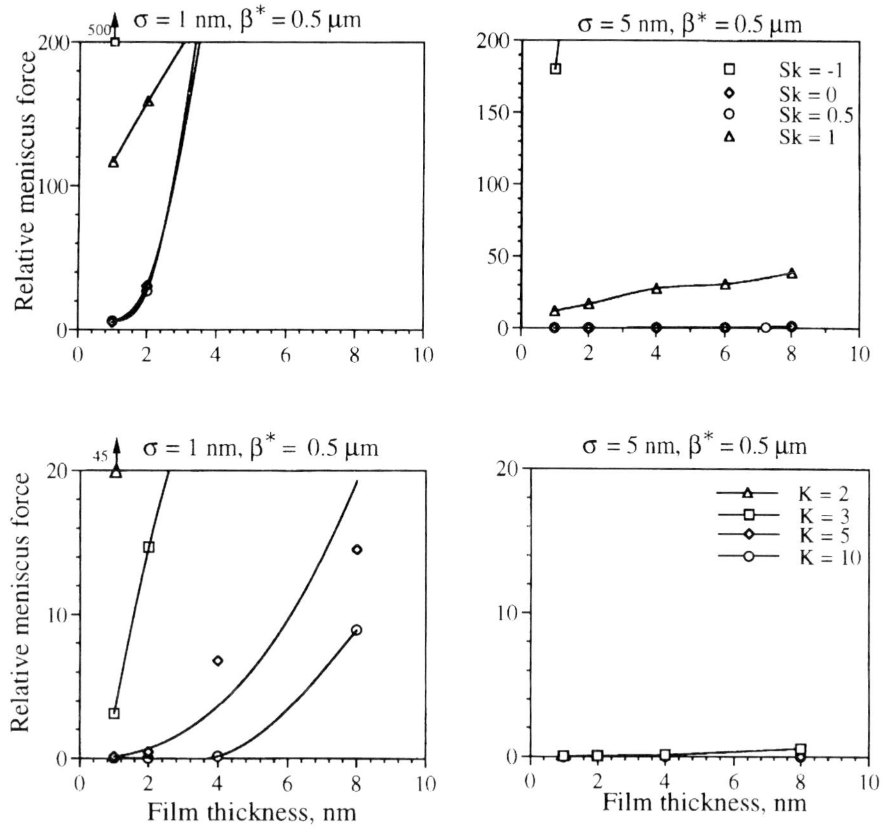

Fig. 10 Effect of σ on relative meniscus force as a function of film thickness

ACKNOWLEDGEMENT

Financial support for this research was provided by the membership of the Computer Microtribology and Contamination Laboratory.

REFERENCES

1 **Bhushan, B.** *Tribology and Mechanics of Magnetic Storage Devices*, 2nd edition, 1996 (Springer-Verlag, New York).
2 **Greenwood, J. A.** and **Williamson, J. B. P.** Contact of nominally flat surfaces. *Proc. R. Soc. (Lond.)*, 1966, **A295**, 300–319.
3 **Onions, R. A.** and **Archards, J. F.** The contact of surfaces having a random structure. *J. Phys. D: Appl. Phys.*, 1973, **6**, 289–304.
4 **Webster, M. N.** and **Sayles, R. S.** A numerical model for the elastic frictionless contact of real rough surfaces. *Trans. ASME, J. Tribology*, 1986, **108**, 314–320.
5 **Tian, X.** and **Bhushan, B.** A numerical three-dimensional model for the contact of rough surfaces by variational principle. *Trans. ASME, J. Tribology*, 1996, **118**, 33–42.
6 **Bhushan, B.** Contact mechanics of rough surfaces in tribology: multiple asperity contact. *Tribology Lett.* (in press).
7 **Thomas, T. R.** *Rough Surfaces*, 1982 (Longman, New York).
8 **Whitehouse, D. J.** *Handbook of Surface Metrology*, 1994 (Institute of Physics Publishing, Bristol).
9 **Merriman, T.** and **Kannel, J.** Analyses of the role of surface roughness on contact stresses between elastic cylinders with and without soft surface coating. *Trans. ASME, J. Tribology*, 1989, **111**, 87–94.
10 **Sayles, R. S.** and **Thomas, T. R.** A stochastic explanation of some structural properties of a ground surface. *Int. Prod. Res.*, 1976, **14**, 641–655.
11 **Mirzamaani, M.** and **Doerner, M. F.** Magnetic performance and tribology of sputter-textured thin film disks. *IEEE Trans. Magn.*, 1996, **32**, 3638–3641.
12 **Ranjan, R., Lambeth, D. N., Tromel, M., Goglia, P.** and **Li, Y.** Laser texturing for low-flying-height media. *J. Appl. Phys.*, 1991, **69**, 5745–5747.
13 **Baumgart, P., Krajnovih, D. J., Nguyen, T. A.** and **Tam, A. C.** A new laser texturing technique for high performance magnetic disk drives. *IEEE Trans. Magn.*, 1995, **31**, 2946–2951.
14 **Kotwal, C. A.** and **Bhushan, B.** Contact analysis of non-Gaussian surfaces for minimum static and kinetic friction and wear. *Tribology Trans.*, 1996, **39**, 890–898.

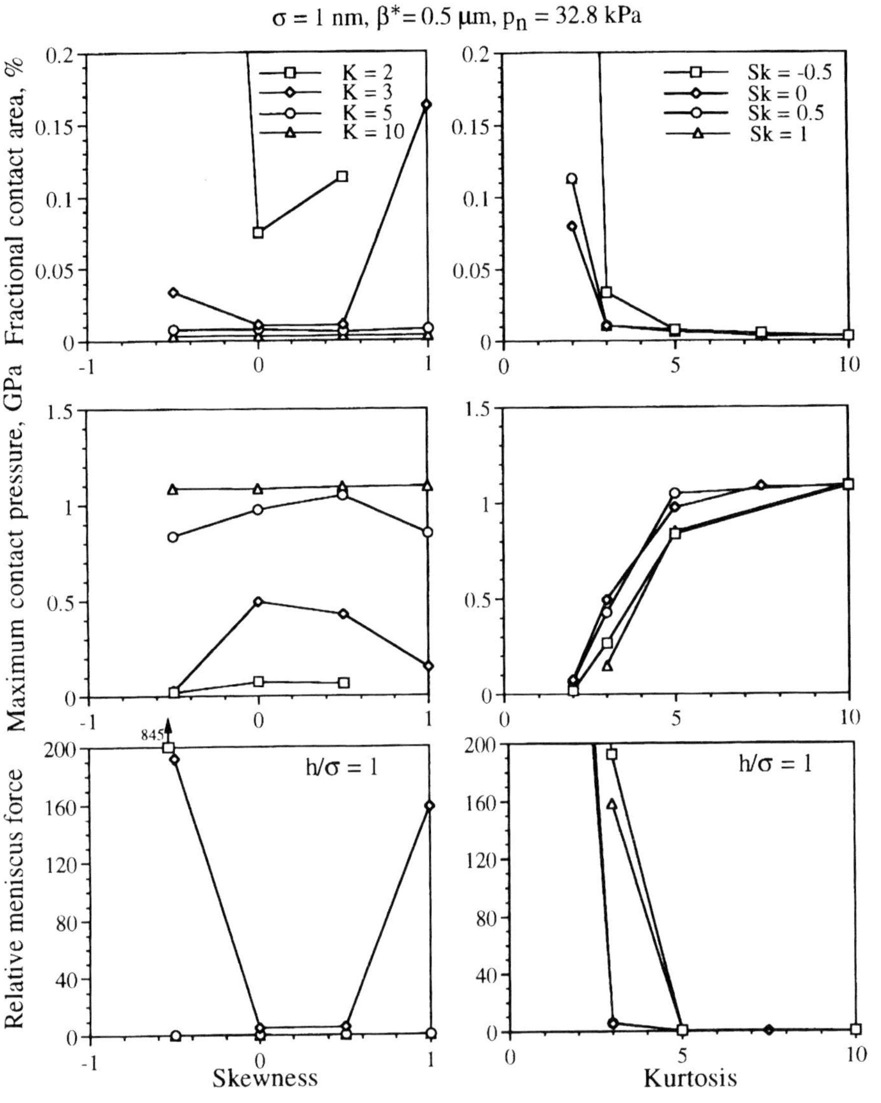

Fig. 11 Interplay of skewness and kurtosis on fractional contact area, maximum contact pressure and relative meniscus force (for film thickness/σ = 1)

15 **Whitehouse, D. J.** and **Archard, J. F.** The properties of random surfaces of significance in their contact. *Proc. R. Soc. (Lond.)*, 1970, **A316**, 97–121.

16 **Hirst, W.** and **Hollander, A. E.** Surface finish and damage in sliding. *Proc. R. Soc. (Lond.)*, 1974, **A337**, 379–394.

17 **Hu, Y. Z.** and **Tonder, K.** Simulation of 3-D random surface by 2-D digital filter and Fourier analysis. *Int. J. Mach. Tool Mf.*, 1992, **32**, 82–90.

18 **Johnson, N. L.** System of frequency curves generated by methods of translation. *Biometrika*, 1949, **36**, 149–167.

19 **Elderton, W. P.** and **Johnson, N. L.** *Systems of Frequency Curves*, 1969 (Cambridge University Press, London).

20 **Tian, X.** and **Bhushan, B.** The micro-meniscus effect of a thin liquid film on the static friction of rough surface contact. *J. Phys. D: Appl. Phys.*, 1996, **29**, 163–178.

21 **Poon, C. Y.** and **Bhushan, B.** Numerical contact and stiction analyses of Gaussian isotropic surfaces for magnetic head slider/disk contact. *Wear*, 1996, **202**, 68–82.

Microtribology

Micro/nanotribology using atomic force microscopy/friction force microscopy: state of the art

B Bhushan
Computer Microtribology and Contamination Laboratory, Department of Mechanical Engineering, The Ohio State University, Columbus, Ohio, USA

Abstract: Atomic force microscopy/friction force microscopy (AFM/FFM) techniques are increasingly used for tribological studies of engineering surfaces at scales ranging from atomic and molecular to microscales. These techniques have been used to study surface roughness, adhesion, friction, scratching/-wear, indentation, detection of material transfer and boundary lubrication and for nanofabrication/nanomachining purposes.

Micro/nanotribological studies of materials of scientific and engineering interest have been conducted. Commonly measured roughness parameters are found to be scale dependent, requiring the need of scale-independent fractal parameters to characterize surface roughness. Measurement of atomic-scale friction of a freshly cleaved highly orientated pyrolytic graphite exhibited the same periodicity as that of corresponding topography. However, the peaks in friction and those in corresponding topography were displaced relative to each other. Variations in atomic-scale friction and the observed displacement have been explained by the variations in interatomic forces in the normal and lateral directions. Local variation in microscale friction is found to correspond to the local slope, suggesting that a ratchet mechanism is responsible for this variation. Directionality in the friction is observed on both micro- and macroscales which results from the surface preparation and anisotropy in surface roughness. Microscale friction is generally found to be smaller than macroscale friction as there is less ploughing contribution in microscale measurements. Microscale friction is load dependent and friction values increase with an increase in the normal load, approaching the macrofriction at contact stresses higher than the hardness of the softer material. The wear rate for single-crystal silicon is negligible below 20 μN and is much higher and remains approximately constant at higher loads. Elastic deformation at low loads is responsible for negligible wear. The mechanism of material removal on a microscale is studied. At the loads used in the study, material is removed by the ploughing mode in a brittle manner without much plastic deformation. Most of the wear debris is loose. Evolution of the wear has also been studied using AFM. Wear is found to be initiated at nanoscratches. AFM has been modified to obtain load–displacement curves and for measurement of nanoindentation hardness and Young's modulus of elasticity, with the depth of indentation as low as 1 nm. Hardness of ceramics on the nanoscale is found to be higher than that on the microscale. Ceramics exhibit significant plasticity and creep on the nanoscale. Scratching and indentation on nanoscales are powerful ways to screen for adhesion and resistance to deformation of ultra-thin films. Detection of material transfer on the nanoscale is possible with AFM. Boundary lubrication studies and measurement of lubricant-film thickness with a lateral resolution on a nanoscale have been conducted using AFM. Self-assembled monolayers and chemically bonded lubricant films with a mobile fraction are superior in wear resistance.

Friction and wear on micro- and nanoscales at low loads have been found to be generally smaller compared to that at macroscales. Therefore, micro/nanotribological studies may help define the regimes for ultra-low friction and near-zero wear.

Keywords: atomic force microscopy, friction force microscopy, micro/nanotribology, surface roughness, deformation, wear

Presented as an invited paper at the World Tribology Congress held in London on 8–12 September 1997. The MS was received on 13 January 1997 and was accepted for publication on 3 July 1997.

1 INTRODUCTION

The recent emergence and proliferation of proximal probes, in particular tip-based microscopies [scanning tunnelling microscopy and atomic force microscopy (**1**–**3**)], have allowed the study of surface topography, adhesion, friction, wear, lubrication and measurement of mechanical properties, all on a micro- to nanometre scale, and the imaging of lubricant molecules. The availability of supercomputers to conduct atomic scale simulations has led to the development of a new field referred to as microtribology or nanotribology (**4**–**7**). This field is concerned with experimental and theoretical investigations of processes ranging from atomic and molecular scales to the microscale, occurring during adhesion, friction, wear and thin-film lubrication at sliding surfaces. At most solid–solid interfaces of technological relevance, contact occurs at numerous asperities; a sharp atomic force microscope tip sliding on a surface simulates just one such contact.

The micro/nanotribological studies are needed to develop fundamental understanding of interfacial phenomena on a small scale and to study interfacial phenomena in micro- and nanostructures used in magnetic storage systems, microelectromechanical systems (MEMS) and other industrial applications (**8**–**12**). Friction and wear of lightly loaded micro/nanocomponents are highly dependent on the surface interactions (few atomic layers). These structures are generally lubricated with molecularly thin films. Micro- and nanotribological studies are also valuable in gaining a fundamental understanding of interfacial phenomena in macrostructures to provide a bridge between science and engineering (**9**, **13**–**15**).

Atomic force microscopy (AFM) can be used for the measurement of all engineering surfaces that may be either electrically conducting or insulating. AFM has become a popular surface profiler for topographic measurements on the micro- to nanoscale. AFM has been modified to measure both normal and friction forces and this instrument is generally called the 'friction force microscope' (FFM) or 'lateral force microscope' (LFM). By using a standard or a sharp diamond tip mounted on a stiff cantilever beam, researchers have used AFM for scratching, wear and measurements of elastic/plastic mechanical properties (such as load–displacement curves, indentation hardness and modulus of elasticity). Boundary lubrication studies of molecularly thick lubricant films can be conducted using FFMs.

Surface roughness, adhesion, friction, wear and lubrication at the interface between two solids with and without liquid films have been studied using AFM and the FFM. AFM and its modifications have also been used for nanomechanical characterization. The status of current understanding of micro/nanotribology of engineering interfaces follows.

2 EXPERIMENTAL TECHNIQUES

2.1 AFM/FFM

The atomic force microscope relies on a scanning technique to produce very high resolution, three-dimensional images of sample surfaces. AFM measures ultra-small forces (less than 1 nN) present between the AFM tip surface and a sample surface. These small forces are measured by measuring the motion of a very flexible cantilever beam having an ultra-small mass. The deflection can be measured to within 0.02 nm, so for a typical cantilever force constant of 10 N/m, a force as low as 0.2 nN can be detected. AFM is capable of investigating surfaces of both conductors and insulators on an atomic scale. In the operation of high-resolution AFM, the sample is generally scanned; however, AFMs are now available where the tip is scanned and the sample is stationary (**5**). To obtain atomic resolution with AFM, the spring constant of the cantilever should be weaker than the equivalent spring between atoms. A cantilever beam with a spring constant of about 1 N/m or lower is desirable. Tips have to be as sharp as possible. Tips with a radius ranging from 10 to 100 nm are commonly available.

A commercial AFM/FFM commonly used to conduct studies of friction, scratching, wear, indentation and lubrication from micro- to atomic scales and nanofabrication/nanomachining is shown in Fig. 1 (**5**). Simultaneous measurements of surface roughness and friction force can be made with this instrument. In AFM/FFM, the sample is mounted on a PZT tube scanner which consists of separate electrodes to scan precisely the sample in the *XY* plane in a raster pattern and to move the sample in the vertical (*Z*) direction. A sharp tip at the end of a flexible cantilever is brought in contact with the sample. Normal and frictional forces applied at the tip–sample interface are measured using a laser beam deflection technique. A laser beam from a diode laser is directed by a prism on to the back of a cantilever near its free end, tilted downward at about 10° with respect to a horizontal plane. The reflected beam from the vertex of the cantilever is directed through a mirror on to a quad photodetector (split photodetector with four quadrants). The differential signal from the top and bottom photodiodes provides the AFM signal, which is a sensitive measure of the cantilever vertical deflection. Topographic features of the sample cause the tip to deflect in the vertical direction as the sample is scanned under the tip. This tip deflection will change the direction of the reflected laser beam, changing the intensity difference between the top and bottom photodetector (AFM signal). In the AFM operating mode of the 'height mode', for topographic imaging or for any other operation in which the applied normal force is to be kept constant, a feedback circuit is used to modulate the voltage applied to the PZT scanner to adjust the height of the PZT, so that the cantilever vertical deflection (given by the intensity difference between the top and bottom detectors) will remain almost constant

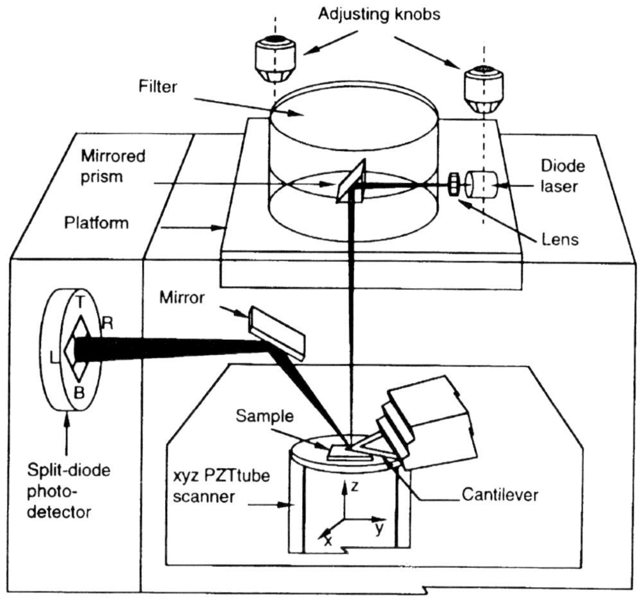

Fig. 1 Schematic of a commercial atomic force microscope/friction force microscope (AFM/FFM) using the laser-beam deflection method

during scanning. The PZT height variation is thus a direct measure of surface roughness of the sample.

For measurement of friction force applied at the tip surface during sliding, the other two (left and right) quadrants of the photodetector (arranged horizontally) are used. In the 'friction mode', the sample is scanned back and forth in a direction orthogonal to the long axis of the cantilever beam. Friction force between the sample and the tip will produce a twisting of the cantilever. As a result, the laser beam will be reflected out of the plane defined by the incident beam, with the beam reflected vertically from an untwisted cantilever. This produces an intensity difference of the laser beam received in the left and right quadrants of the photodetector. The intensity difference between the left and right detectors (FFM signal) is directly related to the degree of twisting and hence to the magnitude of friction force. One problem associated with this method is that any misalignment between the laser beam and the photodetector axis would introduce error in the measurement. However, following the procedures developed by Ruan and Bhushan (**16**), the average FFM signal for the sample scanned in two opposite directions is subtracted from the friction profiles of each of the two scans to eliminate the misalignment effect. This method provides three-dimensional maps of the friction force. By following the friction force calibration procedures developed by Ruan and Bhushan (**16**) and Bhushan (**5**), voltages corresponding to friction forces can be converted to force units.

Topographic measurements are typically made using a sharp tip on a cantilever beam with normal stiffness of about 0.58 N/m at a normal load of about 10 nN. Friction measurements are carried out in the load range of 10–150 nN. The tip is scanned in such a way that its trajectory on the sample forms a triangular pattern. Scanning speeds in the fast and slow scan directions depend on the scan area and scan frequency. A maximum scan size of 125 μm × 125 μm and scan rate of 122 Hz typically can be used. Higher scan rates are used for small scan lengths.

For nanoscale boundary lubrication studies, the samples are typically scanned over an area of 1 μm × 1 μm at a normal force of about 300 nN, in a direction orthogonal to the long axis of the cantilever beam. The samples are generally scanned with a scan rate of 1 Hz and a scanning speed of 2 μm/s. The coefficient of friction is monitored during scanning for a desired number of cycles. After the scanning test, a larger area of 2 μm × 2 μm is scanned at a normal force of 40 nN to observe any wear scar.

For microscale scratching, microscale wear and nanoscale indentation hardness measurements, a three-sided pyramidal single-crystal natural diamond tip with an apex angle of 80° and a radius of about 100 nm mounted on a stainless steel cantilever beam with normal stiffness of about 25 N/m is used at relatively higher loads (1–150 mN). For scratching and wear studies, the sample is generally scanned in a direction orthogonal to the long axis of the cantilever beam (typically at a rate of 0.5 Hz) so that friction can be measured during scratching and wear. The tip is mounted on the beam such that one of its edges is orthogonal to the long axis of the beam; therefore, wear during scanning along the beam axis is higher (about 2–3 times) than that during scanning orthogonal to the beam axis. For wear studies, typically an area of 2 μm × 2 μm is scanned at various normal loads (ranging from 1 to 100 mN) for a selected number of cycles.

For nanoindentation hardness measurements the scan size is set to zero and then normal load is applied to make the indents. During this procedure the diamond tip is continuously pressed against the sample surface for about 2 s at various indentation loads. The sample surface is scanned before and after the scratching, wear or indentation to obtain the initial and the final surface topography, at a low normal load of about 0.3 mN using the same diamond tip. An area larger than the indentation region is scanned to observe the indentation marks. Nanohardness is calculated by dividing the indentation load by the projected residual area of the indents.

For measurements of surface roughness, friction force, nanoscale scratching and wear, a microfabricated square-pyramidal Si_3N_4 or silicon tip with a tip radius ranging from 10 to 50 nm (Fig. 2a) is generally used at loads ranging from 10 to 150 nN. For measurements of micro-scale scratching and wear and for nanoindentation hardness measurements and nanofabrication, a three-sided pyramidal single-crystal natural diamond tip with a tip radius of about 100 nm (Fig. 2b) is generally used at relatively high loads ranging from 10 to 150 mN.

(a) Square pyramidal Si_3N_4 tip

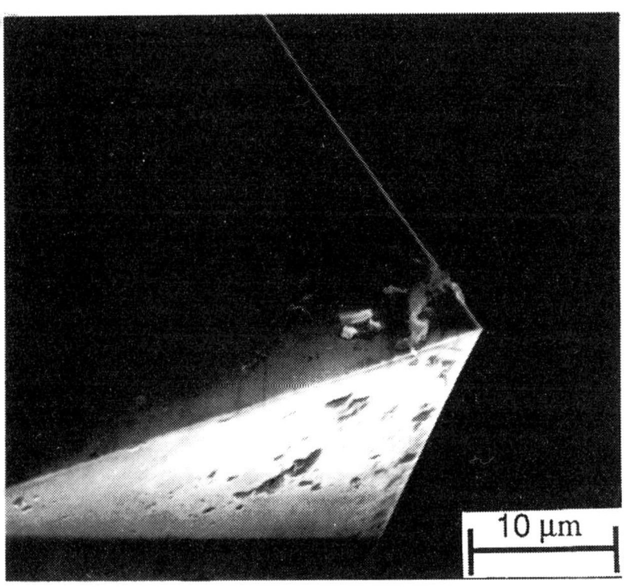

(b) Three-sided pyramidal (natural) diamond tip

Fig. 2 SEM micrographs of a PECVD Si_3N_4 cantilever beam with a tip and a stainless steel cantilever beam with a diamond tip

2.2 Nano/picoindenter

As described earlier, conventional AFMs have been used for indentation studies on nanometre-scale depths. In these studies the hardness value is based on the projected residual area after imaging the indent. Direct imaging of the indent allows the piling up of ductile material around the indenter to be quantified. However, it becomes difficult to identify the boundary of the indentation mark with great accuracy. This makes the direct measurement of contact area somewhat inaccurate. A technique with the dual capability of depth-sensing as well as *in situ* imaging, which is most appropriate in nanomechanical property studies, has been developed (**17**–**20**). This indentation system is used to make load–displacement measurements and subsequently carry out *in situ* imaging of the indent. A schematic of the nano/picoindenter system used is shown in Fig. 3. The indentation system consists of a three-plate transducer with electrostatic actuation hardware used for direct application of normal load and a capacitive sensor used for the measurement of vertical displacement. The AFM head is replaced with this transducer assembly while the specimen is mounted on the PZT scanner which remains stationary during indentation experiments. The transducer consists of a three-plate (Be–Cu) capacitive structure which provides a high-sensitivity, large dynamic range and a linear output signal with respect to load or displacement. The tip is mounted on the centre plate. The upper and lower plates serve as drive electrodes. Load is applied as appropriate voltage to the drive electrodes, thereby generating an electrostatic force between the centre plate and the drive electrodes. Vertical displacement of the tip (indentation depth) is found by measuring the displacement of the centre plate relative to the two outer electrodes using a capacitance technique. The load resolution is 100 nN or better and the displacement resolution is 0.1 nm. At present, a load range of 1–10 mN can be employed. Loading rates can be varied by changing the load/unload period. The AFM functions as a platform providing an *in situ* image of the indent with a lateral resolution of 1 nm and a vertical resolution of 0.2 nm. The load–displacement data can be acquired and displayed on the display monitor. The hardness value can be obtained from the load–displacement data as well as from direct measurement of the projected residual area of the indent after imaging. Young's modulus of elasticity is obtained from the slope of the unloading curve.

A three-sided Berkovich indenter with a tip radius of about 0.1 mm has been generally used for the measurements (**5**). Sharper diamond tips with an included angle of 60–90° and tip radii of 30–60 nm are sometimes employed

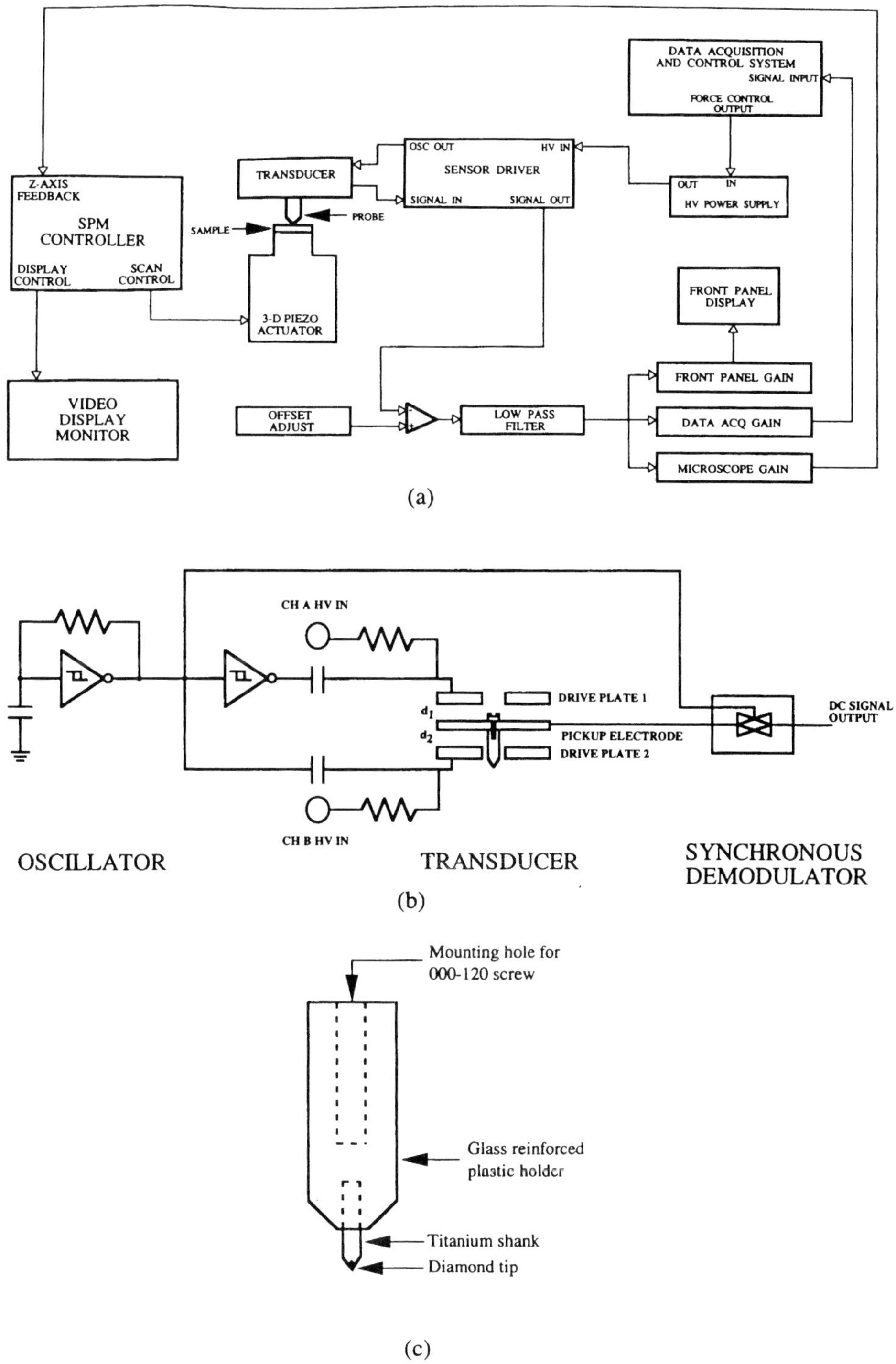

Fig. 3 Schematics of (a) a three-plate transducer with (b) electrostatic actuation hardware and (c) a capacitance sensor (**17**)

for shallower indentations (on the order of 1 nm). In order to obtain an accurate relation between the indentation depth and the projected contact area, tip shape calibration needs to be done (**5**). Also for surfaces with r.m.s. roughness on the order of the indentation depth, the original (unindented) profile is subtracted from the indented profile (**21**).

In a typical indentation experiment the tip is lowered close to the sample (ideally $< 100\ \mu m$) and the scan size and scan rate are selected. The tip is engaged to the sample surface by a stepper motor with a set point of 1 nA (about 1 mN). A desired image area is captured prior to indentation. The feedback is set to zero to disable the scanner and the scan size is set to zero so that the indenter will be positioned at the centre of the image. An appropriate set point for the preload condition is selected. The indentation rate can be varied by changing the load/unload period.

3 SURFACE ROUGHNESS

Solid surfaces, irrespective of the method of formation, contain surface irregularities or deviations from the prescribed geometrical form. When two nominally flat surfaces are placed in contact, surface roughness causes contact to occur at diskrete contact points. Deformation occurs in these points, and may be either elastic or plastic, depending on the nominal stress, surface roughness and material properties. The sum of the areas of all the contact points constitutes the real area that would be in contact, and for most materials at normal loads, this will be only a small fraction of the area of contact if the surfaces were perfectly smooth. In general, the real area of contact must be minimized in order to minimize adhesion, friction and wear (**9**, **14**).

Characterizing surface roughness is therefore important for predicting and understanding the tribological properties of solids in contact. AFM has been used to measure surface roughness on length scales from nanometres to micrometres. Roughness plots of a glass–ceramic disk measured using AFM (lateral resolution ~ 15 nm), a non-contact optical profiler (lateral resolution ~ 1 μm) and a stylus profiler (lateral resolution ~ 0.2 μm) are shown in Fig. 4a. Figure 4b compares the profiles of the disk obtained with different instruments at a common scale. The figures show that roughness is found at scales ranging from millimetre to nanometre scales. The measured roughness profile is dependent on the lateral and normal resolutions of the measuring instrument (**22**–**25**). Instruments with different lateral resolutions measure features with different scale lengths. It can be concluded that a surface is composed of a large number of length scales of roughness that are superimposed on each other.

Surface roughness is most commonly characterized by the standard deviation of surface heights, which is the square root of the arithmetic average of squares of the vertical deviation of a surface profile from its mean plane. Due to the multiscale nature of surfaces, it is found that the variances of surface height and its derivatives and other roughness parameters depend strongly on the resolution of the roughness-measuring instrument or any other form of filter, and hence are not unique for a surface (**24**–**26**) (see, for example, Fig. 5). Therefore, a rough surface should be characterized in such a way that the structural information of roughness at all scales is retained. It is necessary to quantify the multiscale nature of surface roughness.

A unique property of rough surfaces is that if a surface is repeatedly magnified, increasing details of roughness are observed right down to the nanoscale. In addition, the roughnesses at all magnifications appear quite similar in structure, as qualitatively shown in Fig. 6. That statistical self-affinity is due to similarity in appearance of a profile under different magnifications. Such a behaviour can be characterized by fractal analysis (**24**, **27**). The main conclusion from these studies is that a fractal characterization of surface roughness is *scale independent* and provides information of the roughness structure at all length scales that exhibit the fractal behaviour.

The structure function and power spectrum of a self-affine fractal surface follow a power law and can be written as (the Ganti and Bhushan model)

$$S(\tau) = C\eta^{(2D-3)}\tau^{(4-2D)} \tag{1}$$

$$P(\omega) = \frac{c_1\eta^{(2D-3)}}{\omega^{(5-2D)}} \tag{2a}$$

and

$$c_1 = \frac{\Gamma(5-2D)\sin[\pi(2-D)]}{2\pi}C \tag{2b}$$

The fractal analysis allows the characterization of surface roughness by two parameters D and C, which are instrument-independent and unique for each surface. D (ranging from 1 to 2 for the surface profile) primarily relates to relative power of the frequency contents and C to the amplitude of all frequencies. η is the lateral resolution of the measuring instrument, τ is the size of the increment (distance) and ω is the frequency of the roughness. Note that if $S(\tau)$ or $P(\omega)$ are plotted as a function of τ or ω respectively, on a log–log plot, then the power law behaviour would result in a straight line. The slope of the line is related to D and the location of the spectrum along the power axis is related to C.

Figure 7 presents the structure function of a thin-film rigid disk measured using AFM, a non-contact optical profiler and a stylus profiler. Horizontal shift in the structure functions from one scan to another arises from the change in the lateral resolution. D and C values for various scan lengths are listed in Table 1. It is noted that the fractal dimension of the various scans is fairly constant (1.26–1.33); however, C increases/decreases monotonically with σ for the AFM data. The error in estimation of η is believed to be responsible for variations in C. These data show that the disk surface follows a fractal structure for three decades of length scales.

Majumdar and Bhushan (**28**, **29**) developed a new fractal theory of contact between two rough surfaces. This model

Fig. 4 Surface roughness plots of a glass–ceramic disk (a) measured using an atomic force microscope (lateral resolution ~ 15 nm), a non-contact optical profiler (NOP) (lateral resolution ~ 1 μm) and a stylus profiler (SP) with a stylus tip of 0.2 μm radius (lateral resolution ~ 0.2 μm) and (b) measured using an AFM (~ 150 nm), SP (~ 0.2 μm) and NOP (~ 1 μm) and plotted on a common scale (**25**)

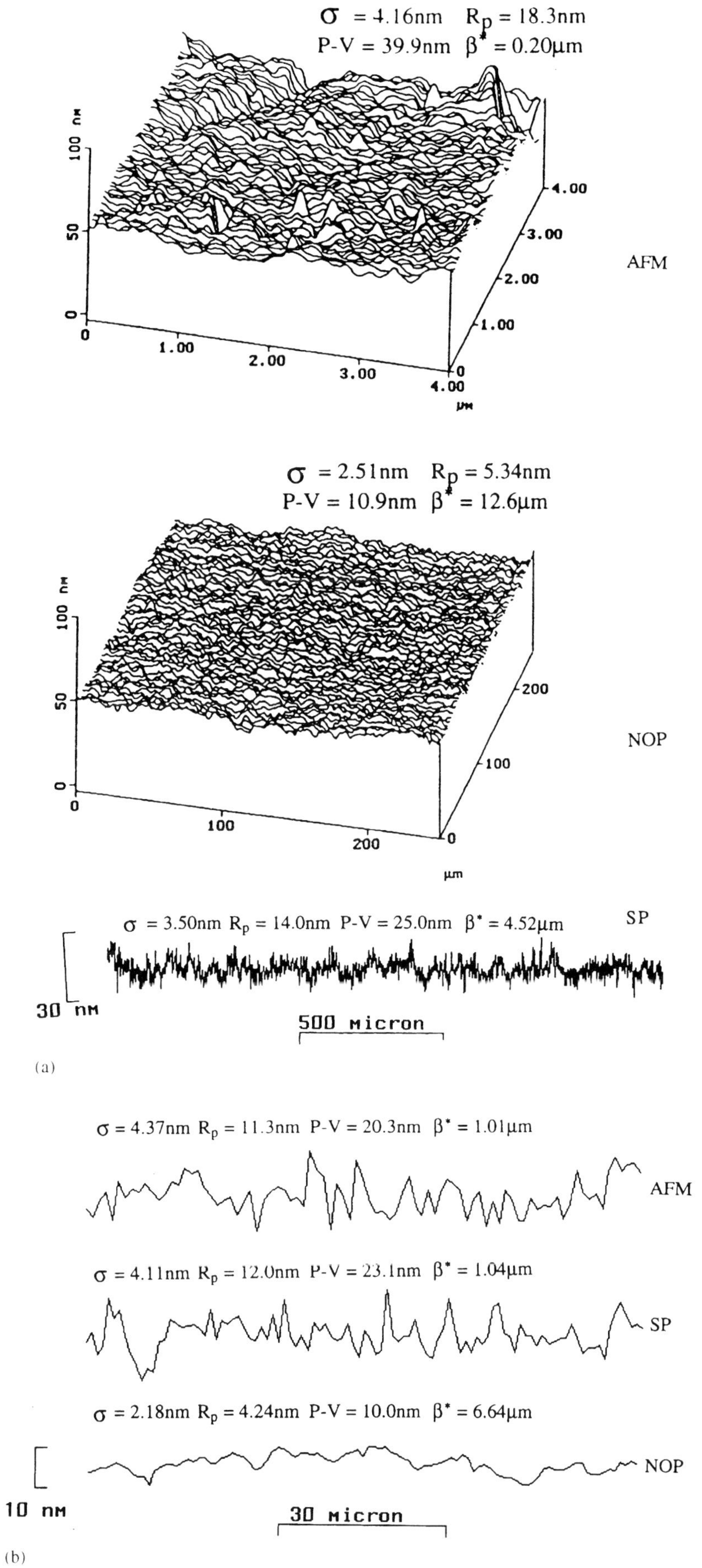
σ = 4.16nm Rp = 18.3nm
P-V = 39.9nm β* = 0.20μm
AFM
σ = 2.51nm Rp = 5.34nm
P-V = 10.9nm β* = 12.6μm
NOP
σ = 3.50nm Rp = 14.0nm P-V = 25.0nm β* = 4.52μm
SP
30 nm
500 micron
(a)
σ = 4.37nm Rp = 11.3nm P-V = 20.3nm β* = 1.01μm
AFM
σ = 4.11nm Rp = 12.0nm P-V = 23.1nm β* = 1.04μm
SP
σ = 2.18nm Rp = 4.24nm P-V = 10.0nm β* = 6.64μm
NOP
10 nm
30 micron
(b)

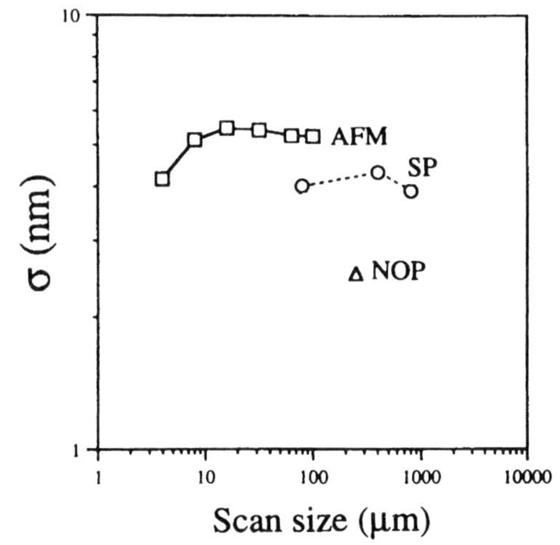

Fig. 5 Scale dependence of standard deviation of surface heights for a glass–ceramic disk measured using an atomic force microscope (AFM), stylus profiler (SP) and non-contact optical profiler (NOP)

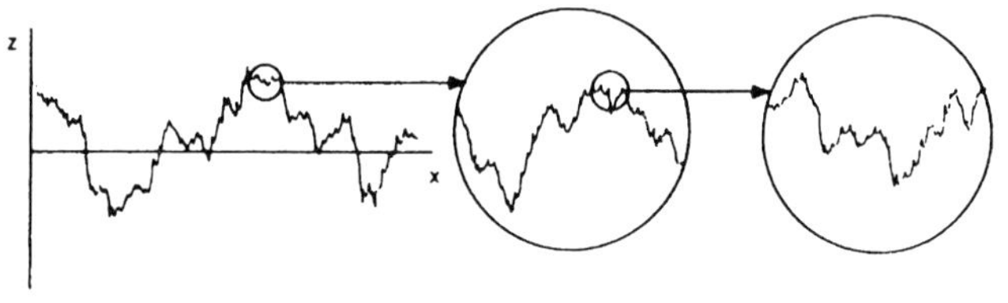

Fig. 6 Qualitative description of statistical self-affinity for a surface profile

Table 1 Surface roughness parameters for a polished thin-film rigid disk (**24**)

Scan size (μm × μm)	σ(nm)	D	C(nm)
1 (AFM)*	0.7	1.33	9.8×10^{-4}
10 (AFM)	2.1	1.31	7.6×10^{-3}
50 (AFM)	4.8	1.26	1.7×10^{-2}
100 (AFM)	5.6	1.30	1.4×10^{-2}
250 (NOP)†	2.4	1.32	2.7×10^{-4}
4000 (NOP)	3.7	1.29	7.9×10^{-5}

*AFM = atomic force microscope.
†NOP = non-contact optical profiler.

has been used to predict whether contacts experience elastic or plastic deformation and to predict the statistical distribution of contact points. For a review of contact models, see Bhushan (**30**, **31**).

4 ADHESION AND FRICTION

To study friction mechanisms on an atomic scale, a well-characterized freshly cleaved surface of highly oriented pyrolytic graphite (HOPG) has been studied by Mate *et al.* (**32**) and Ruan and Bhushan (**33**). The atomic-scale friction force of HOPG exhibited the same periodicity as that of the corresponding topography (Fig. 8a), but the peaks in friction and those in topography were displaced relative to each other (Fig. 8b). A Fourier expansion of the interatomic potential was used to calculate the conservative interatomic

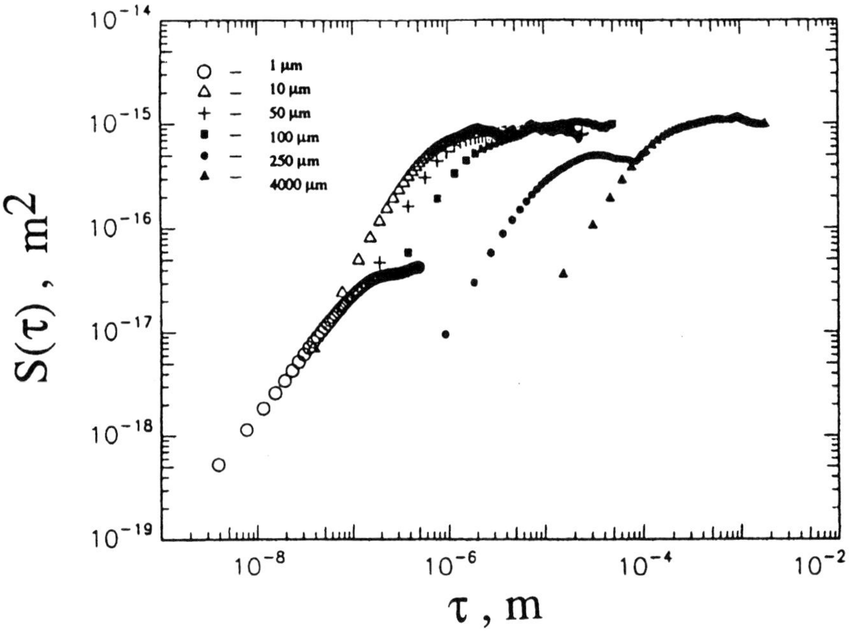

Fig. 7 Structure functions for the roughness data measured at various scan sizes using AFM (scan sizes: 1 μm × 1 μm, 10 μm × 10 μm, 50 μm × 50 μm and 100 μm × 100 μm), NOP (scan size: 250 μm × 250 μm) and SP (scan length: 4000 μm) for a magnetic thin-film rigid disk (**24**)

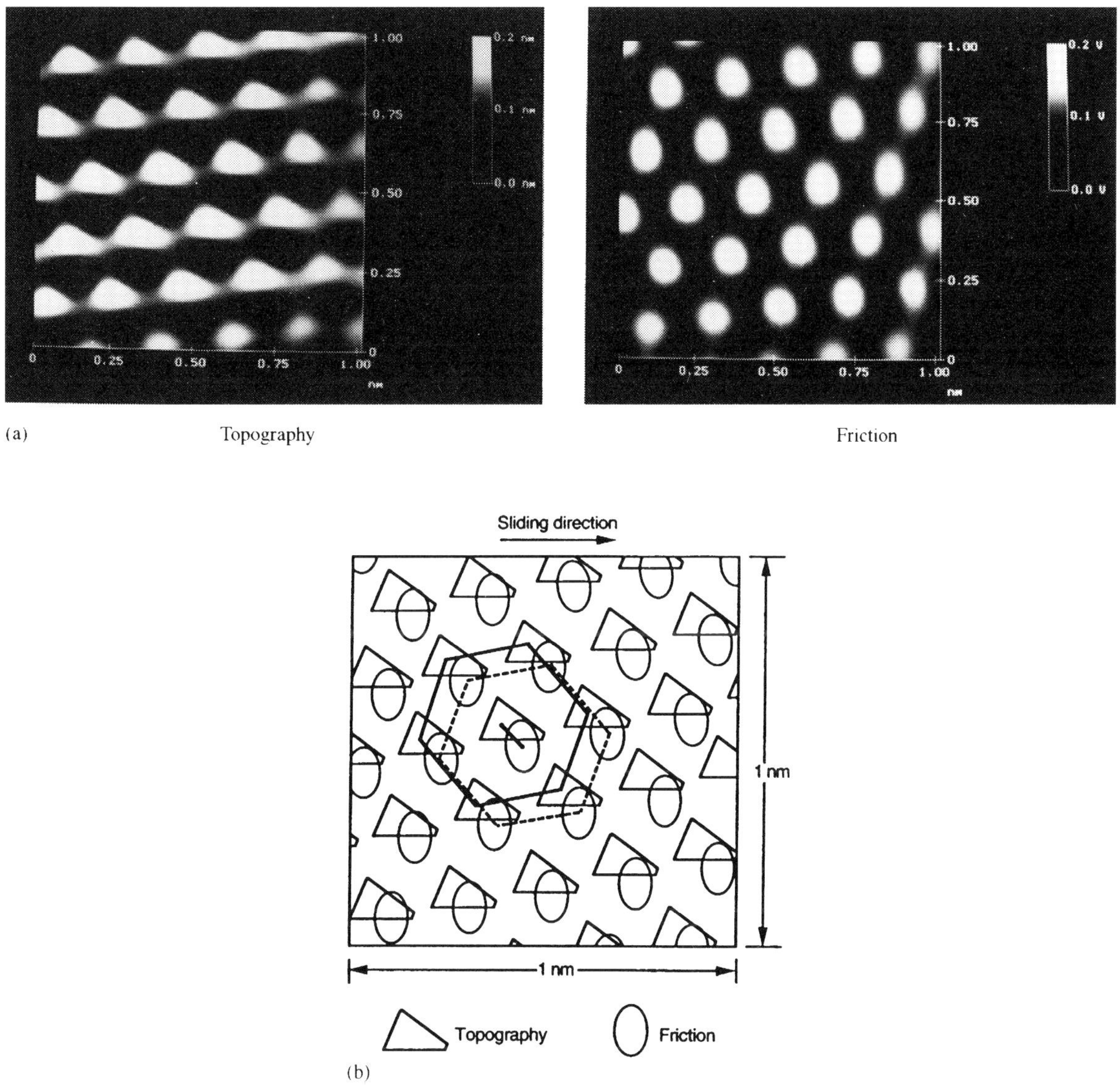

Fig. 8 (a) Grey-scale plots of surface topography (left) and friction profiles (right) of a 1 nm × 1 nm area of freshly cleaved HOPG, showing the atomic-scale variation of topography and friction. (b) Diagram of superimposed topography and friction profiles from (a); the symbols correspond to the maxima. Note the spatial shift between the two profiles (**33**)

forces between atoms of the FFM tip and those of the graphite surface. Maxima in the interatomic forces in the normal and lateral directions do not occur at the same location, which explains the observed shift between the peaks in the lateral force and those in the corresponding topography. Furthermore, the observed local variations in friction force were explained by variations in the intrinsic lateral force between the sample and the FFM tip (**33**). These variations may not necessarily occur as a result of the atomic-scale stick-slip process (**32**), but can be due to variations in the intrinsic lateral force between the sample and the FFM tip.

Friction forces of HOPG have also been studied. Local variations in the microscale friction of cleaved graphite are observed, which arise from structural changes that occur during the cleaving process (**34**). The cleaved HOPG surface is largely atomically smooth but exhibits line-shaped regions in which the coefficient of friction is more than an order of magnitude larger. Transmission electron microscopy indicates that the line-shaped regions consist of graphite planes of a different orientation, as well as of amorphous carbon. Differences in friction have also been observed for multiphase ceramic materials (**35**) and for organic mono- and multilayer films (**36**), which again seems to be the result of structural variations in the surfaces. These measurements suggest that the FFM can be used for structural mapping of the surfaces. FFM measurements can be used to map chemical variations, as indicated by the use of the FFM with a modified probe tip to map the spatial arrangement of chemical functional groups in mixed organic monolayer films (**37**). Here, sample regions that had stronger interactions with the functionalized probe tip exhibited larger friction.

Local variations in the microscale friction of scratched surfaces can be significant, and seem to depend on the local surface slope rather than the surface height distribution (**5, 21, 26**). Directionality in friction is sometimes observed on the macroscale; on the microscale this is the norm (**5, 21**). This is because most engineering surfaces have asymmetric surface asperities so that the interaction of the FFM tip with the surface is dependent on the direction of the tip motion. Moreover, during surface finishing processes material can be transferred preferentially on to one side of the asperities, which also causes asymmetry and directional dependence. Reduction in local variations and in directionality of frictional properties therefore requires careful optimization of surface roughness distributions and of surface-finishing processes.

Table 2 shows the coefficient of friction measured for two surfaces on micro- and macroscales. The coefficient of friction is defined as the ratio of friction force to the normal load. The values on the microscale are much lower than those on the macroscale. Contact stresses at AFM conditions generally do not exceed the sample hardness which minimizes plastic deformation. When measured for the small contact areas and very low loads used in microscale studies, the indentation hardness and modulus of elasticity are higher than at the macroscale. Lack of plastic deformation and improved mechanical properties reduce the degree of wear. In addition, the small apparent areas of contact reduce the number of particles trapped at the interface, and thus minimize the 'ploughing' contribution to the friction force.

At higher loads (with contact stresses exceeding the hardness of the softer material), however, the coefficient of friction for microscale measurements increases towards values comparable with those obtained from macroscale measurements, and surface damage also increases (Fig. 9) (**38**). Thus Amontons' law of friction, which states that the coefficient of friction is independent of apparent contact area and normal load, does not hold for microscale measurements. These findings suggest that microcomponents sliding under lightly loaded conditions should experience very low friction and near-zero wear.

Table 2 Surface roughness (standard deviation of surface heights or σ) and coefficients of friction on micro- and macroscale of various samples

Material	σ (nm)	Microscale coefficient of friction versus Si_3N_4 tip*	Macroscale coefficient of friction versus alumina ball†
Si(111)	0.11	0.03	0.18
C^+-implanted Si	0.33	0.02	0.18

*Tip radius of about 50 nm in the load range of 10–150 nN (2.5–6.1 GPa), a scanning speed of 5 μm/s and scan area of 1 μm × 1 μm.
†Ball radius of 3 mm at a normal load of 0.1 N (0.3 GPa) and average sliding speed of 0.8 mm/s.

5 SCRATCHING, WEAR AND INDENTATION

AFM can be used to investigate how surface materials can be moved or removed on micro- to nanoscales, e.g. in scratching and wear (**15**) (where these things are undesirable) and nanomachining/nanofabrication (where they are desirable). AFM can also be used for measurements of mechanical properties on micro- to nanoscales. Figure 10 shows microscratches made on Si(111) at various loads and a scan velocity of 2 μm/s after 10 cycles (**39**). As expected, the depth of scratch increases with load. Such microscratching measurements can be used to study failure mechanisms on the microscale and to evaluate the mechanical integrity (scratch resistance) of ultra-thin films at low loads.

By scanning the sample in two dimensions using AFM, wear scars are generated on the surface. Figure 11 shows the effect of normal load on the wear rate. It is noted that the wear rate is very small below 20 μN of normal force. A normal force of 20 μN corresponds to contact stresses comparable to the hardness of silicon. Primarily, elastic deformation at loads below 20 μN is responsible for low wear. A typical wear mark generated at a normal load of 40 μN for one scan cycle and imaged using AFM at 300 nN load is shown in Fig. 12a. The inverted line plot of the wear mark is given in Fig. 12b showing the uniform material removal at the bottom of the wear mark. Next the mechanism of material removal at light loads is examined on the microscale in AFM wear experiments (**40**). Figure 13 shows a secondary electron image of the wear mark and associated wear particles. The specimen used for SEM (scanning electron microscopy) was not scanned after initial wear, to retain wear debris in the wear region. Wear debris is clearly observed. The AFM image of the wear mark shows small debris at the edges, swiped during AFM scanning. Thus the debris is 'loose' (not sticky) and can be removed during AFM scanning. SEM micrographs show both cutting type and ribbon-like debris. TEM (transmission electron microscopy) studies were performed to enable the material removal process to be understood. The TEM micrograph of the worn region in Fig. 14 shows evidence of bend contours passing through the wear mark. The bend contours around and inside the wear mark suggest that there are some residual stresses around and inside the wear mark region. No major surface dislocation activity, phase transformation or crack formation is observed in the plan view of TEM. The dislocation activity and/or cracking probably occurs at the subsurface. TEM studies suggest that for loads of 40 μN, material is mostly removed in a brittle manner or is chipped, although the presence of some ribbon-like debris in SEM studies suggests some plastic deformation as well.

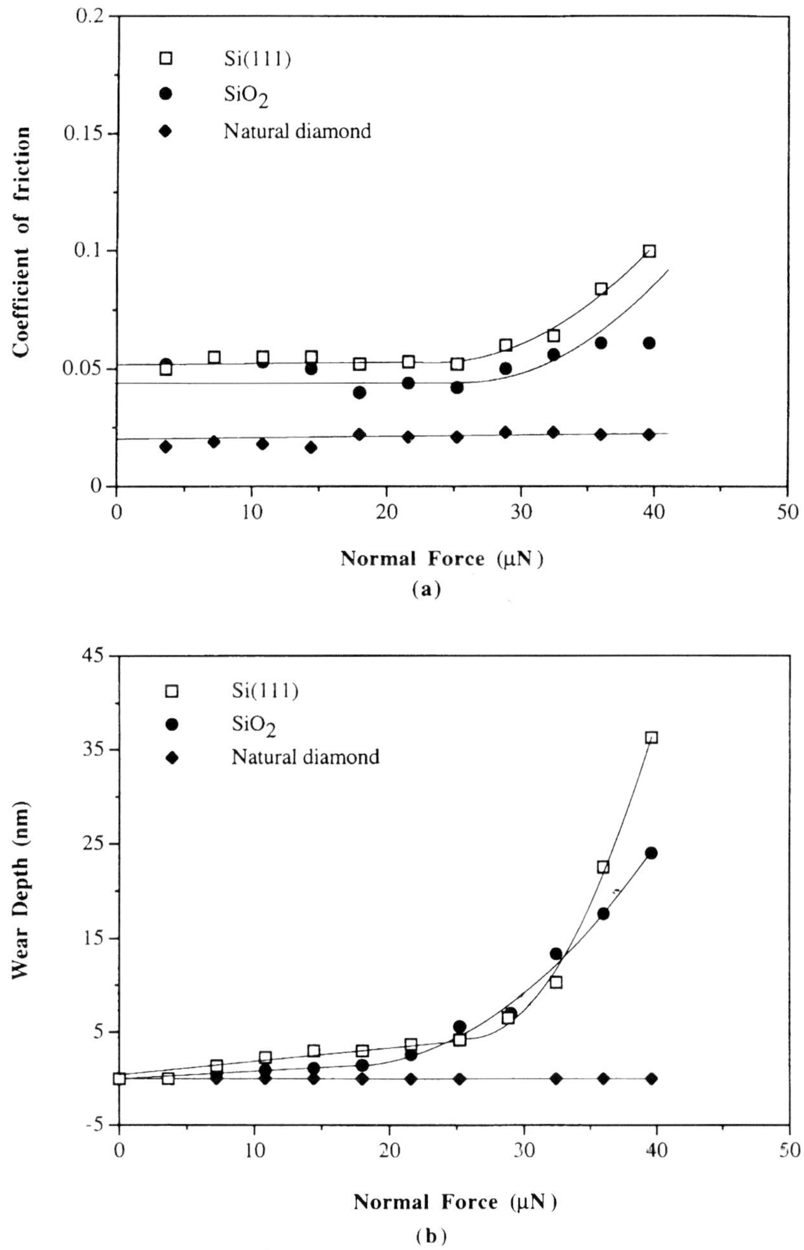

Fig. 9 (a) Coefficient of friction as a function of normal force and (b) corresponding wear depth as a function of normal force for silicon, SiO_2 coating and natural diamond. Inflections in the curves for silicon and SiO_2 correspond to the hardnesses of these materials (**38**)

The evolution of wear of a diamond-like carbon coating on a polished aluminium substrate is shown in Fig. 15, which illustrates how the microwear profile for a load of 20 μN develops as a function of the number of scanning cycles (**21**). Wear is not uniform, but is initiated at the nanoscratches, indicating that surface defects (with high surface energy) act as initiation sites. Thus, scratch-free surfaces will be relatively resistant to wear.

Mechanical properties such as hardness and Young's modulus of elasticity can be determined on micro- to picoscales using AFM (**5, 41**) and a new nano/picoindentation system used in conjunction with AFM (**17**–**20, 42**). Indentability on the scale of subnanometres can be studied by monitoring the slope of cantilever deflection as a function of sample travelling distance after the tip is engaged and the sample is pushed against the tip. For a rigid sample, cantilever deflection equals the sample travelling distance, but the former quantity is smaller if the tip indents the sample. Figure 16 shows the load–displacement curves at different peak loads for Si(100). Load–displacement data at residual depths as low as about 1 nm can be obtained. Loading/unloading curves are not smooth, but exhibit sharp diskontinuities, particularly at high loads (shown by arrows in the figure). Any diskontinuities in the

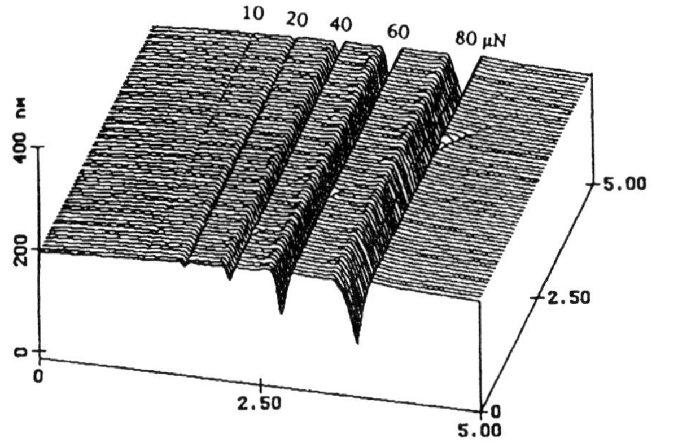

Fig. 10 Surface plots of Si(111) scratched at various loads and a scanning velocity of 2 μm/s. Note that x and y axes are in μm and z axis is in nm (**39**)

loading part of the curve probably result from slip. The sharp diskontinuities in the unloading part of the curves are believed to be due to formation of lateral cracks which form at the base of the median crack, which results in the surface of the specimen being thrust upwards.

The indentation hardness of surface films with an indentation depth of as small as about 1 nm has been measured for Si(111) (**17**, **41**). Triangular indentations are observed for shallow penetration depths. The hardness of silicon on the nanoscale is found to be higher than on the microscale (Fig. 17). This decrease in hardness with an increase in indentation depth can be rationalized on the basis that as the volume of deformed material increases there is a higher probability of encountering material defects. Bhushan and Koinkar (**41**) have used AFM measurements to show that ion implantation of silicon surfaces increases their hardness and thus their wear resistance. The formation of surface alloy films with improved mechanical properties by ion

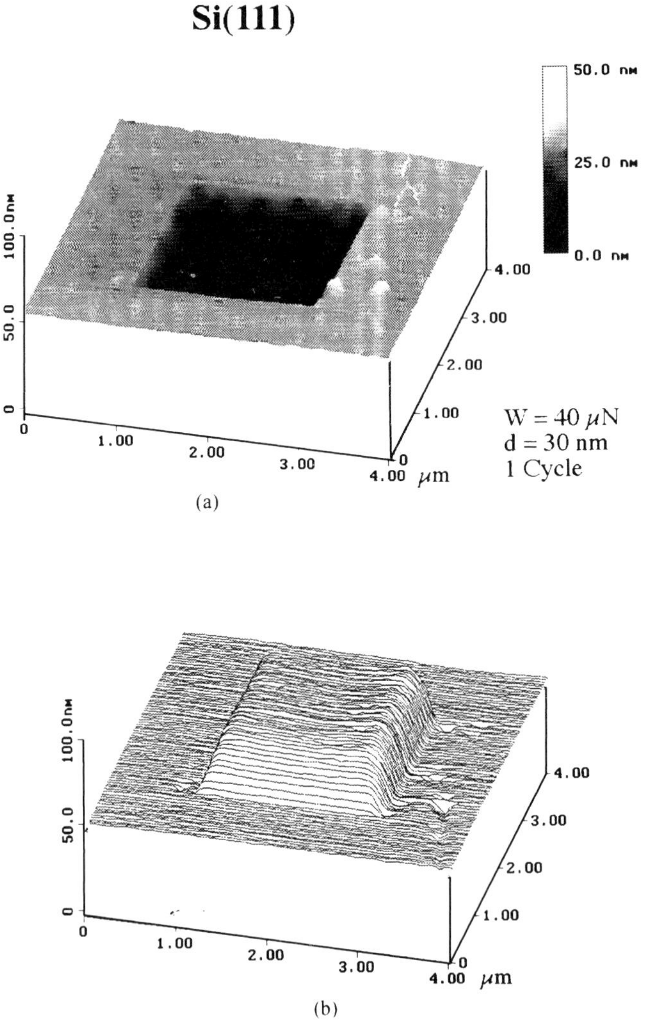

Fig. 12 (a) Typical grey scale and (b) inverted AFM images of wear mark created using a diamond tip at a normal force of 40 μN and one scan cycle on the Si(111) surface (**40**)

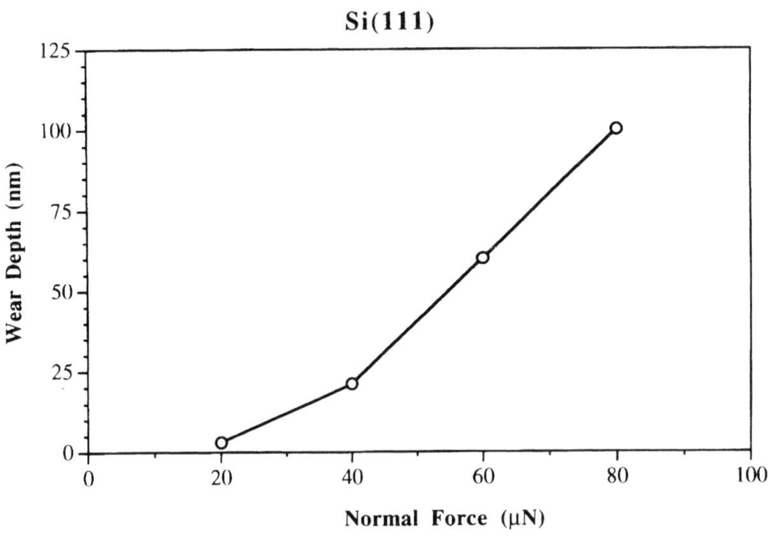

Fig. 11 Wear depth as a function of normal force for Si after one cycle (**40**)

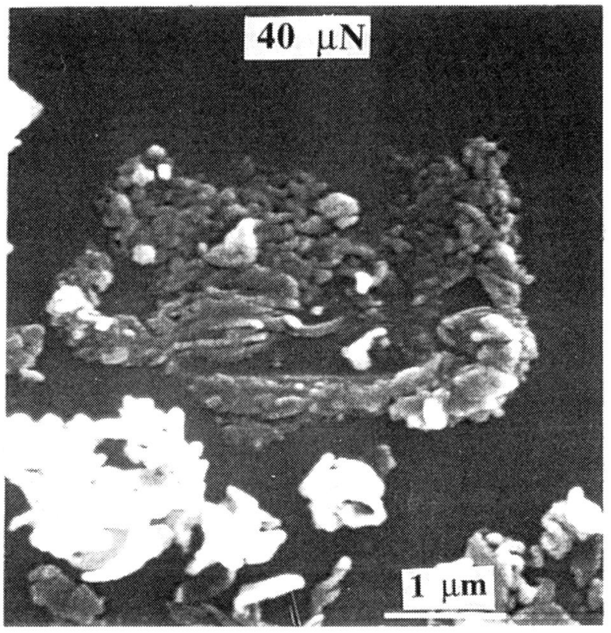

Fig. 13 Secondary electron image of wear mark and debris particles for Si produced at a normal force of 40 μN and one scan cycle (**40**)

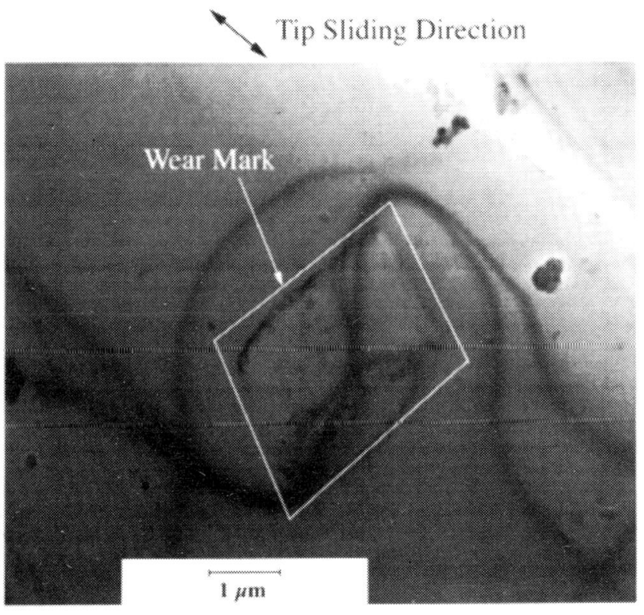

Fig. 14 Bright field TEM micrograph showing wear mark and bend contour around and inside wear mark in Si produced at a normal force of 40 μN and one scan cycle (**40**)

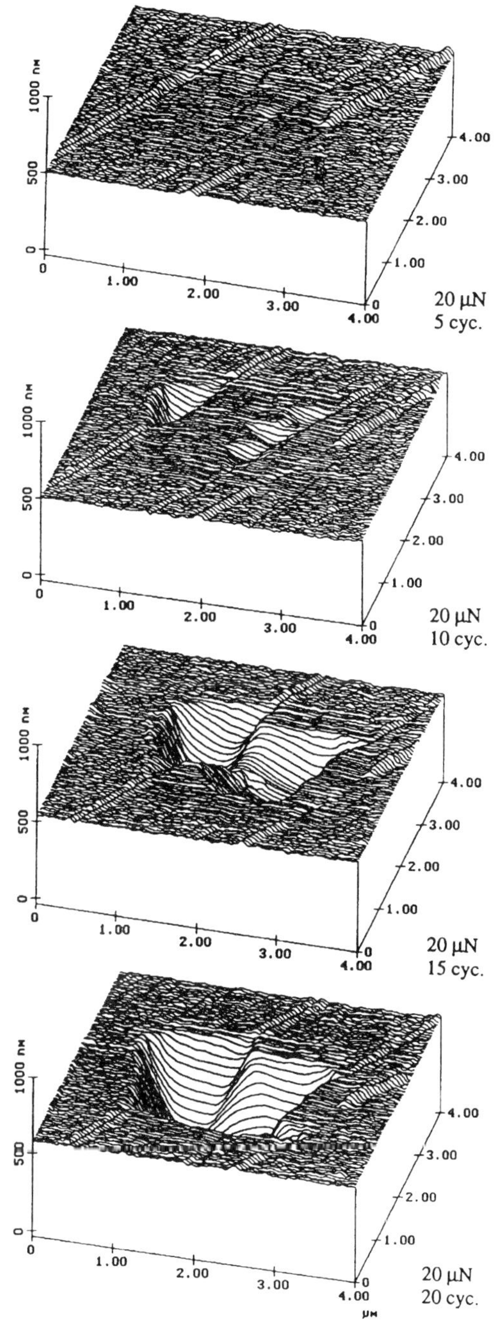

Fig. 15 Surface plots of diamond-like carbon-coated thin-film disk showing the worn region; the normal load and number of test cycles are indicated (**21**)

implantation is growing in technological importance as a means of improving the mechanical properties of materials. The hardness of 20 nm thick diamond-like carbon films has been measured by Kulkarni and Bhushan (**20**).

Young's modulus of elasticity is calculated from the slope of the indentation curve during unloading (**5**, **17**–**20**). Maivald *et al.* (**43**) used AFM in the 'force modulation mode' to measure surface elasticities. The AFM tip is scanned over the modulated sample surface, with the feedback loop keeping the average force constant. For the same applied force, a soft area deforms more, and thus causes less cantilever deflection, than a hard area. The ratio of modulation amplitude to the local tip deflection is then used to create a force modulation image. The force modulation mode makes it easier to identify soft areas on hard substrates.

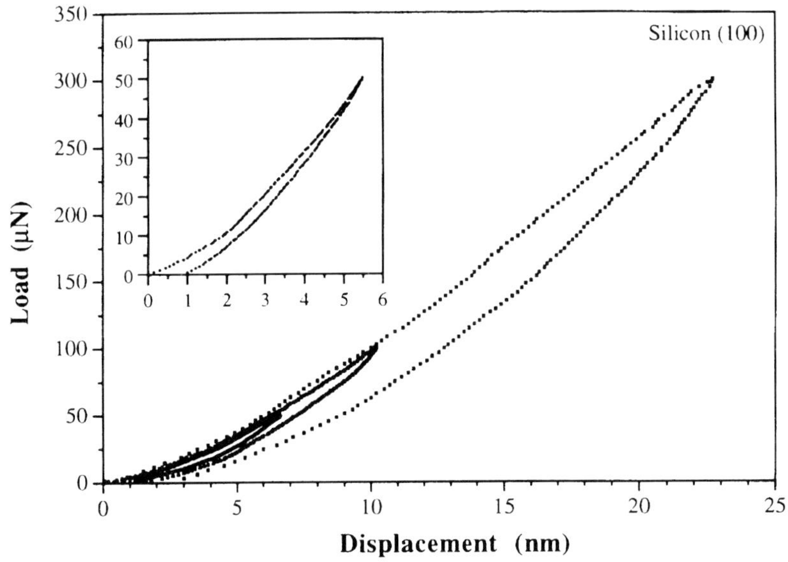

Fig. 16 Load–displacement curves at various peak loads for Si(100) (**17**)

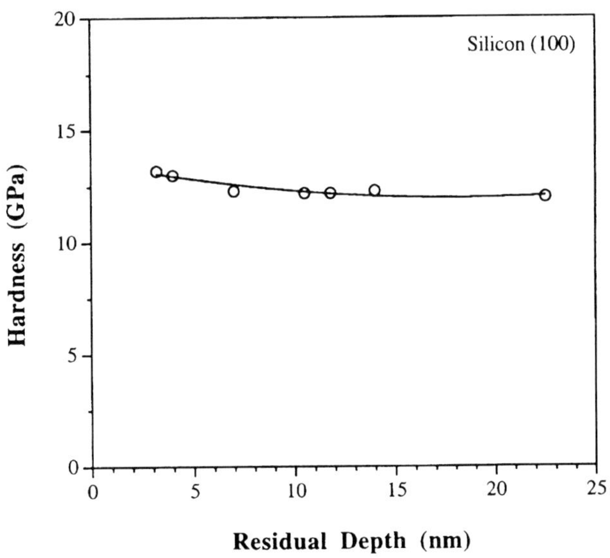

Fig. 17 Indentation hardness as a function of residual indentation depth for Si(100) (**17**)

The nano/picoindentation system has been used to study the creep and strain-rate effects of ceramics. Bhushan *et al.* (**17**) and Kulkarni and Bhushan (**19**, **20**) have reported that ceramics exhibit significant plasticity and creep on the nanoscale.

Detection of transfer of material on the nanoscale is possible with AFM. Indentation of C60-rich fullerene films with an AFM tip has been shown (**44**) to result in the transfer of fullerene molecules to the AFM tip, as indicated by diskontinuities in the cantilever deflection as a function of the sample travelling distance in subsequent indentation studies.

6 BOUNDARY LUBRICATION

The classical approach to lubrication uses freely supported multimolecular layers of liquid lubricants (**9**, **13**). The liquid lubricants are chemically bonded to improve their wear resistance (**5**, **9**). To study depletion of boundary layers, microscale friction measurements were made as a function of the number of cycles of virgin Si(100) surface and silicon surface lubricated with Z-15 and Z-Dol PFPE lubricants (Fig. 18) (**45**, **46**). Z-Dol is a PFPE (perfluoropolyether) lubricant with hydroxyl end groups. Its lubricant film was thermally bonded at 150 °C for 30 min (BUW—bonded, unwashed) and, in some cases, the unbonded fraction was washed off with a solvent to provide a chemically bonded layer of the lubricant (BW—bonded, washed) film. In Fig. 18a, the unlubricated silicon sample shows a slight increase in friction force followed by a drop to a lower steady state value after 20 cycles. Depletion of native oxide and possible roughening of the silicon sample are believed to be responsible for the decrease in friction force after 20 cycles. The initial friction force for the Z-15 lubricated sample is lower than that of unlubricated silicon and increases gradually to the friction force value, which is comparable to that of silicon after 20 cycles. This suggests that depletion of the Z-15 lubricant occurs in the wear track. In the case of the Z-Dol coated silicon sample, the friction force starts out to be very low and remains low during the 100 cycle test, which suggests that Z-Dol does not become displaced/depleted as readily as Z-15. The nanowear results for BW and BUW/Z-Dol samples with different thicknesses are shown in Fig. 18b. The BW with a thickness of 2.3 nm exhibits an initial decrease in the friction force in the first few cycles and then remains steady for more than 100 cycles. The decrease in friction force possibly arises from the alignment of any free liquid

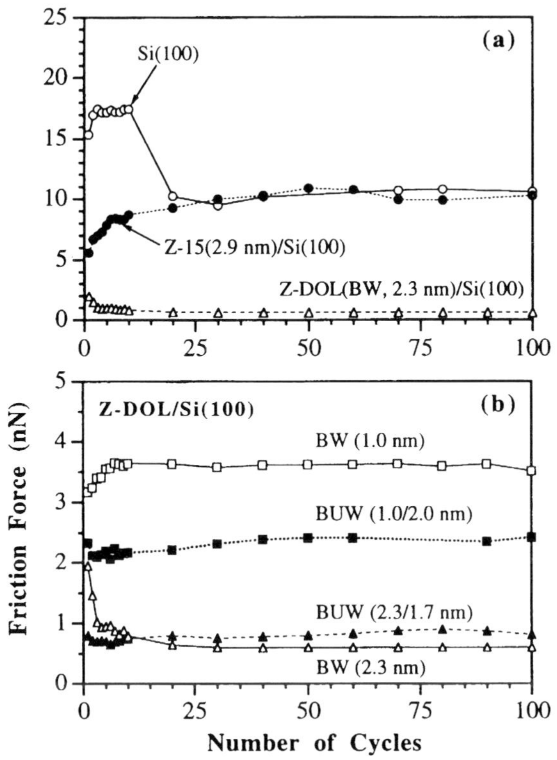

Fig. 18 Friction force as a function of number of cycles using a silicon nitride tip at a normal force of 300 nN for (a) unlubricated silicon, Z-15 and bonded washed (BW) Z-Dol and (b) bonded Z-Dol before washing (BUW) and after washing (BW) (**45**)

lubricant present over the bonded lubricant layer. BUW with a thickness of 4.0 nm exhibits a behaviour similar to BW (2.3 nm). Lubricated BW and BUW samples with thinner films exhibit a higher value of coefficient of friction. Among the BW and BUW samples, BUW samples show the lower friction because of an extra unbonded fraction of the lubricant.

The effect of the operating environment on the coefficient of friction of unlubricated and lubricated samples is shown in Fig. 19. Silicon (100) samples were lubricated with 2.9 nm thick Z-15 and 2.3 nm thick Z-Dol bonded and washed (BW) lubricants. The coefficient of friction in a dry environment is lower than that in a high-humidity environment. The author believes that in the humid environment, the condensed water from the humid environment competes with the liquid film present on the sample surface, and interaction of the liquid film (water for the unlubricated sample and polymer lubricant for the lubricated sample) to the substrate is weakened and the boundary layer of the liquid forms puddles. This dewetting results in a poorer lubrication performance, resulting in high friction. Since Z-Dol is a bonded lubricant with superior frictional properties, the dewetting effect in a humid environment for Z-Dol is more pronounced than for Z-15.

The effect of scanning speed on the coefficient of friction of unlubricated and lubricated samples is shown in Fig. 20. The coefficient of friction for the unlubricated silicon sample and lubricated sample with Z-15 decreases with the logarithm increase in the scanning velocity in an ambient environment. These samples are insensitive to scanning velocity in dry environments. Samples lubricated with Z-Dol do not show any effect of scanning velocity on the friction. Alignment of liquid molecules (shear thinning) is believed to be responsible for the drop in friction, with an increase in scanning velocity for samples with a mobile film and exposed to ambient environment.

For lubrication of microdevices, a more effective approach involves the deposition of organized, dense molecular layers of long-chain molecules on the surface contact. Such monolayers and thin films are commonly produced by Langmuir–Blodgett (L–B) deposition and by chemical grafting of molecules into self-assembled monolayers (SAMs). Based on the measurements, SAMs of octodecyl (C18) compounds based on aminosilanes on an oxidized silicon exhibited a lower coefficient of friction of 0.018 and greater durability than L–B films of zinc arachidate

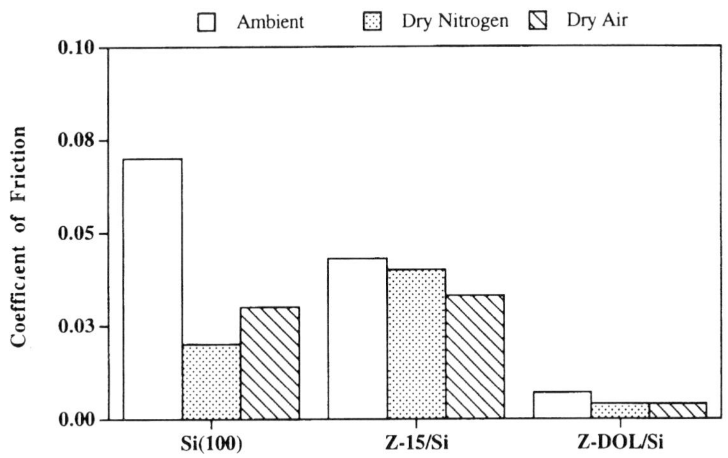

Fig. 19 Coefficient of friction for unlubricated and lubricated Si(100) samples in ambient (~ 50% RH), dry nitrogen (~ 5% RH) and dry air (~ 5% RH) (**46**)

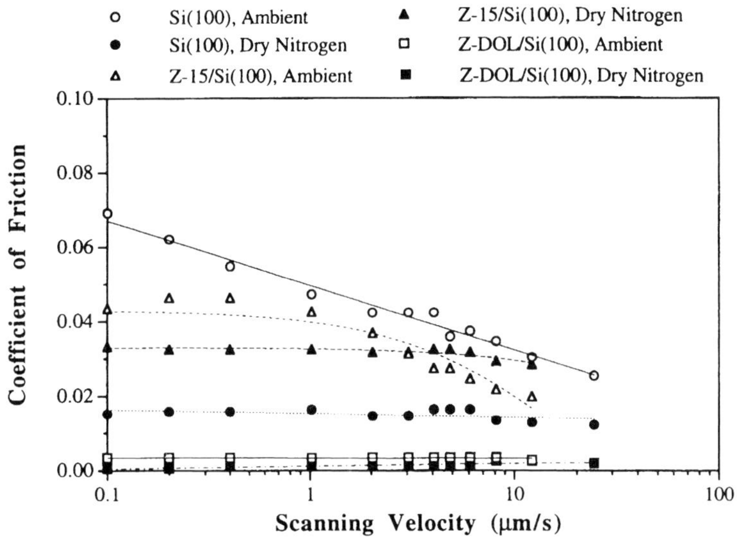

Fig. 20 Coefficient of friction as a function of scanning velocity for unlubricated and lubricated Si(100) samples in ambient (~ 50% RH), dry nitrogen (~ 5% RH) and dry air (~ 5% RH) (**46**)

adsorbed on a gold surface coated with octadecylthiol (ODT) (with a coefficient of friction of 0.03) (Fig. 21) (**47**). L–B films are bonded to the substrate by weak van der Waals attraction, whereas SAMs are chemically bound via covalent bonds. Because of the choice of chain length and terminal linking group that SAMs offer, they hold great promise for boundary lubrication of microdevices.

Measurement of ultra-thin lubricant films with nanometre lateral resolution can be made with AFM (**5**). The lubricant thickness is obtained by measuring the force on the tip as it approaches, contacts and pushes through the liquid film and ultimately contacts the substrate. The distance between the sharp 'snap-in' (owing to the formation of a liquid whose meniscus is between the film and the tip) at the liquid surface and the hard repulsion at the substrate surface is a measure of the liquid film thickness. This technique is now used routinely in the information-storage industry for thickness measurements (with nanoscale spatial resolution) of lubricant films a few nanometres thick, in rigid magnetic disks.

Lubricant film thickness can also be measured using friction force microscopy (**45**). Figure 22 shows grey scale plots of the surface topography and friction force obtained simultaneously for unbonded Dennum-type perfluoropolyether lubricant film on silicon. The friction force plot shows well-distinguished low and high friction regions roughly corresponding to high and low regions in surface topography (thick and thin lubricant regions). A uniformly

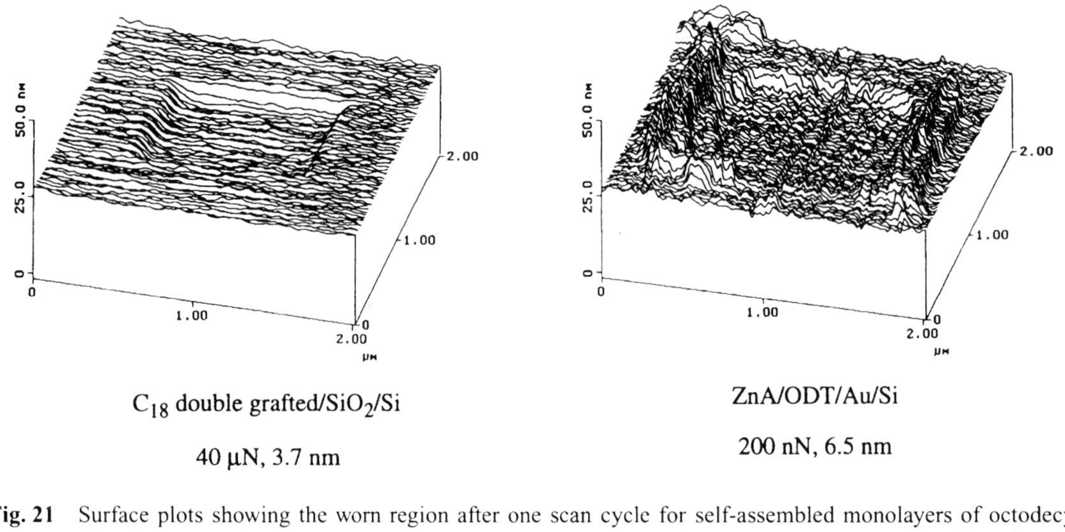

Fig. 21 Surface plots showing the worn region after one scan cycle for self-assembled monolayers of octodecyl silanol (C18) (left) and zinc arachidate (ZnA) (right). Normal force and wear depths are indicated. Note that wear of ZnA occurs at only 200 nN as compared to 40 mN for C18 film (**47**)

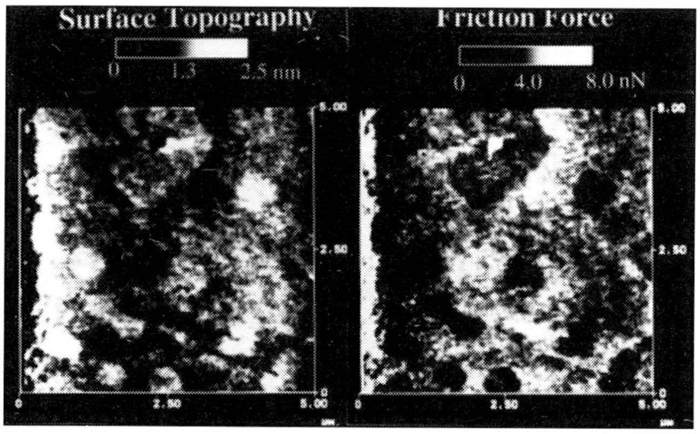

Fig. 22 Grey-scale plots of the surface topography and friction force obtained simultaneously for unbonded perfluoropolyether lubricant film on silicon (**45**)

lubricant sample does not show such a variation in the friction. Friction force imaging can thus be used to measure the lubricant uniformity on the sample surface where identification cannot only be by surface topography.

7 CONCLUSIONS AND FUTURE OUTLOOK

AFM/FFM has been developed as a versatile instrument to conduct a variety of micro/nanotribological studies. These studies are crucial in the development of a fundamental understanding on micro- to nanoscales. These studies are of significant interest in magnetic storage and MEMS industries. Micro/nanotribology is expected to grow rapidly and dominate the tribology research arena.

ACKNOWLEDGEMENTS

Financial support for this research was provided by the Office of Naval Research, Department of the Navy (Contract N00014-96-1-10292). The information herein does not necessarily reflect the position or policy of the government and no official endorsement should be inferred.

REFERENCES

1 **Binnig, G., Rohrer, H., Gerber, Ch.** and **Weibel, E.** Surface studies by scanning tunnelling microscopy. *Phys. Rev. Lett.*, 1982, **49**, 57–61.

2 **Binnig, G., Quate, C. F.** and **Gerber, Ch.** Atomic force microscopy. *Phys. Rev. Lett.*, 1986, **56**, 930–933.

3 **Binnig, G., Gerber, Ch., Stoll, E., Albrecht, T. R.** and **Quate, C. F.** Atomic resolution with atomic force microscope. *Europhys. Lett.*, 1987, **3**, 1281–1286.

4 **Singer, I. L.** and **Pollock, H. M.** *Fundamentals of Friction: Macroscopic and Microscopic Processes*, 1992 (Kluwer Academic, Dordrecht, Netherlands).

5 **Bhushan, B.** *Handbook of Micro/nanotribology*, 1995 (CRC Press, Boca Raton, Florida).

6 **Bhushan, B., Israelachvili, J. N.** and **Landman, U.** Nanotribology: friction, wear and lubrication at the atomic scale. *Nature*, 1995, **374**, 607–616.

7 **Bhushan, B.** *Micro/nanotribology and Its Applications*, NATO ASI Series E: Applied Sciences, Vol. 330, 1997, pp. 169–191 (Kluwer Academic, Dordrecht, Netherlands).

8 **Bhushan, B.** Micro/nanotribology and its application to magnetic storage devices and MEMS. *Tribology Int.*, 1995, **28**, 85–95.

9 **Bhushan, B.** *Tribology and Mechanics of Magnetic Storage Devices*, 2nd edition, 1996 (Springer-Verlag, New York).

10 **Bhushan, B.** Nanotribology and nanomechanics of MEMS devices. In Proceedings of MEMS '96, 1996, pp. 91–98 (Institute of Electrical and Electronics Engineers, New York).

11 **Bhushan, B.** and **Koinkar, V. N.** Microtribological studies of doped single-crystal silicon and polysilicon films for MEMS devices. *Sensors and Actuators A*, 1997, **57**, 91–102.

12 **Bhushan, B.** and **Li, X.** Micromechanical and tribological characterization of doped single-crystal silicon and polysilicon films for microelectromechanical systems devices. *J. Mater. Res.*, 1997, **12**, 54–63.

13 **Bowden, F. P.** and **Tabor, D.** *The Friction and Lubrication of Solids*, Parts I and II, 1950 and 1964 (Clarendon, Oxford).

14 **Bhushan, B.** and **Gupta, B. K.** *Handbook of Tribology: Materials, Coatings and Surface Treatments*, 1991 (McGraw-Hill, New York); 1997 (Krieger, Malabar, Florida).

15 **Bhushan, B.** *Mechanics and Reliability of Flexible Magnetic Media*, 1992 (Springer-Verlag, New York).

16 **Ruan, J.** and **Bhushan, B.** Atomic-scale friction measurements using friction force microscopy: part I—general principles and new measurement techniques. *Trans. ASME, J. Tribology*, 1994, **116**, 378–388.

17 **Bhushan, B., Kulkarni, A. V., Bonin, W.** and **Wyrobek, J. T.** Nano/picoindentation measurement using a capacitive transducer system in atomic force microscopy. *Phil. Mag.*, 1996, **74**, 1117–1128.

18 **Kulkarni, A. V.** and **Bhushan, B.** Nanoscale mechanical property measurements using modified atomic force microscopy. *Thin Solid Films*, 1996, **290–291**, 206–210.

19 **Kulkarni, A. V.** and **Bhushan, B.** Nano/picoindentation measurements on single-crystal aluminum using modified atomic force microscopy. *Mater. Lett.*, 1996, **29**, 221–227.

20 **Kulkarni, A. V.** and **Bhushan, B.** Nanoindentation measurement of amorphous carbon coatings. *J. Mater. Res.*, 1997, **12** (in press).

21 **Bhushan, B., Koinkar, V. N.** and **Ruan, J.** Microtribology of magnetic media. *Proc. Instn Mech. Engrs, Part J, Journal of Engineering Tribology*, 1994, **208**(J1), 17–29.

22 **Bhushan, B.** and **Blackman, G. S.** Atomic force microscopy of magnetic rigid disks and sliders and its applications to tribology. *Trans. ASME, J. Tribology*, 1991, **113**, 452–458.

23 **Oden, P. I., Majumdar, A., Bhushan, B., Padmanabhan, A.** and **Graham, J. J.** AFM imaging, roughness analysis and contact mechanics of magnetic tape and head surfaces. *Trans. ASME, J. Tribology*, 1992, **114**, 666–674.

24 **Ganti, S** and **Bhushan, B.** Generalized fractal analysis and its applications to engineering surfaces. *Wear*, 1995, **180**, 17–34.

25 **Poon, C. Y.** and **Bhushan, B.** Comparison of surface roughness measurements by stylus profiler, AFM and non-contact optical profiler. *Wear*, 1995, **190**, 76–88.

26 **Koinkar, V. N.** and **Bhushan, B.** Effect of scan size and surface roughness on microscale friction measurements. *J. Appl. Phys.*, 1997, **81**, 2472–2479.

27 **Majumdar, A.** and **Bhushan, B.** Role of fractal geometry in roughness characterization and contact mechanics of surfaces. *Trans. ASME, J. Tribology*, 1990, **112**, 205–216.

28 **Majumdar, A.** and **Bhushan, B.** Fractal model of elastic–plastic contact between rough surfaces. *Trans. ASME, J. Tribology*, 1991, **113**, 1–11.

29 **Bhushan, B.** and **Majumdar, A.** Elastic–plastic contact model for bifractal surfaces. *Wear*, 1992, **153**, 53–64.

30 **Bhushan, B.** Contact mechanics of rough surfaces in tribology: single asperity contact. *Appl. Mech. Rev.*, 1996, **49**, 275–298.

31 **Bhushan, B.** Contact mechanics of rough surfaces in tribology: multiple asperity contact. *Tribology Lett.* (in press).

32 **Mate, C. M., McClelland, G. M., Erlandsson, R.** and **Chiang, S.** Atomic-scale friction of a tungsten tip on a graphite surface. *Phys. Rev. Lett.*, 1987, **59**, 1942–1945.

33 **Ruan, J.** and **Bhushan, B.** Atomic-scale and microscale friction of graphite and diamond using friction force microscopy. *J. Appl. Phys.*, 1994, **76**, 5022–5035.

34 **Ruan, J.** and **Bhushan, B.** Frictional behaviour of highly oriented pyrolytic graphite. *J. Appl. Phys.*, 1994, **76**, 8117–8120.

35 **Koinkar, V. N.** and **Bhushan, B.** Microtribological studies of Al_2O_3, Al_2O_3–TiC, polycrystalline and single-crystal Mn–Zn ferrite and SiC head slider materials. *Wear*, 1996, **202**, 110–122.

36 **Meyer, E., Overney, R., Luthi, R., Brodbeck, D., Howald, L., Frommer, J., Guntherodt, H.-J., Wolter, O., Fujihira, M., Takano, T.** and **Gotoh, Y.** Friction force microscopy of mixed Langmuir–Blodgett films. *Thin Solid Films*, 1992, **220**, 132–137.

37 **Frisbie, C. D., Rozsnyai, L. F., Noy, A., Wrighton, M. S.** and **Lieber, C. M.** Functional group imaging by chemical force microscopy. *Science*, 1994, **265**, 2071–2074.

38 **Bhushan, B.** and **Kulkarni, A. V.** Effect of normal load on microscale friction measurements. *Thin Solid Films*, 1996, **278**, 49–56.

39 **Bhushan, B.** and **Koinkar, V. N.** Tribological studies of silicon for magnetic recording applications. *J. Appl. Phys.*, 1994, **75**, 5741–5746.

40 **Koinkar, V. N.** and **Bhushan, B.** Scanning and transmission electron microscopies of single-crystal silicon microworn/machined using atomic force microscopy. *J. Mater. Res.*, 1997, **12** (in press).

41 **Bhushan, B.** and **Koinkar, V. N.** Nanoindentation hardness measurements using atomic force microscopy. *Appl. Phys. Lett.*, 1994, **64**, 1653–1655.

42 **Koinkar, V. N.** and **Bhushan, B.** Microtribological properties of hard amorphous carbon protective coatings for thin film magnetic disks and heads. *Proc. Instn Mech. Engrs, Part J, Journal of Engineering Tribology*, 1997, **211**(J4), 365–372.

43 **Maivald, P., Butt, H. J., Gould, S. A. C., Prater, C. B., Drake, B., Gurley, J. A., Elings, V. B.** and **Hansma, P. K.** Using force modulation to image surface elasticities with the atomic force microscope. *Nanotechnology*, 1991, **2**, 103–106.

44 **Ruan, J.** and **Bhushan, B.** Nanoindentation studies of fullerene films using atomic force microscopy. *J. Mater. Res.*, 1993, **8**, 3019–3022.

45 **Koinkar, V. N.** and **Bhushan, B.** Micro/nanoscale studies of boundary layers of liquid lubricants for magnetic disks. *J. Appl. Phys.*, 1996, **79**, 8071–8075.

46 **Koinkar, V. N.** and **Bhushan, B.** Microtribological studies of unlubricated and lubricated surfaces using atomic force/friction force microscopy. *J. Vac. Sci. Technol.*, 1996, **A14**, 2378–2391.

47 **Bhushan, B., Kulkarni, A. V., Koinkar, V. N., Boehm, M., Odoni, L., Martelet, C.** and **Belin, M.** Microtribological characterization of self-assembled and Langmuir–Blodgett monolayers by atomic and friction force microscopy. *Langmuir*, 1995, **11**, 3189–3198.

Microtribological properties of hard amorphous carbon protective coatings for thin-film magnetic disks and heads

V N Koinkar and **B Bhushan**
Computer Microtribology and Contamination Laboratory, Department of Mechanical Engineering, The Ohio State University, Columbus, Ohio, USA

Abstract: For long durability of magnetic media and head sliders, protective overcoats of hydrogenated amorphous carbon (a-C:H) are generally used. In this study, microtribological studies of hydrogenated amorphous carbon coatings deposited on a single-crystal silicon using three different deposition techniques—sputtering, ion beam and cathodic arc—were studied using atomic force/friction force microscopy (AFM/FFM). Roughnesses of all coatings at two scan sizes of 1 μm × 1 μm and 10 μm × 10 μm are comparable. Surface topography of sputtered carbon coating shows some particulates on the surface. Cathodic arc carbon coating exhibits the lowest coefficient of friction value followed by ion beam and sputtered carbon coatings. Microscratch and wear resistance and nanohardness of cathodic arc carbon coating are superior to those of ion beam and sputtered carbon coatings. Cathodic arc deposited carbon coatings are potential candidates for magnetic disks and heads.

Keywords: microtribology, hard carbon coatings, magnetic disks, magnetic heads

1 INTRODUCTION

In order to achieve ever increasing high recording densities in magnetic rigid-disk drives, it is necessary to continue to reduce the effective magnetic spacing (ems) between magnetic surfaces of heads and disks (**1**, **2**). Read/write heads are designed such that during normal operation of the magnetic rigid-disk drives, air bearing (flying height) forms between the head and thin-film magnetic rigid disk. However, during contact start/stop (CSS) of the magnetic rigid-disk drives physical contact occurs between the head and the disk. It is very important to protect the magnetic disk and head air-bearing surfaces (ABS) from wear and corrosion for longer durability. Better protective coatings with almost near zero wear are essential to increase the durability of magnetic storage components (**2**, **3**).

The dimensions and load used in the magnetic storage devices are continuously decreasing and mechanical and tribological properties of protective coatings are becoming extremely important on a microscale (**3**, **4**). Thicknesses of protective layers need to be decreased to reduce the ems. Hydrogenated amorphous carbon (a-C:H), also referred to as a diamond-like carbon (DLC), and silicon oxide coatings are almost exclusively used to protect the magnetic disk. A liquid lubricant layer a couple of monolayers thick is applied over these protective coatings to improve the friction and wear characteristics. Hydrogenated amorphous carbon coatings on ABS of rigid-disk head sliders deposited by d.c. or r.f. magnetron sputtering are commercially used on thin-film magnetoresistive (MR) rigid-disk heads (**5**–**7**). Carbon coatings should have a high degree of sp^3 bonding. Previous investigations suggest that in order to achieve a higher degree of sp^3 bonding content, kinetic energy of depositing species should be on the order of 100 eV or higher. Various deposition techniques are employed to deposit hard amorphous carbon films. For details on deposition techniques, kinetic energies and deposition rate to deposit hard a-C:H, see reference (**8**).

The objective of this research work is to investigate the tribological properties of amorphous carbon coatings deposited by various techniques on small scale and light loads using atomic force/friction force microscopy (AFM/FFM). These studies will help to provide fundamental understanding of friction and wear mechanisms. They will also provide guidance in selection of better protective coatings for future magnetic storage

The MS was received on 3 October 1996 and was accepted for publication on 19 April 1997.

devices with ultra low friction and almost near zero wear.

2 EXPERIMENTAL

2.1 Test specimens

Amorphous carbon coatings with a thickness of 20 nm were deposited on single-crystal silicon (100) wafer using d.c. magnetron sputtering, direct ion beam and cathodic arc techniques. Carbon coatings deposited by d.c. magnetron sputtering, direct ion beam and cathodic arc are referred to as sputtered, ion beam and cathodic arc respectively throughout this article. Single-crystal silicon (100) specimens were also used to compare microtribological performance of different carbon coatings. Sputtered hydrogenated carbon coating was deposited from Ar–7%H_2 at a deposition rate of 2.1 nm/s. Deposition conditions for ion beam and cathodic arc carbon coatings are presented by Gupta and Bhushan (**8**).

2.2 Atomic force/friction force microscope (AFM/FFM) and test procedures

A modified AFM/FFM (Nanoscope III, Digital Instruments, Santa Barbara, California) was used to study microtribological behaviour of a-C:H coatings deposited by different deposition techniques. A square pyramidal silicon nitride (Si_3N_4) tip with a radius of about 30–50 nm and spring constant of about 0.6 N/m was used to measure surface topography and microscale friction. A normal force ranging from 10 to 200 nN was used for microscale friction measurements. Surface roughness and microscale friction measurements were performed at scan sizes of 1 μm × 1 μm and 10 μm × 10 μm and at a scan rate of 2 Hz. The variation in the microscale friction is about ±10 per cent.

A three-sided pyramidal natural diamond tip (radius ~ 100 nm) mounted on a gold coated stainless steel cantilever beam (spring constant = 40 N/m) was used for microscratch and microwear studies. Normal force in the range of 0.3–150 μN could be applied using these cantilever beams. Microscratch tests were carried out for five different normal forces and for 10 scan cycles at a scanning velocity of about 5 μm/s. For microwear tests, an area of 2 μm × 2 μm was scanned at a normal force of 60 μN for different numbers of scan cycles at a scanning velocity of 4.0 μm/s. The average wear depth was measured after completion of a given number of scan cycles by scanning a larger area at a light load. A larger specimen area was scanned after microscratch and wear tests to observe scratch and wear marks at a normal force of 0.3 μN. Average scratch and wear depths obtained from several measurements are reported in this article. Specimens were scanned orthogonally to the long axis of cantilever beam during surface topography, microscale friction, microscratch and wear measurements.

Nanoindentation measurements were made using a depth-sensing nano/picoindenter indentation system (Hysitron Inc.) (**9**). Hardness measurements were made on 100 nm thick coatings. Load–displacement curves were obtained using a Berkovitch indenter mounted in AFM at a peak load of 700 μN. Nanohardness values were calculated from the load versus displacement data. All measurements were performed in the ambient atmosphere (room temperature = 22 ± 1 °C, relative humidity = 45 ± 5° RH and class 10 000).

3 RESULTS AND DISCUSSION

3.1 Surface roughness and microfriction

Surface height and friction force plots were obtained simultaneously at two different scan sizes 1 μm × 1 μm and 10 μm × 10 μm. Si_3N_4 tips were used for surface roughness and friction measurements. A summary of surface roughness [root mean square (r.m.s.) of surface

Table 1 Summary of roughness and microscale friction measurements at two different scan sizes, microscratch/wear and nanohardness data

	Scan size								
	1 μm × 1 μm			10 μm × 10 μm					
Specimen[‡]	R.m.s. (nm)	P-V* (nm)	$\mu^{\dagger}$	R.m.s. (nm)	P-V* (nm)	$\mu^{\dagger}$	Scratch depth[‡] (nm)	Wear depth[§] (nm)	Nanohardness[•] (GPa)
Si(100)	0.08	1.2	0.05	—	—	—	3.0	47	13.0
Sputtered carbon	0.22	1.8	0.06	0.42	4.7	0.09	1.0	2.0	17.2
Ion beam carbon	0.12	0.78	0.06	0.63	7.1	0.07	1.3	3.3	15.2
Cathodic arc carbon	0.07	0.60	0.04	0.41	14.2	0.05	0.4	0.7	24.9

*Peak-to-valley distance.
†Coefficient of friction.
‡Normal force = 20 μN, 10 scan cycles.
§Normal force = 60 μN, 1 cycle.
•Measured at a normal force of 700 μN.

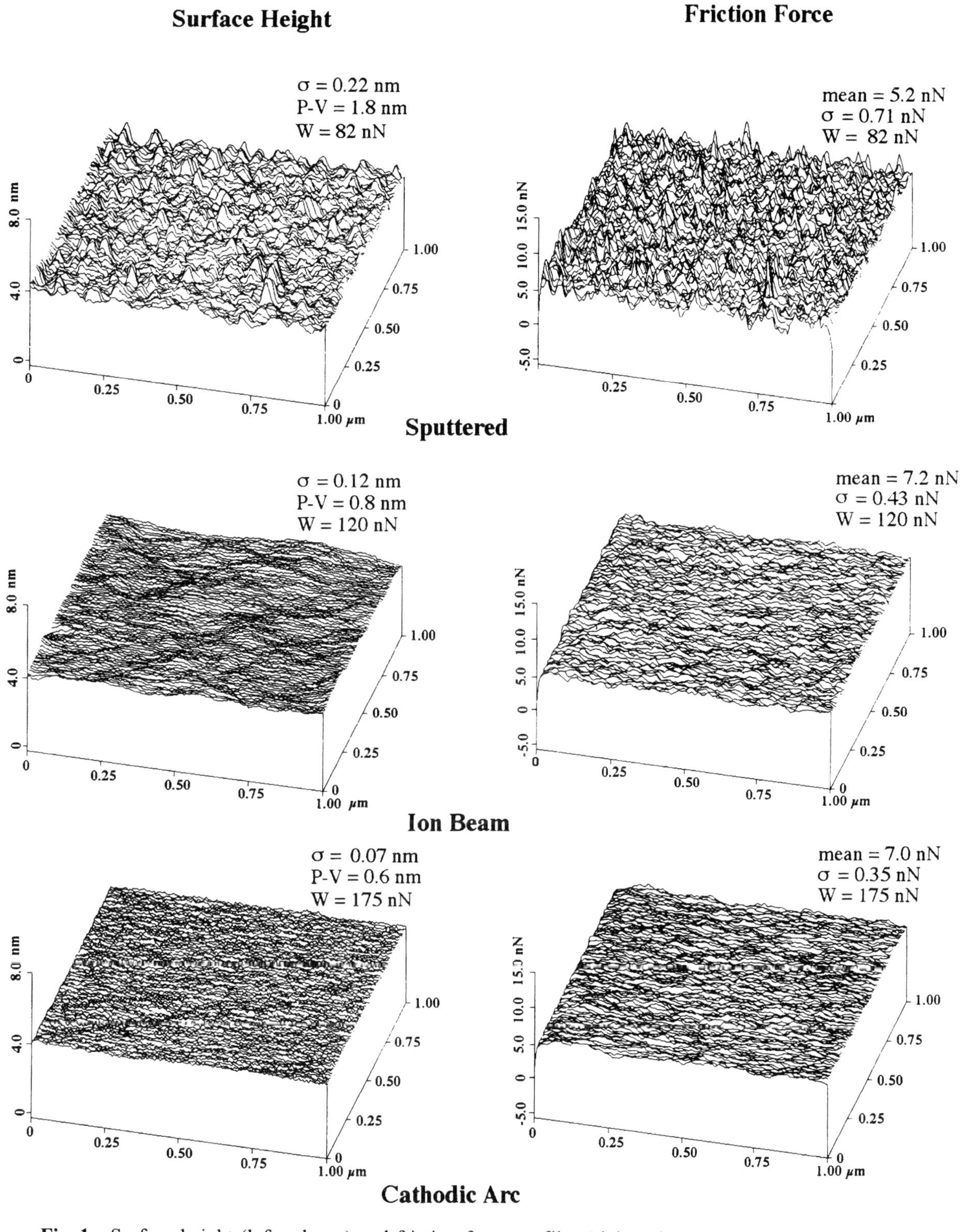

Fig. 1 Surface height (left column) and friction force profiles (right column) (three-dimensional) for sputtered, ion beam and cathodic arc carbon coatings. Standard deviations of surface heights (σ) or r.m.s., peak-to-valley (P-V) distance, mean and normal force used are given in the figure

heights] measurements is given in Table 1. Three-dimensional profiles of surface height and friction force obtained simultaneously for sputtered, ion beam and cathodic arc carbon are shown in Fig. 1. Roughness of ion beam and cathodic carbon is comparable. Surface topography of sputtered carbon shows particulates on the surface and P-V (peak-to-valley) roughness of about 1.8 nm at a scan size of 1 μm × 1 μm. Roughness of sputtered carbon is twice that of the ion beam and cathodic arc carbon coatings at the scan size of 1 μm × 1 μm. Uncoated single-crystal silicon exhibits the lowest roughness. At the scan size of 10 μm × 10 μm roughness values for all carbon coatings used here are comparable.

The friction force increases linearly with an increase in normal force for all carbon coatings studied here. Cathodic arc carbon exhibits the lowest coefficient of microscale friction at scan sizes of 1 μm × 1 μm and 10 μm × 10 μm. The coefficients of friction for ion beam and sputtered carbon coatings are comparable at a scan size of 1 μm × 1 μm. At a scan size of 10 μm ×10 μm, sputtered carbon exhibits a higher coefficient of friction than that of ion beam carbon. The particulates present on the surface of sputtered carbon, which contribute to the microscale friction, result in higher values of overall friction as compared to other carbon coatings. It was observed that the coefficients of microscale friction values for carbon coatings are higher than those of the uncoated single-crystal silicon.

Figure 2 shows the grey scale plots of surface height, surface slope and friction force obtained simultaneously for sputtered carbon coating. If the surface height and friction force plots are compared there is some correlation between both. Particulate in the surface height exhibits lower friction. To understand the mechanism for local variation in microfriction, the slope of the surface roughness profile is obtained along the tip sliding direction. Figure 2 shows the two-dimensional grey scale plots of the slope of the surface height profile as well. Surface slope is calculated by taking the first derivative of the surface height profile in the same tip sliding direction. Comparing the surface slope and friction force profiles, a correlation is not observed between the two as had been observed in the case of other materials (**3**). High points (also high slopes) on the coating surface correspond to low friction. It suggests that the nodules exhibit low friction which may arise from a change in structure of these nodules.

3.2 Microscratch and microwear studies

Surface profiles (three-dimensional) of scratch marks generated at different normal forces for 10 scan cycles for uncoated single-crystal silicon, sputtered, ion beam and cathodic arc carbon are shown in Fig. 3. A plot of the average scratch depth as a function of the normal force is shown in Fig. 4. Single-crystal silicon could be scratched at a normal force of 10 μN and 10 scan cycles. The scratch depth increases almost linearly with an increase in the normal force for single-crystal silicon. Sputtered carbon exhibits better scratch resistance up to a normal force of 60 μN. Once the normal force exceeds the value of 60 μN sputtered carbon coating breaks catastrophically without any warning where the tip penetrates into the underlying single-crystal silicon substrate. The coating thickness is marked in Fig. 4. Many debris particles are visible around the scratch mark generated at 80 μN for sputtered carbon coating. The scratch depth for ion beam carbon coating increases linearly with an increase in the normal force. However, the maximum average depth obtained at a normal force of 80 μN is about 5 nm. Cathodic arc carbon coating shows excellent scratch resistance as compared to the other two carbon coatings and uncoated single-crystal silicon. Ion beam and cathodic arc carbon coatings do not break at

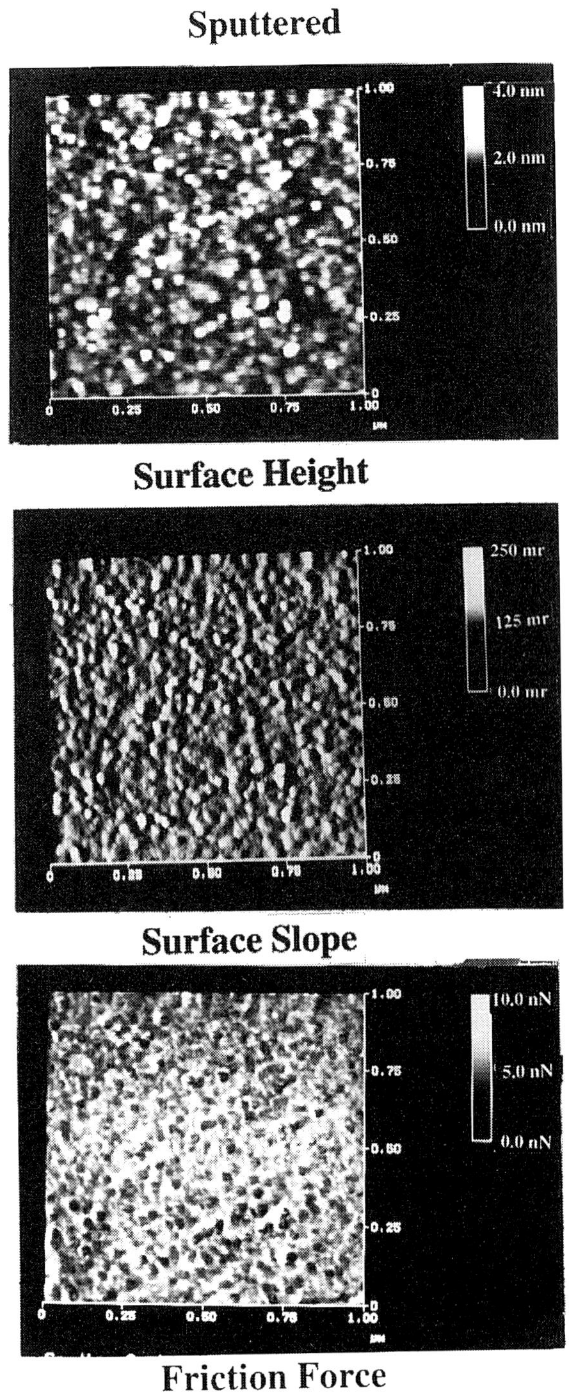

Fig. 2 Two-dimensional grey scale plots of surface height, slope of the surface height profile and friction force profile for sputtered carbon

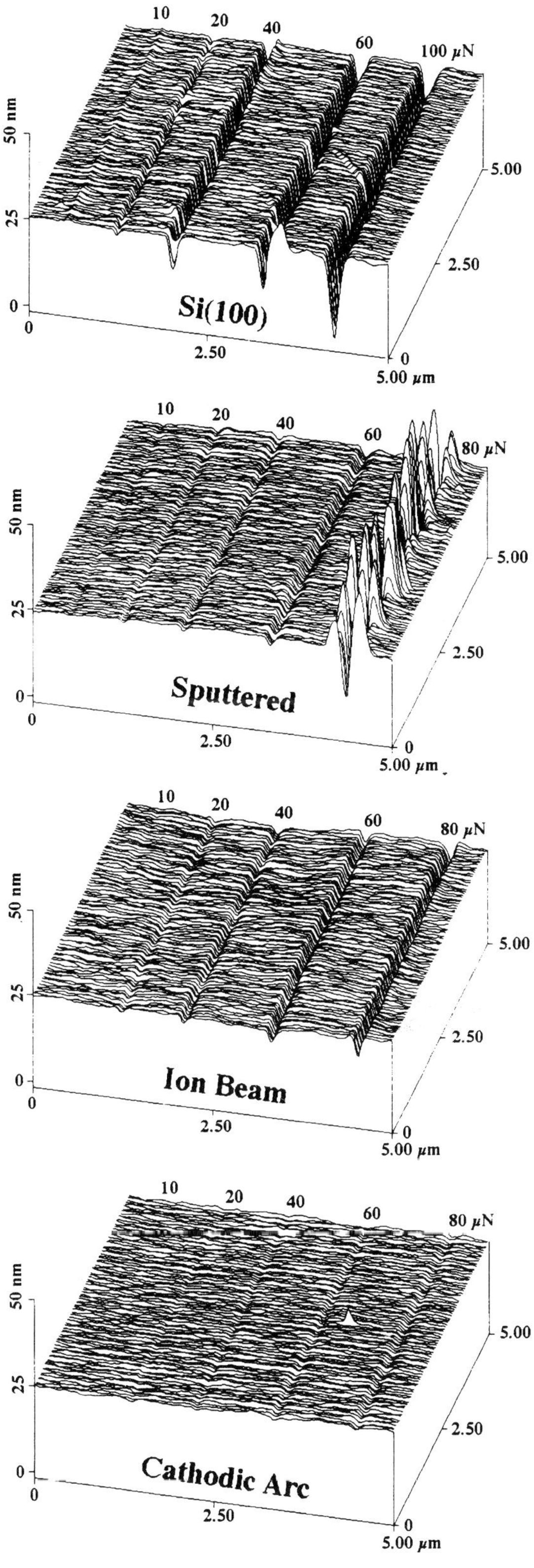

Fig. 3 Surface profiles (three-dimensional) of scratches generated at different normal forces and 10 scan cycles for uncoated single-crystal silicon, sputtered, ion beam and cathodic arc carbon coatings. Normal forces used for scratching the surfaces are given in the figure

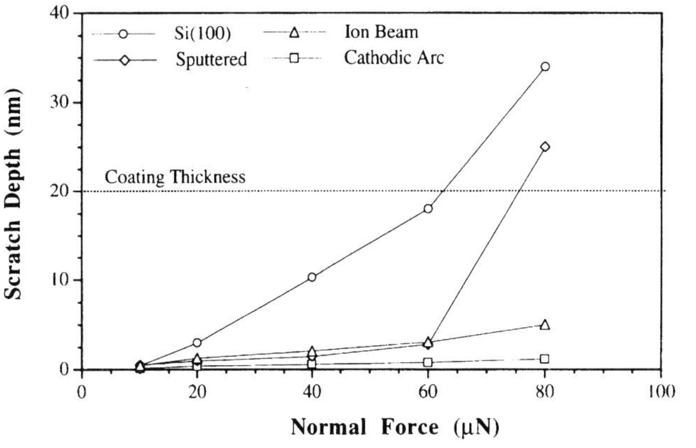

Fig. 4 Scratch depth as a function of normal force after 10 scan cycles for uncoated and carbon-coated single-crystal silicon

the maximum normal force of 80 μN used in the scratch test.

Microwear studies were performed at a normal force of 60 μN for different numbers of scan cycles. Figure 5 shows the surface profiles of wear marks (centre region of 2 μm × 2 μm) for uncoated single-crystal silicon, sputtered, ion beam and cathodic arc carbon coatings. Carbon-coated silicon specimens exhibit better wear resistance than that of uncoated single-crystal silicon for one scan cycle. Debris particles generated during the first scan cycle of wear are loose and could be moved out of the scan area with a light load.

Figure 6 shows the wear behaviour for an increase in the number of cycles for uncoated and coated single-crystal silicon. Uncoated single-crystal silicon exhibits poor wear resistance, as expected from the scratch test. Ion beam carbon coating shows a linear increase in the wear depth with an increase in the number of scan cycles up to about seven cycles and then the tip penetrates through the coating into underlying substrate. The wear resistance of sputtered carbon coating is better for only one scan cycle. At the second scan cycle, sputtered carbon coating breaks catastrophically and the tip penetrates into the underlying single-crystal silicon substrate. Wear evolution of the sputtered coating is shown in Fig. 7a. As observed previously, wear debris generated during the first scan cycle are loose and move out of the scan area. Once the coating is broken, at the edges of the wear mark a lot of sticky debris is visible (see the profiles after 2 and 3 scan cycles in Fig. 7a). The wear depth for cathodic arc carbon coating increases slowly with an increase in the number of scan cycles up to 75 cycles. Up to the 75th scan cycle, a very small volume of material is removed non-uniformally from the wear track. As soon as about 75 scan cycles are exceeded the wear rate increases and the coating fails at a scan cycle of about 120.

Microscratch and microwear studies indicate that cathodic arc carbon coating exhibits excellent resistance to

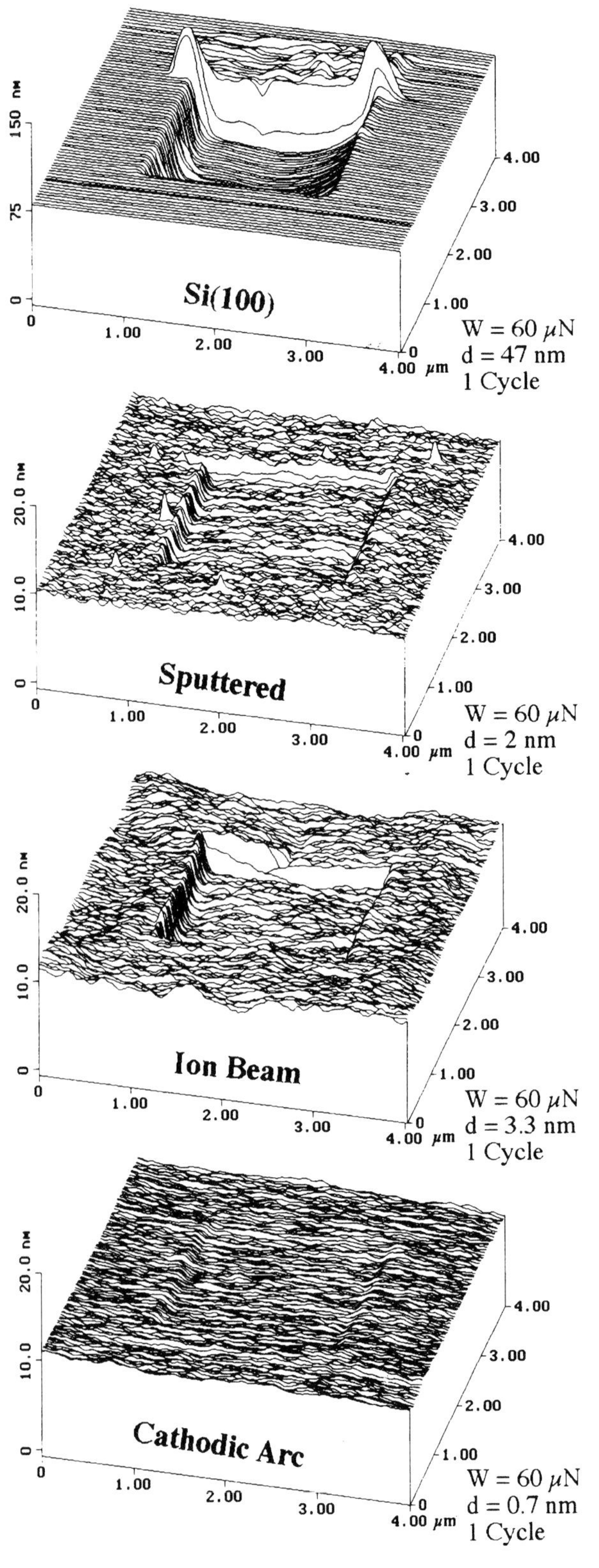

Fig. 5 Surface profiles showing the worn region (centre of 2 μm × 2 μm) after one scan cycle at a normal force of 60 μN for uncoated single-crystal silicon, sputtered, ion beam and cathodic arc carbon coatings. The normal force, average wear depth and number of scan cycles are indicated in the figure

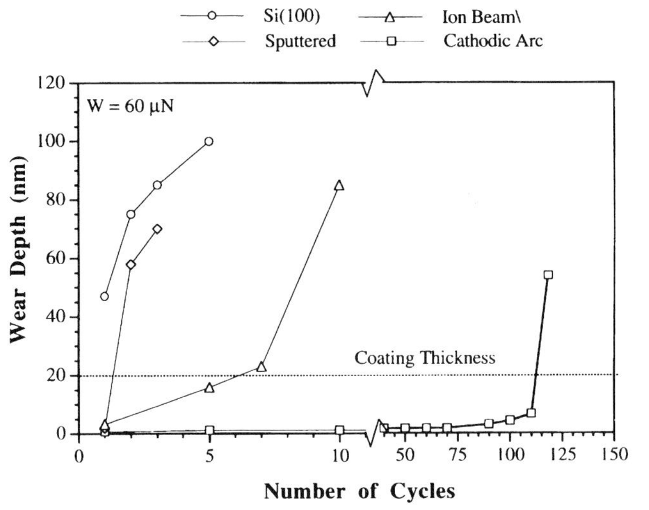

Fig. 6 Wear depth as a function of number of cycles at a normal force of 60 μN for uncoated single-crystal silicon, sputtered, ion beam and cathodic arc carbon coatings

wear and scratch followed by ion beam and sputtered carbon coatings. These results are summarized in Fig. 8, showing better microtribological performance than that of the carbon coatings studied here.

3.3 Nanoindentation hardness studies

Nanoindentation hardness was measured at a normal force of 700 μN. A summary of hardness values is presented in Table 1. Cathodic arc carbon exhibits the highest hardness of about 24.9 GPa. Hardness values for sputtered and ion beam carbon are 17.2 and 15.2 GPa respectively. Uncoated single-crystal silicon shows a lower value of hardness of about 13.0 GPa. The highest hardness of cathodic arc carbon is believed to be responsible for excellent tribological properties.

Sputtered carbon is harder than ion beam carbon. However, sputtered carbon exhibits poor scratch and wear resistance, probably due to its poor adhesion to the substrate. Kinetic energies of deposition species play a major role in the adhesion between coating and substrate. Kinetic energies of species in d.c./r.f. sputtering deposition are low, in the range of 1–10 eV, as compared to 50–500 eV in direct ion beam deposition and 100–2500 eV in cathodic arc deposition (**10–12**). Studies performed here clearly show the effect of kinetic energies of deposition species on microtribological performance. Amorphous carbon coatings deposited with high kinetic energies are much more durable to scratch and wear and are also harder as compared to carbon coatings deposited using techniques with low kinetic energies of deposition species.

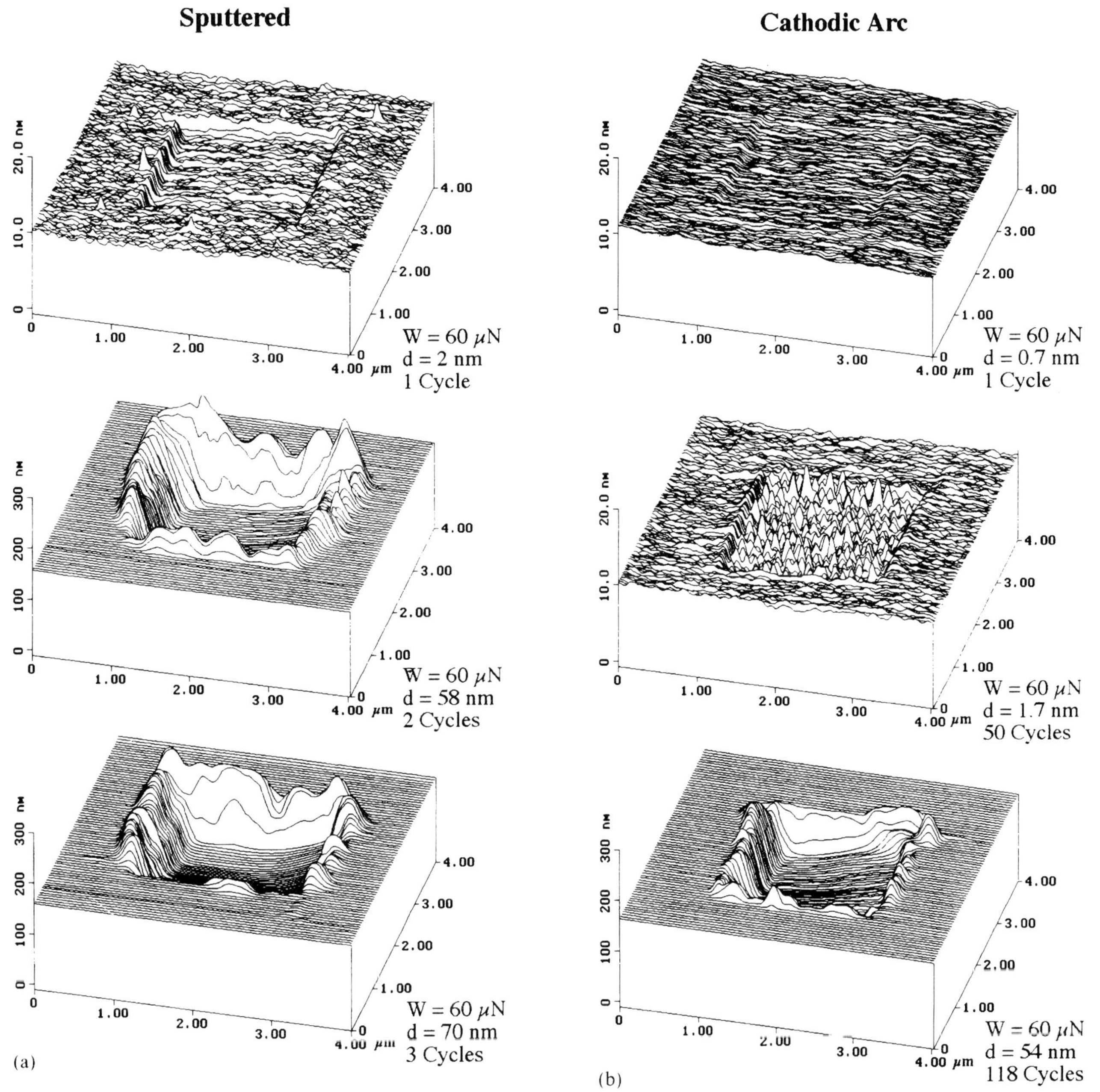

Fig. 7 Surface profiles showing the worn region (centre 2 μm × 2 μm) after different numbers of cycles at a normal force of 60 μN for (a) sputtered and (b) cathodic arc carbon coatings. The normal force, average wear depth and number of scan cycles are indicated in the figure

4 CONCLUSIONS

Microtribological studies were performed on hydrogenated amorphous carbon coatings deposited by three deposition techniques on single-crystal silicon. Roughness and microscale friction, scratching, wear and indentation behaviour are studied in detail. Cathodic arc carbon exhibits excellent microtribological and mechanical performance followed by ion beam and sputtered carbon coatings. Amorphous coatings deposited on high kinetic energy species exhibit excellent microtribological performance compared to those of coatings deposited on low kinetic energy species.

ACKNOWLEDGEMENTS

The authors are grateful to Dr Rajiv Ranjan and Dr Caroline Ross of Komag for sputtered carbon, Paul Reader of Ion Tech. Inc. for ion beam carbon and Simone Anders

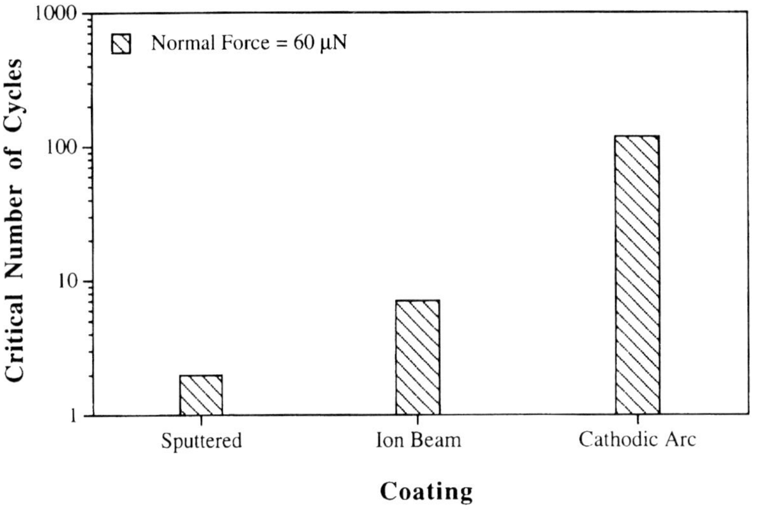

Fig. 8 Plot of the critical number of scan cycles required to break the carbon coating in a microwear test at a normal force of 60 μN

of Lawrence Berkeley Laboratory for cathodic arc carbon. This research was supported by the Office of Naval Research, Department of Navy (Contract N00014-96-1-10292). The content of this information does not necessarily reflect the position or policy of the US Government, and no official endorsement should be inferred.

REFERENCES

1 **Bhushan, B.** and **Koinkar, V. N.** Microscale mechanical and tribological characterization of hard amorphous carbon coatings as thin as 5 nm for magnetic disks. *Surf. Coatings Technol.*, 1995, **76–77**, 655–669.

2 **Bhushan, B.** *Tribology and Mechanics of Magnetic Storage Devices*, 2nd edition, 1996 (Springer-Verlag, New York).

3 **Bhushan, B.** *Handbook of Micro/nanotribology*, 1995 (CRC Press, Boca Raton, Florida).

4 **Bhushan, B., Koinkar, V. N.** and **Ruan, J.** Microtribology of magnetic media. *Proc. Instn Mech. Engrs, Part J*, 1994, **208**(J1), 17–29.

5 **Ganapathi, S. K.** and **Riener, R. A.** Slider overcoats for enhanced interface durability in magnetic recording applications. *Trans. ASME, J. Tribology*, 1995, **117**, 86–93.

6 **Chang, H. C., Chen, M. M., Horng, C. T.** and **Schwenker, R. O.** Thin film magnetic head having a protective coating and method for making same. US Pat. 5,175,658, 1992.

7 **Grill, A., Horng, C. T., Mayerson, B. S., Patel, V. V.** and **Russak, M. A.** Magnetic head slider having a protective coating thereon. US Pat. 5,159,508, 1992.

8 **Gupta, B. K.** and **Bhushan, B.** Mechanical and tribological properties of hard carbon coatings for magnetic recording heads. *Wear*, 1995, **190**, 110–122.

9 **Bhushan, B., Kulkarni, A. V., Bonin, W.** and **Wyrobek, J. T.** Nano/picoindentation measurements using capacitive transducer system in atomic force microscopy. *Phil. Mag.*, 1996, **A74**, 1117–1128.

10 **Cuomo, J. J., Pappas, D. L., Bruley, J., Doyle, J. P.** and **Saenger, K. L.** Vapor deposition processes for amorphous carbon film with sp^3 fractions approaching diamond. *J. Appl. Phys.*, 1991, **70**, 1706–1711.

11 **Bhushan, B.** and **Gupta, B. K.** *Handbook of Tribology: Materials, Coatings, and Surface Treatments*, 1991 (McGraw-Hill, New York).

12 **Catherine, Y.** Preparation techniques for diamond-like carbon. In *Diamond-Like Films and Coatings* (Eds R. E. Clausing, L. L. Horton, J. C. Angus and P. Koidal), 1991, pp. 193–227 (Plenum, New York).

Subject Index

Author Index